OXFORD MASTER SERIES IN ATOMIC, OPTICAL, AND LASER PHY

OXFORD MASTER SERIES IN PHYSICS

The Oxford Master Series is designed for final year undergraduate and beginning graduate students in physics and related disciplines. It has been driven by a perceived gap in the literature today. While basic undergraduate physics texts often show little or no connection with the huge explosion of research over the last two decades, more advanced and specialized texts tend to be rather daunting for students. In this series, all topics and their consequences are treated at a simple level, while pointers to recent developments are provided at various stages. The emphasis in on clear physical principles like symmetry, quantum mechanics, and electromagnetism which underlie the whole of physics. At the same time, the subjects are related to real measurements and to the experimental techniques and devices currently used by physicists in academe and industry. Books in this series are written as course books, and include ample tutorial material, examples, illustrations, revision points, and problem sets. They can likewise be used as preparation for students starting a doctorate in physics and related fields, or for recent graduates starting research in one of these fields in industry.

CONDENSED MATTER PHYSICS

1. M.T. Dove: *Structure and dynamics: an atomic view of materials*
2. J. Singleton: *Band theory and electronic properties of solids*
3. A.M. Fox: *Optical properties of solids*
4. S.J. Blundell: *Magnetism in condensed matter*
5. J.F. Annett: *Superconductivity, superfluids, and condensates*
6. R.A.L. Jones: *Soft condensed matter*
17. S. Tautz: *Surfaces of condensed matter*
18. H. Bruus: *Theoretical microfluidics*

ATOMIC, OPTICAL, AND LASER PHYSICS

7. C.J. Foot: *Atomic physics*
8. G.A. Brooker: *Modern classical optics*
9. S.M. Hooker, C.E. Webb: *Laser physics*
15. A.M. Fox: *Quantum optics: an introduction*
16. S.M. Barnett: *Quantum information*

PARTICLE PHYSICS, ASTROPHYSICS, AND COSMOLOGY

10. D.H. Perkins: *Particle astrophysics, second edition*
11. Ta-Pei Cheng: *Relativity, gravitation and cosmology*

STATISTICAL, COMPUTATIONAL, AND THEORETICAL PHYSICS

12. M. Maggiore: *A modern introduction to quantum field theory*
13. W. Krauth: *Statistical mechanics: algorithms and computations*
14. J.P. Sethna: *Statistical mechanics: entropy, order parameters, and complexity*

Quantum Information

STEPHEN M. BARNETT

Department of Physics
University of Strathclyde, Glasgow

OXFORD
UNIVERSITY PRESS

OXFORD

UNIVERSITY PRESS

Great Clarendon Street, Oxford OX2 6DP

Oxford University Press is a department of the University of Oxford.
It furthers the University's objective of excellence in research, scholarship,
and education by publishing worldwide in

Oxford New York

Auckland Cape Town Dar es Salaam Hong Kong Karachi
Kuala Lumpur Madrid Melbourne Mexico City Nairobi
New Delhi Shanghai Taipei Toronto

With offices in

Argentina Austria Brazil Chile Czech Republic France Greece
Guatemala Hungary Italy Japan Poland Portugal Singapore
South Korea Switzerland Thailand Turkey Ukraine Vietnam

Oxford is a registered trade mark of Oxford University Press
in the UK and in certain other countries

Published in the United States
by Oxford University Press Inc., New York

© Oxford University Press 2009

British Library Cataloguing in Publication Data

Data available

Library of Congress Cataloging in Publication Data

Data available

Printed in Great Britain
on acid-free paper by
Clays Ltd, St Ives plc

ISBN: 978–0–19–852762–6
ISBN: 978–0–19–852763–3

10 9 8 7 6 5 4

For Claire, for Reuben, and for Amy.
Love always.

And in memory of beloved mother, Rita.

Preface

The last ten years have witnessed an explosion of interest in the idea that quantum phenomena might have a vital role to play in the development of future information technology. In communications, the development of quantum cryptography has emerged as a possible long-term solution to the problem of information security. At the same time, the demand for improved computational ability and the move towards ever smaller components has resulted in a race to make the world's first quantum computer. These were linked, dramatically, by Shor's demonstration that a quantum computer could challenge the security of established and currently favoured methods of secure communication, and this has added to the excitement.

This book is an introduction to the field of quantum information. It is aimed at readers who are new to the field and also at those who wish to make sense of the already bewildering and extensive literature. I have aimed to cover, in an introductory manner, what seem to me to be all of the most fundamental ideas in the field. The emphasis, throughout, is on theoretical aspects of the subject, not because these are the most important, but because it is only by understanding these that the true significance of practical developments can be appreciated.

I have included a large number of exercises at the end of each chapter. I would certainly not expect readers to attempt all of these, but my aim was to provide problems on every aspect of the text, so that there are examples to attempt on whatever catches the reader's imagination. A very selective bibliography is given at the end of each chapter, which includes relevant books, review articles, and a few papers. This is obviously in no sense an exhaustive list and I have not attempted to represent the already vast literature on the subject. The suggestions for further reading include only the texts that I have found especially useful in writing the book or that provide further material, and are intended only as a starting point in exploring the wider literature. This means that I have not, for the most part, cited the original papers in which important developments were made, although these are readily available through the suggestions for further reading. This transition is inevitable, of course, as the field becomes better established (who cites *Principia* in mechanics lectures?). I hope that authors who find their work described but not cited in this way will accept the implied and intended compliment.

I would like to record my gratitude to Alison Yao, who not only produced the figures for this book but also carefully proofread the entire text and made many helpful suggestions. Naturally, the responsibility for any remaining errors or residual lack of clarity is mine alone. This book builds upon knowledge obtained by working with and listening to a great many very talented people. It would be futile to attempt to list all of them here, but I would like to thank explicitly the former students: Thomas Brougham, Tony Chefles, Sarah Croke, Kieran Hunter, Norbert Lütkenhaus, and Lee Phillips, from whom I learnt so much about quantum information. Especial thanks must also go to Simon Phoenix, with whom, more than twenty years ago, I first started to think about applying quantum theory to optical communications. (Those were the days!) Chapter 2 is based on the first chapter of *Methods in Theoretical Quantum Optics*, Oxford University Press, Oxford (1997). I would like to thank OUP for their kind permission for this, and my colleague and co-author on *Methods*, Paul Radmore, for enthusiastically encouraging me to do this. I must thank also, for their patience and continued encouragment, Sönke Adlung and his colleagues at OUP. I hope that the resulting text justifies the wait.

Glasgow   November 2008

Contents

Probability and information

1.1   Introduction

The science of information theory begins with the observation that there is a fundamental link between probabilities and information. As early as the mid eighteenth century, Bayes recognized that probabilities depend on what we know; if we acquire additional information then this modifies the probability. For example, the probability that it is raining when I leave for work in the morning is about 0.2, but if I look out of the window ten minutes before leaving and see that it is raining then this additional information adjusts the probability to in excess of 0.9.

Information is a function of probabilities: it is the entropy associated with the probability distribution. This conclusion grew out of investigations into the physical nature of entropy by Boltzmann and his followers. The full power of entropy as the quantity of information was revealed by Shannon in his mathematical theory of communication. This work laid the foundations for the development of information and communications theory by proving two powerful theorems which limit our ability to communicate information.

The link between probability and information has far wider application than just communications. Indeed, we can expect the ideas of information theory to be applicable to any statistical or probabilistic problem. Quantum mechanics is a probabilistic theory and so it was inevitable that a quantum information theory would be developed. In quantum theory, probabilities are secondary quantities calculated by taking the squared modulus of probability amplitudes and this gives rise to interference effects. Consider, for example, the famous two-slit experiment depicted in Fig. 1.1. A single particle launched at the slits can arrive at a point P on the screen by passing either through slit 1 or through slit 2. In classical statistical mechanics this leads to a probability

$$P = P_1 + P_2, \tag{1.1}$$

where P_1 and P_2 are, respectively, the probabilities that the particle passed through slit 1 or slit 2 and went on to arrive at the point P. In quantum theory, however, we associate a complex probability amplitude with each of the slits, ψ_1 and ψ_2, and write

$$
\begin{aligned}
P &= |\psi_1 + \psi_2|^2 \\
&= |\psi_1|^2 + |\psi_2|^2 + \psi_1\psi_2^* + \psi_1^*\psi_2 \\
&= P_1 + P_2 + 2\sqrt{P_1 P_2}\cos[\arg(\psi_1\psi_2^*)],
\end{aligned}
\tag{1.2}
$$

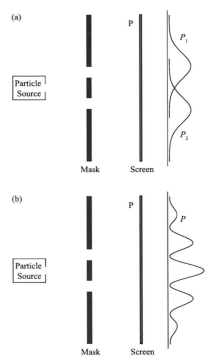

Fig. 1.1 The two-slit experiment and the (a) classical and (b) quantum probabilities for the position at which a single particle is detected on the screen.

Which-way information Equation 1.2 assumes that it is not possible to determine through which slit the particle has passed. Monitoring the slit through which the particle passes gives back the classical result in eqn 1.1.

where $P_1 = |\psi_1|^2$ and $P_2 = |\psi_2|^2$. Clearly, this quantity can be greater or less than its classical counterpart (eqn 1.1) depending on the phase of the complex quantity $\psi_1\psi_2^*$, and this phase depends on the distances between P and both of the slits. This is the signature of quantum interference. The role of probability amplitudes makes quantum theory very different from classical statistical mechanics. The fundamental link between probability and information then leads one to expect that quantum information will behave in a fundamentally different way from its classical counterpart, and this does indeed turn out to be the case.

Quantum information combines the fields of information science and quantum theory. For this reason we begin by treating first information theory and then, in Chapter 2, elements of quantum theory. The remainder of the book is devoted to the application of quantum phenomena to communications and information-processing tasks. Quantum information theory is a new and rapidly developing discipline and it is too soon to attempt a complete presentation of it. For this reason I have included what seem to me to be the most fundamental and important elements of the field. My aim is to be complete and self-contained but to avoid, as far as possible, lengthy and formal mathematical proofs.

1.2 Conditional probabilities

Consider an event, such as a measurement or the result of a game of chance, which can have a number of possible outcomes. We label this event A and denote the set of outcomes as $\{a_i\}$. The probability that a_i occurs ($A = a_i$) is $P(a_i)$, which is necessarily greater than or equal to zero and less than or equal to one. If the set $\{a_i\}$ contains all possible outcomes then the probabilities sum to unity:

$$\sum_i P(a_i) = 1. \tag{1.3}$$

For a single event, this set of probabilities provides a complete statistical description. If we introduce a second event B, with outcomes $\{b_j\}$, then the probabilities $P(a_i)$ and $P(b_j)$ do not tell us all we need to know. A more complete description is provided by the joint probabilities $\{P(a_i, b_j)\}$. The comma in $P(a_i, b_j)$ denotes *and*, so that we read $P(a_i, b_j)$ as 'the probability that $A = a_i$ *and* that $B = b_j$'. If A and B are independent, uncorrelated events then $P(a_i, b_j) = P(a_i)P(b_j)$, but more generally $P(a_i, b_j)$ can be greater or less than $P(a_i)P(b_j)$. We can construct the single-event probabilities from the joint probabilities by summing over all the outcomes for the other event so that

$$P(a_i) = \sum_j P(a_i, b_j),$$
$$P(b_j) = \sum_i P(a_i, b_j). \tag{1.4}$$

These express the simple and natural conclusion that the probability that $A = a_i$ is equal to the probability that $A = a_i$ and that B takes one of its allowed values.

If we know the value of A, what does this tell us about the possible values of B? It is clear that the information gained by learning the value of A can change the probabilities for each of the values of B, but we would like to be able to quantify this change and to derive the values of the new probabilities. Let us suppose that we discover that $A = a_0$; the quantities of interest are then the *conditional* probabilities $\{P(b_j|a_0)\}$. The vertical line in $P(b_j|a_0)$ denotes *given that*, so that we read $P(b_j|a_0)$ as 'the probability that $B = b_j$ *given that* $A = a_0$'. The conditional probability $P(b_j|a_0)$ relates to occurrences for which $B = b_j$ and $A = a_0$ and it is clear, therefore, that it should be proportional to the joint probability $P(a_i, b_j)$:

$$P(b_j|a_0) = K(a_0)P(a_0, b_j). \tag{1.5}$$

We can find the constant of proportionality, $K(a_0)$, by summing this equation over the set of outcomes $\{b_j\}$. The sum over j of $P(b_j|a_0)$ must be unity as these are a complete set of probabilities for the outcome B. The sum over j of the $P(a_0, b_j)$ is, by eqn 1.4, simply $P(a_0)$ and it follows that $K(a_0) = [P(a_0)]^{-1}$. Hence we can relate the conditional and joint probabilities by the equation

$$P(a_0, b_j) = P(b_j|a_0)P(a_0). \tag{1.6}$$

There is nothing special about the outcome a_0, so in general we can write

$$P(a_i, b_j) = P(b_j|a_i)P(a_i). \tag{1.7}$$

We have not introduced or relied on any concept of cause and effect, and so we could equally well have asked for the conditional probability $P(a_i|b_j)$, that is, the probability that $A = a_i$ given that $B = b_j$. Repeating the preceding analysis for $P(a_i|b_j)$ then tells us that

$$P(a_i, b_j) = P(a_i|b_j)P(b_j). \tag{1.8}$$

It is sometimes helpful to arrange the probabilities in the form of a diagram: a probability tree. An example is given in Fig. 1.2. The leftmost set of lines correspond to the set of possible values of A and the remaining lines to the values of B. The weights of the lines are probabilities associated with events and are read from left to right. Hence the probabilities for the three values of A are $P(a_1) = \frac{1}{2}$, $P(a_2) = \frac{1}{3}$, and $P(a_3) = \frac{1}{6}$. Continuing to the right, the probability for taking a second path is given by its weight so that, for example, the paths emerging from the node a_1 are associated with the conditional probabilities $P(b_1|a_1) = \frac{1}{4}$, $P(b_2|a_1) = \frac{1}{4}$, and $P(b_3|a_1) = \frac{1}{2}$. The probability of traversing any given path through the diagram, associated with a given set of outcomes, can be calculated by multiplying the probabilities labelling the paths. This means, for example, that $P(a_1, b_1) = \frac{1}{2} \times \frac{1}{4} = \frac{1}{8}$, $P(a_1, b_2) = \frac{1}{2} \times \frac{1}{4} = \frac{1}{8}$, and $P(a_1, b_3) = \frac{1}{2} \times \frac{1}{2} = \frac{1}{4}$. We are not required to place events A before the events B, and we can replace the probability

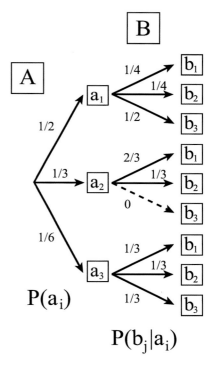

Fig. 1.2 A probability tree for the two events A and B. Note that the dashed line corresponds to a conditional probability that is zero. It is normal practice to omit such zero-weighted lines.

tree in Fig. 1.2 by an equivalent one in which the paths are weighted by the probabilities $P(b_j)$ and then $P(a_i|b_j)$.

Bayes' theorem combines eqns 1.7 and 1.8 to obtain a relation between the two sets of conditional probabilities:

$$P(a_i|b_j) = \frac{P(b_j|a_i)P(a_i)}{P(b_j)}. \tag{1.9}$$

The utility of this theorem can be illustrated by a simple problem. Each morning I walk to work, choosing between a long and a short route. If I choose the short route then I always arrive on time, but if I take the long route then I arrive on time with probability $\frac{3}{4}$. The long route is scenic, so I risk taking it one day in four. If you see me arriving on time, what probability would you infer for me having taken the long route? To solve this problem, we shall use Bayes' theorem and our first task is to identify the two events: these are the choice of route (A) and arriving on time or late (B). Let a_l and a_s denote taking the long route and the short route, respectively, and let b_O and b_L denote arriving on time and late. The information contained in the problem then corresponds to the probabilities

$$P(b_O|a_s) = 1,$$
$$P(b_O|a_l) = \frac{3}{4},$$
$$P(a_l) = \frac{1}{4}. \tag{1.10}$$

The last of these immediately tells us that $P(a_s) = \frac{3}{4}$. In order to find the probability that I took the long route, using Bayes' theorem, we need the probability $P(b_O)$. This is

$$P(b_O) = P(b_O|a_s)P(a_s) + P(b_O|a_l)P(a_l) = \frac{15}{16}. \tag{1.11}$$

The probability that I took the long route, given that I arrived on time, is then

$$P(a_l|b_O) = \frac{P(b_O|a_l)P(a_l)}{P(b_O)} = \frac{1}{5}. \tag{1.12}$$

As a check we can calculate the probability that I took the short route given that I arrived on time:

$$P(a_s|b_O) = \frac{P(b_O|a_s)P(a_s)}{P(b_O)} = \frac{4}{5}, \tag{1.13}$$

which equals $1 - P(a_l|b_O)$ as it should.

Bayes' theorem is often written as a proportionality in the form

$$P(a_i|b_j) \propto P(b_j|a_i)P(a_i). \tag{1.14}$$

This form suffices if we are interested only in the relative size of a set of conditional probabilities. It is also possible to use Bayes' theorem in this form and to normalize at the end of the calculation, as $\sum_i P(a_i|b_j) = 1$, so that

$$P(a_i|b_j) = \frac{P(b_j|a_i)P(a_i)}{\sum_k P(b_j|a_k)P(a_k)}. \tag{1.15}$$

As an example, we repeat the exercise of the preceding paragraph:

$$P(a_1|b_O) \propto P(b_O|a_1)P(a_1) = \frac{3}{16},$$

$$P(a_s|b_O) \propto P(b_O|a_s)P(a_s) = \frac{3}{4}. \tag{1.16}$$

These tell us that $P(a_s|b_O)$ is four times as big as $P(a_1|b_O)$ and hence normalization tells us that $P(a_s|b_O) = \frac{4}{5}$ and $P(a_1|b_O) = \frac{1}{5}$.

Conditional probabilities are not limited to just a pair of events. If we supplement the events A and B by a third event C, with possible outcomes $\{c_k\}$, then we can define a set of joint probabilities for all three events, $\{P(a_i, b_j, c_k)\}$. Each of the commas represents an *and*, so that $P(a_i, b_j, c_k)$ is the probability that $A = a_i$ *and* that $B = b_j$ *and* that $C = c_k$. We can also write conditional probabilities, but need to take care with the notation. In particular, $P(a_i|b_j, c_k)$ is the probability that $A = a_i$ given that $B = b_j$ and that $C = c_k$. The conditional probability $P(a_i, b_j|c_k)$, however, is the probability that $A = a_i$ and that $B = b_j$ given that $C = c_k$. The formulae presented above for a pair of outcomes can readily be extended to treat three (or more) events. For example, the relationship 1.7 tells us that

$$\begin{aligned} P(a_i, b_j, c_k) &= P(a_i|b_j, c_k)P(b_j, c_k) \\ &= P(a_i|b_j, c_k)P(b_j|c_k)P(c_k) \end{aligned} \tag{1.17}$$

and that

$$\begin{aligned} P(a_i, b_j, c_k) &= P(a_i, b_j|c_k)P(c_k) \\ &= P(a_i|b_j, c_k)P(b_j|c_k)P(c_k). \end{aligned} \tag{1.18}$$

Bayes' theorem can also be extended to apply to three or more events. For example,

$$P(a_i|b_j, c_k) = \frac{P(b_j, c_k|a_i)P(a_i)}{P(b_j, c_k)} \tag{1.19}$$

or, alternatively, we can use the form

$$P(a_i|b_j, c_k) \propto P(b_j, c_k|a_i)P(a_i). \tag{1.20}$$

We have not, as yet, examined how learning about one event changes the probability for a second one. How, for example, does learning that $B = b_j$ affect the probability that $A = a_i$? The answer is most elegantly stated in terms of Fisher's likelihood function. If we discover that $B = b_j$ then the relevant probability will be $P(a_i|b_j)$, which is related to the prior probability, $P(a_i)$, by

$$P(a_i|b_j) = \ell(a_i|b_j)P(a_i), \tag{1.21}$$

where $\ell(a_i|b_j)$ is the likelihood of a_i given b_j. Bayes' theorem tells us that $\ell(a_i|b_j)$ is proportional to the conditional probability $P(b_j|a_i)$ but that, unlike the conditional probabilities, it is symmetrical in its arguments. We should note also that $P(b_j|a_i)$ is, in a sense, a function of the b_j in that it is the value of B that is unknown. The likelihood $\ell(a_i|b_j)$, however, is a function of the a_i. Note that if learning that $B = b_j$ does not change the probabilities for the event A, then $\ell(a_i|b_j) = 1$. The likelihood can either increase or decrease the probability that $A = a_i$, corresponding to values of $\ell(a_i|b_j)$ that are greater than or less than unity, respectively.

As with Bayes' theorem, it is often convenient to treat eqn 1.21 as a proportionality by defining ℓ only up to a multiplicative constant and normalizing only at the end of the calculation. This is especially useful when we have a sequence of events modifying the probability. Suppose, as above, that we wish to determine how learning that $B = b_j$ affects the probability that $A = a_i$. We write the associated conditional probability as

$$P(a_i|b_j) \propto \ell(a_i|b_j)P(a_i). \tag{1.22}$$

If we then learn that $C = c_k$, we further modify the probability that $A = a_i$ using the likelihood $\ell(a_i|c_k)$:

$$\begin{aligned} P(a_i|b_j, c_k) &\propto \ell(a_i|c_k)P(a_i|b_j) \\ &\propto \ell(a_i|c_k)\ell(a_i|b_j)P(a_i). \end{aligned} \tag{1.23}$$

Here we have simply applied the likelihood formula but with $P(a_i|b_j)$ acting as the prior probability (prior, that is, to learning the value of C). Note that obtaining $P(a_i|b_j, c_k)$ from $P(a_i)$ requires us merely to multiply by the likelihoods $\ell(a_i|b_j)$ and $\ell(a_i|c_k)$ and then to normalize by ensuring that $\sum_i P(a_i|b_j, c_k) = 1$. Additional events can be accounted for by multiplying by their associated likelihoods.

We can illustrate the use of the likelihood by means of a classic problem in genetics, described by Fisher. Consider mice that can be either black or brown. Black, B, is the dominant gene and brown, b, is the recessive gene. This means that the two genes in brown mice must be bb but that black mice can be either BB or Bb. (Similar properties are to be found throughout nature: in humans, for example, B may denote brown hair or brown eyes, with the corresponding b's being blonde hair and blue eyes.) If we mate a black mouse with a brown one then how does the colour of the resulting offspring modify the probabilities that the mouse is BB or Bb? We note that there are three possible genetic patterns for our black test mouse, namely BB, or Bb with either gene B. In the absence of any information about the ancestry of the mouse, we can only make these possibilities equiprobable. Hence we assign the prior probabilities

$$P(BB) = \frac{1}{3}, \qquad P(Bb) = \frac{2}{3}. \tag{1.24}$$

If any of the offspring are brown then the test mouse must be Bb. If they are all black, however, then the probability that the mouse is BB

increases with each birth. We let x_i denote the colour of the ith mouse to be born, and then the likelihoods for the test mouse to be BB or Bb are

$$\ell(BB|x_i = \text{black}) \propto P(x_i = \text{black}|BB)$$
$$= 1,$$
$$\ell(Bb|x_i = \text{black}) \propto P(x_i = \text{black}|Bb)$$
$$= \frac{1}{2}. \tag{1.25}$$

It follows that after the birth of one black mouse, the probabilities in eqn 1.24 are modified according to

$$P(BB|x_1 = \text{black}) \propto \ell(BB|x_1 = \text{black})P(BB)$$
$$\propto 1 \times \frac{1}{3},$$
$$P(Bb|x_1 = \text{black}) \propto \ell(Bb|x_1 = \text{black})P(Bb)$$
$$\propto \frac{1}{2} \times \frac{2}{3}. \tag{1.26}$$

Normalizing then gives the probabilities

$$P(BB|x_1 = \text{black}) = \frac{1}{2},$$
$$P(Bb|x_1 = \text{black}) = \frac{1}{2}. \tag{1.27}$$

If the second mouse born is also black then

$$P(BB|x_1, x_2 = \text{black}) \propto \ell(BB|x_1 = \text{black})\ell(BB|x_2 = \text{black})P(BB)$$
$$\propto 1 \times 1 \times \frac{1}{3},$$
$$P(Bb|x_1, x_2 = \text{black}) \propto \ell(Bb|x_1 = \text{black})\ell(Bb|x_2 = \text{black})P(Bb)$$
$$\propto \frac{1}{2} \times \frac{1}{2} \times \frac{2}{3}, \tag{1.28}$$

which on normalizing gives

$$P(BB|x_1, x_2 = \text{black}) = \frac{2}{3},$$
$$P(Bb|x_1, x_2 = \text{black}) = \frac{1}{3}. \tag{1.29}$$

The birth of two black mice interchanges the values of the prior probabilities in eqn 1.24 and the birth of additional black mice further inceases the probability that the test mouse is BB. The birth of a single brown mouse, however, makes it certain that the test mouse is Bb. An example of such a mouse family and associated genetic probabilities is given in Fig. 1.3.

1.3 Entropy and information

We have seen how acquiring information about events leads us to modify the probabilities for other as yet undetermined events. It remains,

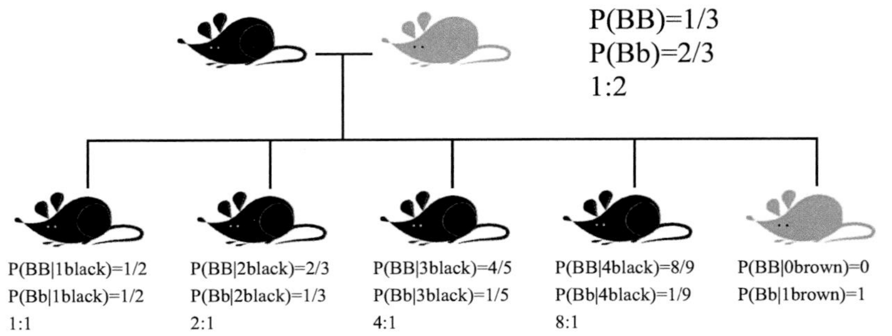

P(BB)=1/3
P(Bb)=2/3
1:2

P(BB|1black)=1/2   P(BB|2black)=2/3   P(BB|3black)=4/5   P(BB|4black)=8/9   P(BB|0brown)=0
P(Bb|1black)=1/2   P(Bb|2black)=1/3   P(Bb|3black)=1/5   P(Bb|4black)=1/9   P(Bb|1brown)=1
1:1   2:1   4:1   8:1

Fig. 1.3 An example of a mouse family. Note the way in which the birth of each mouse modifies the probabilities for the genes of the parent black mouse.

however, for us to deal with the important question of quantifying information itself. It turns out to be both natural and useful to define the quantity of information as the entropy of the associated probability distribution for the event. A formal demonstration of this is given in Appendix A, but we present here a simple argument which makes this plausible.

Consider again an event A with possible outcomes $\{a_i\}$. If one of these events, a_0 say, is certain to occur, so that $P(a_0) = 1$, then learning the value of A tells us nothing and we can acquire no information by measuring A. In other words, observing A tells us nothing that we did not already know. If $P(a_0)$ is close to but not equal to unity, then learning that $A = a_0$ provides some information but not very much. In this case, finding that $A = a_0$ confirms something that we would have confidently guessed. If we find that A takes a very unlikely value, however, then we might be rather surprised and could be said to have acquired a considerably larger quantity of information. The result of this, moreover, might cause us to drastically modify our behaviour. Consider, for example, a fire alarm. The most likely state of the fire alarm is that it makes no sound, and in that state we give it no regard. On the rare occasions on which the alarm does sound, it grabs our attention and leads us to abandon our activities and to leave the building.

The above considerations suggest that learning the value of A provides a quantity of information that *increases* as the corresponding prior probability *decreases*. Let us denote by $h[P(a_i)]$ the information we obtain on learning that $A = a_i$. Suppose that there is a second event B with possible outcomes $\{b_j\}$ and that the two events are independent, so that $P(a_i, b_j) = P(a_i)P(b_j)$. If we know that $A = a_i$ then we think of learning that $B = b_j$ as *adding* to our knowledge or providing *additional* information. Hence it is natural to require that

$$h[P(a_i, b_j)] = h[P(a_i)P(b_j)]$$
$$= h[P(a_i)] + h[P(b_j)]. \qquad (1.30)$$

The fact that the product of probabilities in $h[P(a_i)P(b_j)]$ becomes a sum of terms in the individual probabilities clearly suggests that h is a logarithm:

$$h[P(a_i)] = -K \log P(a_i). \qquad (1.31)$$

We have included a minus sign so that h is a positive quantity, and a positive constant K to be determined or selected later.

It is useful to define information not for individual outcomes, but rather for the event A and its complete set of possible outcomes. We arrive at the information associated with the event A, $H(A)$, by averaging eqn 1.31 over the set of possible outcomes:

$$H(A) = \sum_i P(a_i) h[P(a_i)]$$
$$= -K \sum_i P(a_i) \log P(a_i). \qquad (1.32)$$

We recognize this summation of the products of the probabilities and their logarithms as the entropy, familiar from statistical mechanics.

We have yet to determine the constant K and to select the base for the logarithm. These tasks are closely connected, as

$$\log_b x = \log_b a \log_a x, \qquad (1.33)$$

where the subscript denotes the base of the logarithm. Clearly, we can incorporate the constant K into the base of the logarithms. Equivalently, changing the base of the logarithms merely scales the information. Two choices of base are commonly employed: logarithms to base 2 and the natural base of logarithms, that is, base e. We follow the common convention in information theory by denoting the former as 'log' and the latter as 'ln'. When using base 2, the information is counted in 'bits', and when using base e, it is measured in 'nats'. We denote the information in bits as H and that in nats as H_e, so that

$$H(A) = -\sum_i P(a_i) \log P(a_i) \quad \text{bits} \qquad (1.34)$$

or

$$H_e(A) = -\sum_i P(a_i) \ln P(a_i) \quad \text{nats}. \qquad (1.35)$$

It follows that $H_e = H \ln 2$. The selection of base 2 is a consequence of the prevalence in information technology of the binary system, with digits 0 and 1, and of the use of physical systems with two distinct states to represent these. If the state representing 0 is prepared with a priori probability p (and the state representing 1 with probability $1 - p$) then the associated information is

$$H = -p \log p - (1 - p) \log(1 - p). \qquad (1.36)$$

This quantity, which is plotted in Fig. 1.4, takes its maximum value, of 1 bit, for $p = \frac{1}{2}$. By working with logarithms to base 2, we are led to associate each two-state system and each 0 or 1 with a maximum of one bit of information. It is common to find the term 'bit' also associated with the physical system encoding the 0 or 1 and, indeed, with the binary value itself. The selection of base e is also commonly employed and is

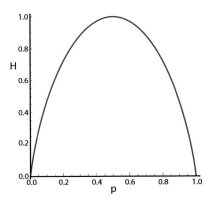

Fig. 1.4 The two-state entropy given in eqn 1.36.

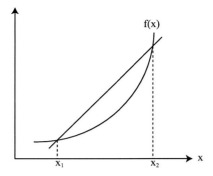

Fig. 1.5 A convex function.

particularly convenient for problems in which analytical methods such as calculus are employed.

The information, or entropy, plays a central role in information theory and a number of important results follow from its properties. As a prelude to describing these, we review the properties of convex and concave functions. A function $f(x)$ is convex if it is continuous and if its value at any point between a pair of points x_1 and x_2 lies below the straight line joining these points, as depicted in Fig. 1.5. More formally, this means that $f(x)$ is convex on an interval (a, b) if, for every pair of points x_1 and x_2 in the interval and for any λ between 0 and 1, we have

$$f[\lambda x_1 + (1 - \lambda)x_2] \leq \lambda f(x_1) + (1 - \lambda)f(x_2). \tag{1.37}$$

It follows that if $f(x)$ is convex then

$$f\left(\sum_{i=1}^{n} p_i x_i\right) \leq \sum_{i=1}^{n} p_i f(x_i) \tag{1.38}$$

for a set of probabilities $\{p_i\}$. A function $f(x)$ is concave if the converse is true or, equivalently, if $-f(x)$ is convex. If the function has a second derivative then a necessary and sufficient condition for it to be convex is that its second derivative is greater than or equal to zero at all points in the interval (a, b). If the second derivative is less than or equal to zero, however, then the function is concave. The importance of this for our study is that $x \log x$ is convex and hence $H(A)$ is a concave function.

The information $H(A)$ is zero if and only if one of the probabilities $P(a_i)$ is unity, with the others being zero. In this case the value of A is already known and so there is no information to be gained by observing it. If A can take n possible values $a_1, a_2, \cdots, a_n$, then $H(A)$ takes its maximum possible value, of $\log n$, if the associated probabilities are all equal: $P(a_i) = \frac{1}{n}$. This is intuitively reasonable, as for these probabilities the value of A is most uncertain, suggesting that the information content is maximum. Any change towards equalization of the probabilities will cause H to increase so that, for example, if $P(a_1) < P(a_2)$ then increasing $P(a_1)$ at the expense of $P(a_2)$ increases the information. More generally, if we replace the probabilities $P(a_i)$ by

$$P'(a_i) = \sum_j \lambda_{ij} P(a_j), \tag{1.39}$$

where $\sum_i \lambda_{ij} = 1 = \sum_j \lambda_{ij}$ and all the λ_{ij} are greater than or equal to zero, then $H(A)$ will increase or remain unchanged: $H' \geq H$. This result is a consequence of the concavity of H.

If we have two events A and B, with associated outcomes $\{a_i\}$ and $\{b_j\}$, then we can write down the information for the two events in terms of the joint probability distribution $P(a_i, b_j)$:

$$H(A, B) = -\sum_{ij} P(a_i, b_j) \log P(a_i, b_j). \tag{1.40}$$

Here we have followed the notation introduced in the preceding section in that the comma in $H(A, B)$ denotes that H is the information associated with the events A *and* B. We can, of course, express the information for the single events A and B in terms of the joint probability distribution using eqn 1.4:

$$H(A) = -\sum_{ij} P(a_i, b_j) \log \sum_k P(a_i, b_k),$$

$$H(B) = -\sum_{ij} P(a_i, b_j) \log \sum_l P(a_l, b_j). \tag{1.41}$$

The values of $H(A)$, $H(B)$, and $H(A, B)$ are constrained by the inequality

$$H(A) + H(B) \geq H(A, B). \tag{1.42}$$

This follows directly from the fact that $H(A) + H(B) - H(A, B)$ can be written as a relative entropy for two possible joint probability distributions $\{P(a_i, b_j)\}$ and $\{P(a_i)P(b_j)\}$:

$$H(A) + H(B) - H(A, B) = \sum_{ij} P(a_i, b_j) \log \left(\frac{P(a_i, b_j)}{P(a_i)P(b_j)} \right)$$

$$= H\left(\{P(a_i, b_j)\} \| \{P(a_i)P(b_j)\}\right). \tag{1.43}$$

It is proven in Appendix B that a relative entropy is always greater than or equal to zero and that it takes the value zero only if the two probability distributions are identical. Hence eqn 1.42 is true and the equality holds if and only if the two events are independent so that $P(a_i, b_j) = P(a_i)P(b_j) \quad \forall i, j$.

The combination in eqn 1.43 plays an important role in communications theory and arises sufficiently often for us to give it a name: the mutual information

$$H(A : B) = H(A) + H(B) - H(A, B) \tag{1.44}$$

is a measure of the correlation between the events A and B. If these correspond, respectively, to the selection and receipt of a signal then $H(A : B)$ is the information transferred by the communication. We can illustrate this idea rather directly by noting that

$$H(A : B) = \sum_{ij} P(a_i, b_j) \log \ell(a_i|b_j). \tag{1.45}$$

Only if learning the value of B changes the probabilities for the values of A will the likelihoods differ from unity, leading to a non-zero mutual information.

The joint information, $H(A, B)$, is bounded from below by the larger of $H(A)$ and $H(B)$. To see this, we write the difference $H(A, B) - H(A)$ in the form

Relative entropy Let $P(a_i)$ and $Q(a_i)$ be distinct probabilities for the event A to correspond to a_i. We define the relative entropy for the two probability distributions as

$$H(P\|Q) = \sum_i P(a_i)$$
$$\times [\log P(a_i) - \log Q(a_i)].$$

Note that the relative entropy is not a symmetric function of the two sets of probabilities:

$$H(P\|Q) \neq H(Q\|P).$$

You will often find the mutual information denoted by $I(A : B)$.

$$H(A, B) - H(A) = -\sum_{ij} P(a_i, b_j) \log \left(\frac{P(a_i, b_j)}{P(a_i)} \right)$$

$$= \sum_{i} P(a_i) \left[\sum_{j} -P(b_j|a_i) \log P(b_j|a_i) \right] , (1.46)$$

which is clearly greater than or equal to zero, as every term in the j-summation is positive or zero because the $P(b_j|a_i)$ are a set of probabilities for the outcomes of the event B. It is zero if the events B are perfectly correlated with the events A so that $P(b_j|a_i) = \delta_{ij}$. A similar treatment for $H(A, B) - H(B)$ establishes that it is also greater than or equal to zero. It follows that the mutual information is bounded from above by the smaller of $H(A)$ and $H(B)$:

$$0 \le H(A : B) \le \text{Inf} \, (H(A), H(B)). \qquad (1.47)$$

We can combine this with eqn 1.42 to place upper and lower bounds on $H(A, B)$ in the form of the inequality

$$\text{Sup} \, (H(A), H(B)) \le H(A, B) \le H(A) + H(B), \qquad (1.48)$$

where $\text{Sup} \, (H(A), H(B))$ denotes the larger of $H(A)$ and $H(B)$.

We see that the positive quantity $H(A, B) - H(A)$ is a function of the conditional probabilities $\{P(b_j|a_i)\}$ and hence it tells us about the information to be gained on learning the value of A. It is useful to define information for such conditional probabilities, and we follow the notation introduced in the preceding section by writing

$$H(B|a_i) = -\sum_{j} P(b_j|a_i) \log P(b_j|a_i). \qquad (1.49)$$

This is the information associated with B given we know that $A = a_i$. If the value of A is known but not specified then the relevant information is the probability-weighted average of eqn 1.49:

$$H(B|A) = \sum_{i} P(a_i) H(B|a_i)$$

$$= H(A, B) - H(A). \qquad (1.50)$$

Positivity of the conditional entropy   The conditional entropy is clearly greater than or equal to zero because $P(b_j|a_i)$ is the probability that $B = b_j$ given that $A = a_i$, so that $H(B|a_i)$ is an entropy.

This quantity is the conditional entropy of B, or the entropy of B conditional on A. We can rewrite eqn 1.50 as

$$H(A, B) = H(A) + H(B|A) \qquad (1.51)$$

which we can think of as the information analogue of Bayes' rule (eqn 1.7) for conditional probabilities. The logarithm occurring in the definition of H means that the multiplication of probabilities in eqn 1.7 is replaced by the addition of the informations in eqn 1.51. This relationship has an appealing interpretation: the entropy or information associated with the joint event, A and B, is the entropy associated with A, plus that for B when A is known.

Bayes' theorem is very much a statement, or perhaps even a definition, of the properties of conditional probabilities and as such it is not in doubt. Its application, however, has caused controversy for two main reasons. (i) How should we interpret probabilities and, in particular, can they be interpreted in terms of frequencies of occurrence? (ii) How should we assign the required prior probabilities? The first of these we shall sidestep by noting that frequencies of occurrence are the way in which probabilities are usually understood in statistical mechanics, in quantum physics and, indeed, in communications theory. For the second we can make use of the information ideas developed above. In assigning prior probabilities (or any other probabilities) on the basis of limited knowledge, it is reasonable to opt to minimize any bias by selecting the most uniform or most uncertain probability distribution that is consistent with what we do know. In other words, if we know nothing then we should set the probabilities to be equal, but if we have some information then we should make the probabilities as nearly equal as is consistent with our prior knowledge. To achieve this we need to maximize the information or entropy associated with the probability distribution given the available information. This approach to assigning prior probabilities is Jaynes's maximum entropy or 'Max Ent' method.

We can illustrate the maximum entropy principle by reference to a simple example. A die is a cube with one of the integers 1, 2, 3, 4, 5, and 6 inscribed on each of its six faces. If it is a true die, then on tossing it we would expect each of these scores to be equally likely to show on the uppermost face. In this case the mean score will be 3.5. What values should we assign for the probabilities p_n for each number n, however, given only that the mean score is, in fact, 3.47? The maximum entropy principle tells us to do this by maximizing the entropy

$$H_e = -\sum_{n=1}^{6} p_n \ln p_n, \tag{1.52}$$

subject to the constraints that the probabilities sum to unity and that the mean score is 3.47. We can find this maximum by using Lagrange's method of undetermined multipliers, as described in Appendix B, and it is for this reason that we work with H_e. We introduce the Lagrange multipliers λ and μ, and vary the quantity

$$\tilde{H} = H_e + \lambda\left(1 - \sum_{n=1}^{6} p_n\right) + \mu\left(3.47 - \sum_{n=1}^{6} np_n\right). \tag{1.53}$$

Variation of the probabilities and setting the variation to zero gives

$$d\tilde{H} = \sum_{n=1}^{6} dp_n\left(-\ln p_n - 1 - \lambda - \mu n\right) = 0, \tag{1.54}$$

the solution of which is

$$p_n = e^{-(1+\lambda)}e^{-\mu n}. \tag{1.55}$$

Properties of information We summarize here the main properties of H.

(i) The entropy associated with an event A is

$$H(A) = -\sum_{i} P(a_i) \log P(a_i)$$
$$\geq 0.$$

It is zero only if one of the probabilities is unity.

(ii) Concavity:

$$H'(A) \geq H(A),$$

where the $P'(a_i)$ are positively weighted averages of the $P(a_i)$.

(iii) The mutual information is

$$H(A:B) = H(A) + H(B)$$
$$\qquad -H(A,B)$$
$$\geq 0.$$

It is zero if and only if A and B are statistically independent.

(iv) The conditional information is

$$H(B|A) = H(A,B) - H(A) \geq 0.$$

It is zero only if each value of A uniquely determines a single value of B.

We can determine the Lagrange multiplier λ by requiring the p_n to sum to unity:

$$p_n = \frac{e^{-\mu n}}{\sum_{k=1}^{6} e^{-\mu k}}. \tag{1.56}$$

The value of μ is fixed by imposing the mean-value constraint and leads, in this case, to the value $\mu = 0.010$.

As a second example of the application of the Max Ent method, we consider the question of whether or not we should treat a pair of variables with known separate statistical properties as correlated or not. For definiteness, we consider two properties of the population of Finland. These are that 90% of the population have blue eyes and that 94% of Finns speak Finnish as their mother tongue. (The remaining 6% have Swedish as their first language.) Given the similarity of the numbers, we might be tempted to associate blue eyes with native Finnish speakers, but the maximum entropy principle suggests otherwise. We present the probabilities conveniently in Fig. 1.6, in which the four entries are the values of the probabilities for the two pairs of properties: blue eyes or not blue eyes and native Finnish or Swedish speakers. The constraints leave only the single parameter x to be fixed in these probabilities, and we do this by maximizing the entropy

Probabilities:	Swedish	Finnish	
Brown eyes	x	0.10 - x	10%
Blue eyes	0.06 - x	0.84 + x	90%
	6%	94%	

Fig. 1.6 A probability table representing the probabilities that a Finn has blue or brown eyes and is a native Finnish or Swedish speaker.

$$H_e = -x \ln x - (0.06 - x) \ln(0.06 - x) - (0.10 - x) \ln(0.10 - x)$$
$$- (0.84 + x) \ln(0.84 + x). \tag{1.57}$$

Differentiating this with respect to x and setting the derivative equal to zero leads to the result $x = 0.006$ or that, in the absence of any additional information, we should use 0.6% as the percentage of Finns who are native Swedish speakers and do not have blue eyes. This is the result we would expect on the basis of uncorrelated traits, that is, if eye colour is independent of mother tongue. This conclusion is quite general, in that the maximum entropy principle will always give uncorrelated probabilities for such problems.

The appearance of the entropy as the quantity of information is highly suggestive of a possible connection between information theory and thermodynamics or statistical mechanics. This connection is, in fact, both useful and profound, and we shall illustrate it with two simple but important examples. Consider first a physical system that can exist in a number of distinct possible configurations. We use n to label the configurations, and associate the energies $\{E_n\}$ with these. One of the tasks in statistical mechanics is to calculate the probability p_n that the system of interest is to be found in its nth configuration given that, on average, the energy is $\bar{E}$. The maximum entropy method tells us to maximize the entropy $H_e = -\sum_n p_n \ln p_n$ subject to the information that is available to us. In this case we know that the probabilities sum to unity, $\sum_n p_n = 1$, and that the mean energy is $\sum_n p_n E_n = \bar{E}$. We can incorporate this information into our problem by introducing the Lagrange multipliers λ and β, and varying the quantity

$$\tilde{H} = H_e + \lambda \left(1 - \sum_n p_n \right) + \beta \left(\bar{E} - \sum_n p_n E_n \right). \qquad (1.58)$$

Varying the probabilities and setting the resulting variation of $\tilde{H}$ to zero gives

$$d\tilde{H} = \sum_n \left(-\ln p_n - 1 - \lambda - \beta E_n \right) dp_n = 0, \qquad (1.59)$$

the solution of which is

$$p_n = e^{-(1+\lambda)} e^{-\beta E_n}. \qquad (1.60)$$

We can eliminate λ by imposing the normalization to give

$$p_n = \frac{e^{-\beta E_n}}{Z(\beta)}, \quad Z(\beta) = \sum_n e^{-\beta E_n}. \qquad (1.61)$$

In principle we can then determine the value of β in terms of $\bar{E}$, but we recognize these probabilities as the Boltzmann distribution and associate β with the inverse temperature, that is $\beta = (k_B T)^{-1}$, and hence associate $Z(\beta)$ with the partition function. Applying the maximum entropy method has led us to an important and fundamental result from thermodynamics.

Our second example, of the connection between information and thermodynamics, introduces the important point that information is stored as an arrangement of physical systems and that these systems, and the relevant arrangements of them, are subject to physical laws. Among these one law, the second law of thermodynamics, deals specifically with entropy. There are a number of statements of the second law. Perhaps the most general of these (due to Clausius) is: *During real physical processes, the entropy of an isolated system always increases. In the state of equilibrium, the entropy attains its maximum value.* The second part of this statement, of course, is equivalent to the information being maximized to obtain the equilibrium probabilities. A closely related statement of the second law (due to Kelvin) is: *A process whose effect is the complete conversion of heat into work cannot occur.*

Szilard showed, by means of a simple model, that information acquisition is intimately connected with (thermodynamic) entropy. He considered an 'intelligent being' operating a heat engine in which a single molecule forms the working fluid. The envisaged scheme is depicted in Fig. 1.7. In Fig. 1.7(a), the molecule is contained within the whole volume, V_0, of its container and is in thermodynamic equilibrium with its surrounding environment, or heat bath, at temperature T. The operation of the heat engine starts, in Fig. 1.7(b), with the insertion of a partition which separates the container into two equal parts, each of volume $V_0/2$. In principle this can be done reversibly, that is with no nett expenditure of energy. Observation by the intelligent being (Fig. 1.7(c)) allows the being to determine in which of the two halves the molecule

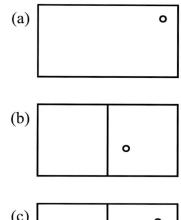

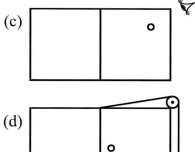

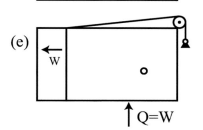

Fig. 1.7 Szilard's illustration of the connection between information and (thermodynamic) entropy.

is contained. The being can then attach a weight to the partition (Fig. 1.7(d)) and so use the isothermal expansion of the one-molecule gas (Fig. 1.7(e)) to extract work by raising the weight. Repeating this sequence allows the being to extract arbitrarily large amounts of work from the surrounding heat bath, apparently in direct violation of the second law of thermodynamics.

It is instructive to quantify the changes associated with the steps in Szilard's thought experiment. We assume that the single molecule behaves like a perfect (classical) gas and so is governed by the (one-molecule) perfect gas law

$$PV = k_B T, \tag{1.62}$$

where P, V, and T are the pressure, volume, and temperature, respectively, of the gas, and $k_B \approx 1.38 \times 10^{-23} J K^{-1}$ is Boltzmann's constant. After the partition is inserted, the volume occupied by the molecule is reduced from V_0 to $V_0/2$, but we do not know on which side of the partition the molecule is to be found. From the point of view of an observer, such as our intelligent being, there is an equal probability that the molecule is to the left or the right of the partition. Information theory tells us to associate one bit or $\ln 2$ nats of information with this situation. Observing in which of the two volumes the molecule is to be found provides this quantity of information. If the partition is attached to a piston or similar mechanical device, represented by a pulley and weight in the figure, then the gradual isothermal expansion of the one-molecule gas can be used to extract an amount of work

$$W = \int_{V_0/2}^{V_0} P \, dV = k_B T \ln 2. \tag{1.63}$$

This energy is supplied, of course, as heat (Q) from the surrounding heat bath: $Q = W$. The expansion of the gas increases the entropy of the gas by the amount

$$\Delta S = \frac{Q}{T} = k_B \ln 2. \tag{1.64}$$

We have extracted $k_B T \ln 2$ of work from the heat bath, and the molecule is back occupying the original volume V_0 and at temperature T. The second law of thermodynamics, as Kelvin formulated it, can be saved by associating the information acquired by our intelligent being with an entropy increase. The process of measuring and recording the position of the molecule is itself necessarily associated with entropy production greater than or equal to $k_B \ln 2$. It has been suggested, with some justification, that the modern science of information began with this observation.

A closely related phenomenon is Landauer's observation that erasing an unknown bit of information requires the dissipation of at least $k_B T \ln 2$ of energy. This is, of course, the amount of energy generated by the isothermal expansion of the single-molecule gas in Szilard's model and we can demonstrate Landauer's result by reference to it. The process of erasing and resetting the bit is depicted in Fig. 1.8. Szilard's

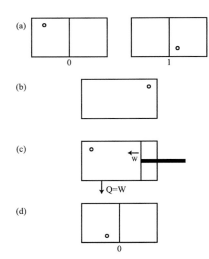

Fig. 1.8 Illustration of Landauer's derivation of the thermodynamic cost of information erasure.

container is divided into two equal volumes and the bit is encoded as the half-volume containing the molecule. We associate (Fig. 1.8(a)) the molecule being positioned to the left of the partition with the bit value 0 and the molecule being in the right half-volume with the bit value 1. Our task is to erase the bit of information by resetting the bit value to zero, whatever the initial bit value. We can achieve this by first removing the partition between the two volumes (Fig. 1.8(b)). This allows the molecule to move freely throughout the volume V_0, which destroys the memory of the bit value. We can reset the bit value to 0 by pushing a new partition (Fig. 1.8(c)) from the right-hand side of the volume to the midpoint (Fig. 1.8(d)). This process constitutes an isothermal compression of the one-molecule gas and requires us to supply the amount of work

$$W = -\int_{V_0}^{V_0/2} P\, dV = k_B T \ln 2. \tag{1.65}$$

The molecule remains in thermodynamic equilibrium with the heat bath and so this energy must, necessarily, be dissipated as heat: $Q = W = k_B T \ln 2$.

1.4   Communications theory

Communication systems exist for the purpose of conveying information between two or more parties. In their simplest manifestation we have a transmitter, who wishes to send a message, and a receiver, who is intended to receive it. In quantum communications, these are universally referred to as Alice, the transmitter, and Bob, the receiver. We shall adopt these useful labels throughout the book, including in this section, in which we treat classical communications theory.

   In order to progress with the mathematical theory of communications it is necessary to appreciate that the operation of any communication system is necessarily probabilistic. The reason for this is that Bob cannot know the intended message prior to receiving it from Alice. If he is already in possession of the message then there is no need for the signal to be sent; the communication is redundant as it can carry no information. In order to convey information, the device must be capable of carrying a signal to Bob associated with any one of a given set of messages. Bob may know the probability for Alice to select each of the possible messages but will not know which was selected until the signal is received.

   Figure 1.9 presents Shannon's schematic diagram of a general communication system. The operation starts with an information source accessible to or operated by Alice. The role of this is to select one of the possible messages for transmission to Bob; indeed, it could be a person, Alice herself, deciding what to send. This message is entered into a transmitter, which prepares a corresponding signal for transmission. Simple examples of such transmitters are telephones and computer keyboards, which act to transform spoken words or keystrokes into electrical signals. The signal leaves Alice's domain and enters the communication

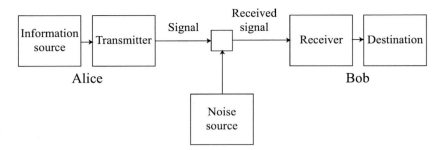

Fig. 1.9 Shannon's model of a general communication system.

channel. Passage through the channel may degrade the quality of the signal and so make it more difficult to read; this possibility is represented in the diagram by the introduction of noise. At the end of the communication channel the noisy signal enters Bob's domain, where it can be decoded and read. The receiver converts the signal into a form that can be read by the destination; for example, a receiving telephone converts the electrical signal back into sound waves that can be heard by the destination, Bob.

We can apply the ideas of the preceding sections to analyse the operation of our model communication system. Let A represent the events occurring in Alice's domain that lead to the preparation of her chosen signal, and let the possible choices of the message and signal be $\{a_i\}$. The probability that Alice selects and prepares the signal a_i is $P(a_i)$ and we suppose this probability is known to Bob, either by prior agreement or by earlier observation of the operation of the channel. The reception event, which we label B, is the receipt of the signal and its decoding to produce a message. The possible values of B, $\{b_j\}$, are the signals that could be received and the associated messages formed from them.

The operation of the communication channel between Alice and Bob is described by specifying the set of conditional probabilities $\{P(b_j|a_i)\}$, that is, the probabilities that any possible message, b_j, will arise given that any possible transmitted message, a_i, is sent. From Bob's perspective, of course, it is the conditional probabilities $\{P(a_i|b_j)\}$ that are important; he knows which message he has received and needs to determine the probabilities that each of the possible original messages was selected given the message he has received. If we have

The Kronecker delta δ_{ij} is the ij-element of the identity matrix. More simply, $\delta_{ij} = 1$ if $i = j$ and is zero otherwise.

$$P(a_i|b_j) = \delta_{ij} \tag{1.66}$$

then each received signal is uniquely decodable and Bob can reconstruct the original message without the possibility of error. More usually, however, noise will tend to induce errors, and it is necessary to devise methods to cope with this.

Shannon's mathematical theory of communication includes two powerful theorems: the noiseless coding theorem and the noisy-channel coding theorem. Each of these deals with redundancy in communication signals and the extent to which it needs to be included. We shall describe these theorems in some detail, but begin by explaining their qualitative features and illustrating their significance. Most messages contain an element of redundancy, by which we mean that not all of the characters are required in order to read them accurately. Try reading, for example, the message

TXT MSSGS SHRTN NGLSH SNTNCS

You probably had little difficulty in restoring the omitted vowels and in reconstructing the sentence. In this sense, the 11 missing vowels were a redundant component of the original message, and removing these has shortened the message without impairing its understandability. It is this principle, of course, that underlies the recently emerged phenomenon of text messaging. The noiseless coding theorem serves to quantify this redundancy and tells us by how much a message can be shortened or compressed and still be decoded without error.

One might wonder why language and other forms of communication include redundancy. The answer, of course, is that redundancy is used to combat errors so that messages affected by noise can still be read. As an example, try reading the message

RQRS BN MK WSAGS NFDBL

It is unlikely that you were able to make sense of this, and the reason for this is that the message has been compressed to remove much of the redundancy and then errors have been introduced on top of this (5 of the characters are incorrect). If the uncompressed message was sent, however, then the received message might be

EQRORS BAN MAKE WESAAGIS UNFEADCBLE

Reading this is possible, and this is true even though some of the redundant letters have themselves been affected by the noise. We build in redundancy in order to combat noise, and the noisy channel theorem tells us how much redundancy we need to introduce in order to faithfully reconstruct the message from the noisy signal. Both coding theorems are statistical in nature and apply strictly in the limit of long messages. They do not tell us how to reach the optimal limits they provide, but rather they give limits on the minimum message length that cannot be improved upon. We shall find that it is the information that sets these limits.

We begin with Shannon's noiseless coding theorem and, in order to appreciate its simplicity, we present a much simplified example. Consider a message comprising a sequence or string of N bits, 0 and 1, and assume that any bit will take the value 0 with probability p and the value 1 with probability $1 - p$. In this simple example we shall assume that this probability is independent of the values taken by the other bits in the string. If the string is N bits in length then there are 2^N possible

If you are struggling, the original messages were 'Text messages shorten English sentences' and 'Errors can make messages unreadable'.

Bits Common usage recognizes two distinct meanings in information theory for the word 'bit': (i) a bit is the unit of information with the logarithms expressed in base 2, and (ii) a bit is a physical system which can be prepared in one of two distinct physical states, and these are associated with the values 0 and 1. Clearly, the maximum information-carrying capacity of a bit (of the second kind) is one bit (of the first kind). Where there is likely to be any confusion we shall use the term 'binary digit' for a two-state system and reserve the term 'bit' for the unit of information: the maximum information that can be carried by one binary digit is one bit.

distinct strings and these can encode 2^N different messages. The probability that any given string is selected having n zeros and $N - n$ ones is $p^n(1-p)^{N-n}$. There are, of course, many possible strings with n zeros, and the probability that the selected string has n zeros is

$$P(n) = \frac{N!}{n!(N-n)!}p^n(1-p)^{N-n}. \tag{1.67}$$

If N is very large, corresponding to a long message, then it is overwhelmingly likely that the number of zeros will be close to Np. This is, of course, the frequency interpretation of probability; in a large sample, the fraction of zeros will be p. In the limit of very large N, therefore, we need only consider *typical* strings, that is, those for which $n \approx pN$. All other possibilities are sufficiently unlikely that they can be omitted from consideration.

If we consider only those messages for which $n = pN$ then this leaves only

$$W = \frac{N!}{n!(N-n)!} \tag{1.68}$$

distinct strings. Taking the logarithm of this and using Stirling's approximation, derived in Appendix C, gives

$$\log W \approx N \log N - n \log n - (N - n) \log(N - n)$$
$$= N\left[-p \log p - (1-p) \log(1-p)\right] = NH(p), \tag{1.69}$$

where $H(p)$ is the entropy associated with the probabilities p and $1 - p$. This gives us an estimate for W, the number of different typical messages that can be sent:

$$W = 2^{NH(p)}. \tag{1.70}$$

Hence it is only necessary to use $NH(p)$ bits rather than N bits to faithfully encode the message. It follows that the original message can be compressed by a factor of $H(p)$, in that the number of bits required can be reduced by this factor. This is Shannon's noiseless coding theorem. It is instructive to note that the single most likely bit string is *not* included in the group of typical messages.

The approximations used in reaching eqn 1.70 are rather drastic and we should consider the effect of these. There are, in particular, many more typical sequences than just those counted in eqn 1.68 in which n differs from pN by only a small number. Including these leads to an increased number of typical messages,

$$W = 2^{N[H(p)+\delta]}, \tag{1.71}$$

where δ tends to zero as $N \to \infty$. It follows that the compressed message will have $N[H(p) + \delta]$ bits and so the required number of bits is reduced by the factor $H(p) + \delta$, and this factor tends to $H(p)$ in the large-N limit. In practice, the required number of bits will exceed $NH(p)$ but the number of additional bits will grow more slowly than linearly in N. The factor by which the number of bits can be reduced cannot exceed $H(p)$

if the message is to be read. More generally, if Alice sends a string of N possible symbols, each of which takes one of the values $\{a_1, a_2, \cdots, a_m\}$, and if the probability for any given symbol is independent of the others, then the number of possible sequences is $2^{N \log m}$. If the probabilities for the m symbols are $\{P(a_i)\}$ then the $N \log m$ bits can be reduced but their number must be greater than or equal to $NH(A)$.

The key idea in compressing a message is to use short sequences of bits to represent commonly occurring symbols and longer sequences for less common ones. In this way, the average length of the bit string is kept as short as possible. An example given by Shannon has a source which produces a sequence of letters chosen from the short alphabet A, B, C, and D. Each symbol is selected independently of the others and the four letters have the probabilities $\frac{1}{2}$, $\frac{1}{4}$, $\frac{1}{8}$, and $\frac{1}{8}$, respectively. The simplest coding scheme would be to use two bits for each of the letters (in the form 00, 01, 10, and 11) so that a sequence on N letters would require $2N$ bits. The entropy associated with the given probabilities is

$$H = - \left(\frac{1}{2} \log \frac{1}{2} + \frac{1}{4} \log \frac{1}{4} + \frac{2}{8} \log \frac{1}{8} \right)$$
$$= \frac{7}{4} \quad \text{bits} \tag{1.72}$$

and this gives the Shannon limit of $\frac{7}{4}N$ bits for a sequence of N letters. We can achieve this limit by encoding the letters as

$$A = 0,$$
$$B = 10,$$
$$C = 110,$$
$$D = 111. \tag{1.73}$$

The average number of bits used to encode a sequence of N letters is then

$$N \left(\frac{1}{2} \times 1 + \frac{1}{4} \times 2 + \frac{2}{8} \times 3 \right) = \frac{7}{4}N, \tag{1.74}$$

which achieves the Shannon limit for noiseless coding. Shannon's theorem tells us, moreover, that this coding sequence is optimal and that there is no hope of devising a more efficient scheme. Note that the zeros and ones occur with equal probability with this coding, so that each binary digit carries its maximum information load of one bit. That the above problem has a simple optimal strategy is somewhat fortuitous. For most situations no optimal strategy is known and the Shannon theorem does not tell us how to construct one. Nevertheless, modern information theory has provided a variety of methods to tackle the problem of efficient coding.

Shannon's noiseless coding theorem tells us how much redundancy exists in a message and limits the extent to which a string can be compressed. His noisy-channel coding theorem tells us how much redundancy we need to incorporate into the message in order to correct all of the errors induced by any noise present in the channel. We begin by

Correlated strings Most messages include correlations between the symbols. For example, in English the presence of the letter 'q' means that it is highly likely that the following letter will be 'u'. Such correlations can be included in Shannon's theory by considering the states of the source $\{c_i\}$ and within these the probabilities for the individual symbols. In this case it is the state of the source that determines the probabilities for the letters, and we need to consider the effects of changes in the state of the source. Shannon's limit for compression still applies but the relevant entropy is now $H(A|C)$.

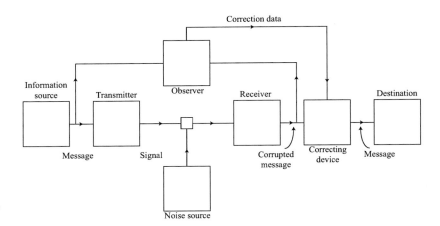

Fig. 1.10 Scheme for correcting the errors on a noisy communication channel.

considering a message comprising N_0 bits and suppose that it has been compressed optimally so that no redundancy remains. This means that all 2^{N_0} messages can be transmitted and that the probability for each of these is 2^{-N_0}. Suppose that the communication channel is noisy so that each of the possible values, 0 or 1, for each bit can be flipped, into 1 or 0, respectively, inducing an error, and let the probability that any single bit value is flipped be q. It follows that an average qN_0 of the bits received by Bob will carry the incorrect value. The arguments employed in obtaining the noiseless coding theorem tell us that for N_0 sufficiently large, the number of errors will be very close to qN_0.

In order to correct the errors, it suffices to tell Bob the positions in the string at which each error has occurred. He can then correct these errors by flipping the bits at these locations. To achieve this, we can postulate the existence of an observer equipped with a separate channel to Bob. This situation is depicted in Fig. 1.10. The number of possible ways in which qN_0 errors can be distributed among the N_0 message bits is

$$E = \frac{N_0!}{(qN_0)!(N_0 - qN_0)!}. \tag{1.75}$$

The correction channel needs to tell Bob where all of the errors are and so needs to comprise at least $\log E$ bits. Using, once again, Stirling's approximation gives

$$\log E \approx N_0 \left[-q \log q - (1 - q) \log(1 - q) \right] = N_0 H(q), \tag{1.76}$$

where $H(q)$ is the entropy associated with the probability for a single bit error. It follows that at least $N_0 \left[1 + H(q) \right]$ bits are required in the combined original signal and correction channels if the corrupted message received by Bob is to be corrected. This quantity is not yet the result we are seeking, as it relies on the existence of a perfect, that is, noiseless, correction channel. If the correction channel is itself noisy, with single-bit errors occurring with probability q, then we can correct the approximately $qN_0 H(q)$ errors on the correction channel with a second observer and correction channel if this second channel carries a minimum of $N_0 H^2(q)$ bits. Correcting for errors on this channel with a third observer and correction channel and then introducing a fourth observer and so on leads to the total number of required bits:

$$N = \frac{N_0}{1 - H(q)}. \tag{1.77}$$

We have chosen each of the correction channels to have the same error probability as the original channel and hence we can replace the entire construction by the *original noisy channel* if we can encode the messages in such a way that Bob can associate each likely, or typical, noisy message with a unique original message. The above analysis shows that for this to be possible we require at least $N = N_0/\left[1 - H(q)\right]$ bits in order to faithfully encode 2^{N_0} messages. Equivalently, N bits can be used to carry faithfully not more than $2^{N[1-H(q)]}$ distinct messages. It follows that, in the large-N limit, each binary digit can carry not more than

$$\frac{\log \left(2^{N[1-H(q)]}\right)}{N} = 1 - H(q) \tag{1.78}$$

bits of information. This result is Shannon's noisy-channel coding theorem. If $q = 0$, so that no errors occur, then each binary digit can carry up to one bit of information. If $q = \frac{1}{2}$, however, then errors occur too frequently to be corrected, and no information can be carried by the channel as $H\left(\frac{1}{2}\right) = 1$. Note that coping with noise, in limiting the information per bit, has reduced the number of possible messages that can be encoded in comparison with a noiseless channel. Here the number of messages that can be reliably encoded in N bits is reduced by at least the factor $2^{NH(q)}$. We can understand this quite simply in terms of the number of readily distinguishable bit sequences. The likely number of errors means that a sequence of N bits will, with very high probability, be transformed into one of about $2^{NH(q)}$ sequences. We can combat efficiently the effects of noise by selecting $2^{N[1-H(q)]}$ messages that are sufficiently different that the likely number of errors will not cause any confusion between them.

Shannon's noisy coding theorem, like its noiseless counterpart, applies to more general situations than the simple binary channel described above. In particular, let Alice's message be encoded in a string of N symbols, with the values $\{a_i\}$, and let the probability for each symbol, $\{P(a_i)\}$, be independent of the others. The symbols received by Bob can take the values $\{b_j\}$ and the channel properties determine the set of conditional probabilities $P(b_j|a_i)$. Alice can produce about $2^{NH(A)}$ different likely messages and, similarly, Bob can receive about $2^{NH(B)}$ likely strings. Each of the likely received strings can be produced by about $2^{NH(A|B)}$ of the likely messages or, equivalently, each message can produce $2^{NH(B|A)}$ different received strings. It follows that the number of messages that can be sent through the channel and, with high probability, be accurately reconstructed by Bob is limited by

$$2^{N[H(B)-H(B|A)]} = 2^{N[H(A)-H(A|B)]} = 2^{NH(A:B)}. \tag{1.79}$$

It is the mutual information, $H(A : B)$, which determines the number of distinct messages that can be sent and hence the quality of the communication channel. A large value of the mutual information indicates a high degree of correlation between A, the symbol selected for transmission, and B, the received symbol. It is convenient to define the *capacity* of the channel, C, as the maximum value of the mutual information:

$$C = \text{Sup } H(A : B). \qquad (1.80)$$

Here the maximization is performed by selecting the probabilities $\{P(a_i)\}$ for the transmitted symbols $\{a_i\}$. The noisy-channel coding theorem then tells us that our N symbols can reliably encode up to 2^{NC} distinct messages and no more than this.

The simplest example of a noisy channel is a binary symmetric channel, in which two input symbols, a_1 and a_2, produce two output symbols, b_1 and b_2. The channel faithfully maps a_1 onto b_1 and a_2 onto b_2 with probability $1 - q$ and induces an error with probability q (as depicted in Fig. 1.11). The corresponding mutual information is

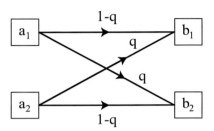

Fig. 1.11 Conditional probabilities $P(b_j|a_i)$ for a binary symmetric channel.

$$
\begin{aligned}
H(A : B) = &- [P(a_1)(1 - q) + P(a_2)q] \log [P(a_1)(1 - q) + P(a_2)q] \\
&- [P(a_2)(1 - q) + P(a_1)q] \log [P(a_2)(1 - q) + P(a_1)q] \\
&+ q \log q + (1 - q) \log(1 - q).
\end{aligned}
\qquad (1.81)
$$

It is straightforward to show that this quantity is maximized by choosing $P(a_1) = \frac{1}{2} = P(a_2)$, to give the channel capacity

$$C = 1 - H(q), \qquad (1.82)$$

which is the value implied by eqn 1.77.

The noisy-channel coding theorem does not tell us how to approach the limit given by the channel capacity, but we can see how to do this in some special cases. One simple example has four possible input symbols, $a_1, a_2, a_3,$ and a_4, and four corresponding output symbols, $b_1, b_2, b_3,$ and b_4. Consider a channel in which each input symbol gives the corresponding output symbol with probability $\frac{1}{2}$:

$$P(b_i|a_i) = \frac{1}{2}, \qquad i = 1, \cdots, 4; \qquad (1.83)$$

the remaining possibilities are that noise can cause $a_1, a_2, a_3,$ and a_4 to be read as $b_2, b_3, b_4,$ and b_1, respectively:

$$P(b_2|a_1) = P(b_3|a_2) = P(b_4|a_3) = P(b_1|a_4) = \frac{1}{2}. \qquad (1.84)$$

The channel conditional probabilities are depicted in Fig. 1.12. We obtain the channel capacity by maximizing the mutual information,

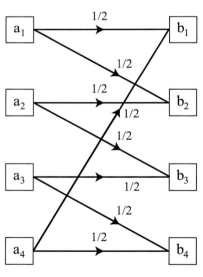

Fig. 1.12 Conditional probabilities $P(b_j|a_i)$ for an example of a noisy channel.

$$
\begin{aligned}
H(A : B) &= H(B) - H(B|A) \\
&= - \sum_{ij} P(a_i)P(b_j|a_i) \log \sum_{k} P(a_k)P(b_j|a_k) \\
&\quad + \sum_{ij} P(a_i)P(b_j|a_i) \log P(b_j|a_i),
\end{aligned}
\qquad (1.85)
$$

by varying the input symbol probabilities $\{P(a_i)\}$. For our problem, this mutual information reduces to

$$H(A:B) = -\frac{1}{2}\left(P(a_1) + P(a_2)\right)\log\left[\frac{1}{2}\left(P(a_1) + P(a_2)\right)\right]$$
$$-\frac{1}{2}\left(P(a_2) + P(a_3)\right)\log\left[\frac{1}{2}\left(P(a_2) + P(a_3)\right)\right]$$
$$-\frac{1}{2}\left(P(a_3) + P(a_4)\right)\log\left[\frac{1}{2}\left(P(a_3) + P(a_4)\right)\right]$$
$$-\frac{1}{2}\left(P(a_4) + P(a_1)\right)\log\left[\frac{1}{2}\left(P(a_4) + P(a_1)\right)\right]$$
$$-1. \tag{1.86}$$

The terms $\frac{1}{2}(P(a_i) + P(a_j))$ are all greater than or equal to zero and sum to unity, and so the first four terms together are mathematically an entropy. It follows that the maximum possible value occurs when all of these terms are equal, and this gives a channel capacity of 1 bit. We can readily see how this one bit per symbol can be realized in this case. If we transmit using only the symbols a_1 and a_3 then the corresponding possible outputs are b_1 and b_2 or b_3 and b_4, respectively, and Bob can determine the selected symbol without error. If Alice encodes the message in such a way that $P(a_1) = \frac{1}{2} = P(a_3)$ then each symbol conveys the limiting value of 1 bit. Alternatively, of course, Alice can use just the symbols a_2 and a_4.

One way of understanding the noisy-channel coding theorem is to suppose that we encode different messages as bit strings, or *codewords*, which are sufficiently different that the expected number of errors is unlikely to introduce any difficulty for Bob in recovering the original message. An efficient coding scheme will be one in which the messages are as close, or similar, to each other as possible but are still distinguishable after passage through the noisy channel. It is helpful to be able to quantify the degree of difference between a pair of bit strings, and this is conveniently expressed as the Hamming distance. If x and y are each strings of N bits then the Hamming distance between the strings, $d_H(x, y)$, is simply the number of bits in the two strings which are different. For example, if $x = 0101$ and $y = 1101$ then $d_H(x, y) = 1$, while if $z = 0110$ then $d_H(x, z) = 2$ and $d_H(y, z) = 3$. The Hamming distance satisfies the natural conditions for a distance and, in particular,

$$d_H(x, y) \geq 0,$$
$$d_H(x, y) = 0 \quad \Leftrightarrow \quad x = y,$$
$$d_H(x, y) = d_H(y, x),$$
$$d_H(x, y) \leq d_H(x, z) + d_H(y, z). \tag{1.87}$$

Bob's ability to detect errors and to correct them is limited by the minimum Hamming distance between any pair of codewords. Consider, for example, a binary [7,4] Hamming code in which one of 2^4 distinct messages (corresponding to 4 bits) is encoded on 7 bits using one of the set of 16 codewords

$$C = \{0000000, 0001011, 0010101, 0011110, 0100111, 0101100,$$
$$0110010, 0111001, 1000110, 1001101, 1010011, 1011000,$$
$$1100001, 1101010, 1110100\ 1111111\}. \qquad (1.88)$$

A casual inspection reveals that the first four bits in each codeword correspond to the 2^4 integers from 0000 to 1111. The remaining three bits are selected to allow for error detection and correction. They are selected so that in each codeword the sum of the first, second, third, and fifth bit values is even, the sum of the first, second, fourth, and sixth bit values is even, and the sum of the second, third, fourth, and seventh bit values is even. If a single error occurs then this coding scheme allows us to correct it without ambiguity. Suppose, for example, that Bob receives the bit string 0100100. This is not one of the codewords eqn 1.88 but Bob can check the sums of the bits mentioned above and find $0+1+0+1 = 2$, $0+1+0+0 = 1$, and $1+0+0+0 = 1$. The assumption of only a single error tells us that it is the fourth bit that is incorrect, and so Bob recovers the codeword 0101100 and from this the protected message string 0101.

Alternative [7,4] Hamming codes
The Hamming code presented here is certainly not unique, and other forms are used. One of these selects the three additional bits such that the sums of bits 2, 3, 4, and 5, of bits 1, 3, 4, and 6, and of bits 1, 2, 4, and 7 are each even. There are, of course, 4! possible distinct [7,4] Hamming codes.

If the minimum Hamming distance for any pair of codewords is d then Bob will be able to detect $d-1$ errors and to correct up to $\frac{1}{2}(d-1)$ errors. For a binary [7,4] Hamming code, we have $d = 3$ and so Bob can detect the presence of up to two errors and, as we have seen, he can correct any single error.

Suggestions for further reading

Box, G. E. P. and Tiao, G. C. (1973). *Bayesian inference in statistical analysis.* Wiley, New York.

Brillouin, L. (1956). *Science and information theory.* Academic Press, New York.

Cover, T. M. and Thomas, J. A. (1991). *Elements of information theory.* Wiley, New York.

Goldie, C. M. and Pinch, R. G. E. (1991). *Communications theory.* Cambridge University Press, Cambridge.

Jaynes, E. T. (2003). *Probability theory: the logic of science.* Cambridge University Press, Cambridge.

Khinchin, A. I. (1957). *Mathematical foundations of information theory.* Dover, New York.

Kullback, S. (1968). *Information theory and statistics.* Dover, New York.

Lee, P. M. (1989). *Bayesian statistics: an introduction.* Edward Arnold, London.

Leff, H. S. and Rex, A. F. (eds) (1990). *Maxwell's demon: entropy, information, computing.* Adam Hilger, Bristol.

Leff, H. S. and Rex, A. F. (eds) (2003). *Maxwell's demon 2: entropy, classical and quantum information, computing.* Institute of Physics Publishing, Bristol.

Loepp, S. and Wootters, W. K. (2006). *Protecting information: from classical error correction to quantum cryptography.* Cambridge University Press, Cambridge.

Mlodinow, L. (2008). *The drunkard's walk: how randomness rules our lives.* Allen Lane, London.

Plenio, M. B. and Vitelli, V. (2001). The physics of forgetting: Landauer's erasure principle and information theory. *Contemporary Physics* **42**, 25.

Shannon, C. E. and Weaver, W. (1963). *The mathematical theory of communication.* University of Illinois Press, Urbana, Il.

Sivia, D. S. (1996). *Data analysis: a Bayesian tutorial.* Oxford University Press, Oxford.

Exercises

(1.1) Show that $P(a_i, b_j) \leq P(a_i), P(b_j)$. What can we infer if $P(a_i, b_j) = P(a_i)$?

(1.2) Two events A and B are strongly correlated so that $P(a_i|b_j) = \delta_{ij}$. Does it follow that $P(b_j|a_i) = \delta_{ij}$?

(1.3) Prepare a probability tree equivalent to that given in Fig. 1.2 but with the first set of lines corresponding to the possible values of B and the second set to the values of A.

(1.4) In Section 1.2, we illustrated Bayes' theorem with an example based on long and short routes to work and the probability of arriving on time. Complete that analysis by using Bayes' theorem to calculate the probabilities that the long and short routes were taken given that I arrived late.

(1.5) A particle counter records counts with an efficiency η. This means that each particle is detected with probability η and missed with probability $1-\eta$. Let N be the number of particles present and n be the number detected. Show that

$$P(n|N) = \frac{N!}{n!(N-n)!}\eta^n(1-\eta)^{N-n}.$$

Calculate $P(N|n)$:

(a) for

$$P(N) = e^{-\bar{N}}\frac{\bar{N}^N}{N!};$$

(b) for all $P(N)$ equally probable;

(c) (trickier) given only that the mean number of particles present is $\bar{N}$.

(1.6) Use Bayes' theorem to express $P(a_i|b_j, c_k)$ in terms of $P(c_k|a_i, b_j)$.

(1.7) Prove that the likelihood $\ell(a_i|b_j)$, as introduced in eqn 1.21, is indeed symmetric in its arguments.

(1.8) For the mouse genetics problem, derive a general expression for the probability $P(BB|x_1, \cdots, x_n =$ black) that the test mouse is BB given that all of n offspring are black.

(1.9) Three friends play a game by each in turn tossing a coin. The winner is the first to throw a head. Amy goes first, then Barbara, and then Claire and they continue playing in this order until there is a winner. What are the probabilities that each of the players will win?

(1.10) On a game show, contestants were presented with three boxes, one of which contained the star prize. They were asked to choose one (and so had a probability of $\frac{1}{3}$ of winning the prize). The host would then open one of the two other boxes, showing the contestant that it did not contain the prize. At this stage, the contestant would be offered the chance to either keep their originally selected box or to exchange it for the one remaining unopened box. In practice, most contestants stuck to their original choice, but what would a good Bayesian do?

(1.11) (a) There are two children, at least one of whom is a boy. What is the probability that they are both boys?
       (b) There are two children, at least one of whom is a boy called Reuben. What is the probability that they are both boys?

       (*Warning:* this subtle problem, posed by Mlodinow, will repay careful thought.)

(1.12) Suppose that there are three people forming a team to play a game. They can discuss their strategy in advance but cannot communicate further once the game has started. Each player is then given a playing card, which is equally likely to be red or black. They each press their card against their forehead so that the other players can see it but they cannot. Each player is then asked to guess whether their own card is red or black by writing down (but not revealing to the other players) either 'red', 'black' or 'pass' (the last of which corresponds to declining to make a guess). The team wins if at least one player makes a correct guess and if no player makes an incorrect guess. Agreeing that one player will make a guess wins with probability $\frac{1}{2}$. There is, however, a strategy that wins with probability $\frac{3}{4}$. Can you find it and explain why it works?

(1.13) Prove the identity in eqn 1.33 and hence show that 1 bit ≈ 0.693 nats.

(1.14) Confirm that the entropy in eqn 1.36 has a single maximum at $p = \frac{1}{2}$ and takes its minimum value at $p = 0$ and at $p = 1$.

(1.15) Prove eqn 1.38 by induction from eqn 1.37.

(1.16) Prove that if $f(x)$ has a second derivative that is greater than or equal to zero everywhere then it is a convex function; that is, it obeys the inequality in eqn 1.37.

       [*Hint:* it might be useful to recall that Taylor's theorem implies that

       $$f(x) = f(x_0) + f'(x_0)(x - x_0) + \frac{f''(\tilde{x})}{2}(x - x_0)^2,$$

       where $\tilde{x}$ lies between x_0 and x.]

(1.17) Prove that

       $$\lim_{x \to 0} x \log x = 0$$

       and hence that $H = 0$ if and only if one of the probabilities is unity. Show that in all other cases H is positive.

(1.18) Prove that if A has n possible values then $H(A)$ takes its maximum value $(\log n)$ when the n associated probabilities are all $\frac{1}{n}$.

(1.19) Prove that changing the probabilities according to eqn 1.39 increases $H(A)$ or leaves it unchanged. Under what conditions will the transformation leave $H(A)$ unchanged?

(1.20) The entropy H can sometimes be difficult to work with, and perhaps for this reason a number of alternative 'entropies' have been introduced. Amongst these are the linear entropy L, the Tsallis entropy T, and the Rényi entropy R. For a set of probabilities $\{p_i\}$, these take the forms

       $$L = \sum_i p_i(1 - p_i),$$

       $$T = \frac{1 - \sum_i p_i^q}{q - 1},$$

       $$R = \frac{\ln\left(\sum_i p_i^q\right)}{1 - q}.$$

       The latter two are defined in terms of a real parameter q, the value of which exceeds unity.

       (a) Show that, like H, each of these is greater than or equal to zero and takes the value zero only if a single probability is unity.
       (b) Show that, also like H, each of these takes its maximum value if all of the probabilities p_i are equal.
       (c) Show that $H_e \geq L$, with the equality holding only when $H_e = 0$.
       (d) Show that

       $$\lim_{q \to 1} T = H_e = \lim_{q \to 1} R.$$

(1.21) If we have three events $A, B,$ and C, show that

       $$H(A) + H(B) + H(C) \geq H(A, B, C).$$

       Under what conditions will the equality hold?

(1.22) Show that if we have three events A, B, and C, then

       $$H(A, B, C) + H(B) \leq H(A, B) + H(B, C).$$

       [*Hint:* you might proceed by first showing that

       $$Q(a_i, b_j, c_k) = \frac{P(a_i, b_j)P(b_j, c_k)}{P(b_j)}$$

       is a mathematically acceptable probability distribution and then use the positivity of the relative entropy.]

(1.23) Show that

       $$H(A : B) = \sum_{ij} P(a_i, b_j) \log \ell(a_i|b_j).$$

(1.24) Suppose that N identical balls are shared between two boxes A and B subject to a given probability distribution. For which probability distribution will $H(A : B)$ take its maximum value, and what is this value? What will be the maximum value of $H(A : B)$ if the balls are distinguishable?

(1.25) An average-die is a six-sided die with the integers 2, 3, 3, 4, 4, and 5 displayed on its faces. We are told that the average score is 3.52. Use the Max Ent method to choose prior probabilities for each of the possible scores.

(1.26) Consider two properties C and D, which can take the values $\{c_i\}$ and $\{d_j\}$, respectively. Given the probabilities $\{P(c_i)\}$ and $\{P(d_j)\}$, use the Max Ent method to suggest values for the joint probabilities $\{P(c_i, d_j)\}$. You should find that the method suggests that C and D should be considered as independent so that $P(c_i, d_j) = P(c_i)P(d_j)$.

(1.27) For the Boltzmann distribution given in eqn 1.61 find the value of the Lagrange multiplier β in terms of the mean energy $\bar{E}$:

(a) for a system with two possible configurations with energies E_0 and E_1

(b) for a quantum harmonic oscillator, which has allowed energies $E_n = (n + \frac{1}{2})\hbar\omega$, $n = 0, 1, 2, \cdots$.

1.28) The thermodynamic entropy is

$$S = -k_B \sum_n p_n \ln p_n,$$

where the probabilities are given in eqn 1.61. Show that the Lagrange multiplier β is indeed $(k_B T)^{-1}$. [*Hint:* It might be useful to recall the thermodynamic relationship

$$\frac{d\bar{E}}{dT} = T\frac{dS}{dT}.]$$

1.29) Repeat the analysis of Szilard's model given in Section 1.3 for the case in which the partition, when initially inserted, divides the volume V_0 into unequal volumes V_1 and $V_0 - V_1$.

1.30) Each bit in a string of N is independent of the others and takes the value 0 with probability p and the value 1 with probability $1 - p$. Show that the fraction of bits taking the value zero will be $p \pm \epsilon$ where ϵ tends to zero as $N \to \infty$.

1.31) The coding scheme in eqn 1.73 uses one string of a single bit, one of two bits and two of three bits in order to encode the four letters A, B, C, and D.

Why can we not use two single bits for the letters A and B, or perhaps just two bits for the letters C and/or D?

(1.32) If the probabilities for the four letters A, B, C, and D in the above problem were 0.49, 0.21, 0.21, and 0.09, respectively, then what would be the Shannon limit for compression of a sequence of N letters? How close to this does the coding scheme in eqn 1.73 get?

(1.33) Suppose that a source produces a sequence of letters A and B with probabilities p and $1 - p$, respectively. If $p \ll 1$ show that $H \approx p\log(e/p)$. Design a coding scheme that provides near-optimal compression.

(1.34) (a) Calculate the entropy for a sequence of letters occurring with same frequencies as in the English language. These frequencies are given in Table 3.1.

(b) Morse code replaces each of the letters by a binary sequence of dots and dashes:

a $= \cdot-$	j $= \cdot---$	s $= \cdots$
b $= -\cdots$	k $= -\cdot-$	t $= -$
c $= -\cdot-\cdot$	l $= \cdot-\cdot\cdot$	u $= \cdots-$
d $= -\cdot\cdot$	m $= --$	v $= \cdots-$
e $= \cdot$	n $= -\cdot$	w $= \cdot--$
f $= \cdots-\cdot$	o $= ---$	x $= -\cdot\cdot-$
g $= --\cdot$	p $= \cdot--\cdot$	y $= -\cdot--$
h $= \cdots\cdot$	q $= ---\cdot-$	z $= --\cdots$
i $= \cdots$	r $= \cdot-\cdot$	

Calculate the average length (in bits) of each letter in Morse code.

(c) Given the answers to parts (a) and (b), how is it that communication by Morse code is possible?

(1.35) Prove that correcting all of the errors by a sequence of observers and correction channels requires, in the limit of long messages, a minimum number of bits given by eqn 1.77.

(1.36) We have seen that the maximum information that can be carried by each bit in a noisy channel is given by eqn 1.78. Why is this zero for $q = \frac{1}{2}$ but positive for higher error probabilities?

(1.37) A measure of the required redundancy for a noisy channel is $N/N_0 - 1$, with N and N_0 related by eqn 1.77. This quantity is the number of extra bits required per message. Plot a graph of this quantity and hence show that efficient coding requires less than about a 10% error rate.

(1.38) Calculate the conditional probabilities $P(a_i|b_j)$ for the binary symmetric channel depicted in Fig. 1.11. Hence confirm that the mutual information is as given in eqn 1.81.

(1.39) A channel has three input symbols (a_1, a_2, a_3) and three output symbols (b_1, b_2, b_3) and the non-zero conditional probabilities are

$$P(b_1|a_1) = 1,$$
$$P(b_2|a_2) = 1 - q = P(b_3|a_3),$$
$$P(b_2|a_3) = q = P(b_3|a_2).$$

(a) Calculate the channel capacity.
(b) Suggest how the channel capacity might be realized for the cases $q = 0$, $q = \frac{1}{2}$, and $q = 1$.

(1.40) It is by no means necessary for a channel to have the same number of input and output symbols. Consider a channel with two input symbols (a_1, a_2) and four output symbols (b_1, b_2, b_3, b_4). The conditional probabilities are

$$P(b_1|a_1) = P(b_2|a_1) = \frac{1}{3},$$
$$P(b_3|a_2) = P(b_4|a_2) = \frac{1}{3},$$
$$P(b_3|a_1) = P(b_4|a_1) = \frac{1}{6},$$
$$P(b_1|a_2) = P(b_2|a_2) = \frac{1}{6}.$$

Calculate the associated channel capacity.

(1.41) A message comprises blocks of seven binary digits. A channel introduces noise in such a way that each block experiences either no errors, which occurs with probability $\frac{1}{8}$, or precisely one error, with each of the seven digits being equally likely to be incorrect (i.e. with probability $\frac{1}{8}$).

(a) Calculate the channel capacity for this system
(b) Can you suggest a coding scheme which realizes this channel capacity?

(1.42) Prove the properties of the Hamming distance, given in eqn 1.87.

(1.43) Recover the original four-bit messages from the following *received* signals protected using the binary [7,4] Hamming code given in eqn 1.88:

(a) 1110000;
(b) 1000101;
(c) 0100110;
(d) 0110010.

(1.44) A set of codewords have a minimum Hamming distance d. Prove that Bob will be able to detect up to $d - 1$ errors and to correct up to $\frac{1}{2}(d - 1)$ errors.

Elements of quantum theory

2

We have seen that there is an intimate relationship between probability and information. The values we assign to probabilities depend on the information available, and information is a function of probabilities. This connection makes it inevitable that information will be an important concept in any statistical theory, including thermodynamics and, of course, quantum physics.

The probabilistic interpretation of quantum theory has probability amplitudes rather than probabilities as the fundamental quantities. This feature, together with the associated superposition principle, is responsible for intrinsically quantum phenomena and gives quantum information theory its distinctive flavour. We shall see that the quantum rules for dynamical evolution and measurement, together with the existence of entangled states, have important implications for quantum information. They also make it possible to perform tasks which are either impractical or impossible within the classical domain. In describing these we shall make extensive use of simple but fundamental ideas in quantum theory. This chapter introduces the mathematical description of quantum physics and the concepts which will be employed in our study of quantum information.

2.1 Basic principles

The state of a physical system in quantum theory is completely specified by its state vector, the ket $|\psi\rangle$. It is common practice to refer to $|\psi\rangle$ as the state of the system. If $|\psi_1\rangle$ and $|\psi_2\rangle$ are possible states then the superposition

$$|\psi\rangle = a_1|\psi_1\rangle + a_2|\psi_2\rangle \tag{2.1}$$

is also a state of the system, where a_1 and a_2 are complex numbers. This superposition principle, which states that any superposition of states is also a state, is perhaps the most fundamental concept in quantum theory. We can trace the existence of probability amplitudes, incompatible observables, and entanglement back to it.

The bra $\langle\psi|$ provides an equivalent representation of the state in eqn 2.1 in the form

$$\langle\psi| = a_1^*\langle\psi_1| + a_2^*\langle\psi_2|, \tag{2.2}$$

where a_1^* and a_2^* are the complex conjugates of a_1 and a_2, respectively. We obtain numbers, including probability amplitudes and probabilities, by forming the overlap between, or inner product of, pairs of states. The overlap between the states $|\psi\rangle$ and $|\phi\rangle$ is the complex number $\langle\psi|\phi\rangle$ or its complex conjugate $\langle\phi|\psi\rangle$, analogous to the scalar or dot product of two vectors. If this overlap is zero, then the states are said to be orthogonal, in analogy with a pair of perpendicular vectors, which have zero scalar product. The inner product of a state with itself is real and strictly positive so that

$$\langle\psi|\psi\rangle > 0. \tag{2.3}$$

If this inner product is unity, so that $\langle\psi|\psi\rangle = 1$, then the state is said to be normalized. In practice, we can always normalize any state by multiplying it by $\langle\psi|\psi\rangle^{-1/2}$.

If the states in eqn 2.1 are orthonormal, that is, both orthogonal ($\langle\psi_1|\psi_2\rangle = 0$) and normalized ($\langle\psi_1|\psi_1\rangle = 1 = \langle\psi_2|\psi_2\rangle$), then the amplitudes a_1 and a_2 are given by the overlaps

$$\langle\psi_1|\psi\rangle = a_1 = \langle\psi|\psi_1\rangle^*,$$
$$\langle\psi_2|\psi\rangle = a_2 = \langle\psi|\psi_2\rangle^*. \tag{2.4}$$

If $|\psi\rangle$ is itself normalized, then $|a_1|^2 + |a_2|^2 = 1$ and we interpret $|a_1|^2$ and $|a_2|^2$ as the probabilities that a suitable measurement will find the system to be in the states $|\psi_1\rangle$ and $|\psi_2\rangle$, respectively. The generalization of eqn 2.1 to many possible states $|\psi_n\rangle$ is

$$|\psi\rangle = \sum_n a_n|\psi_n\rangle, \tag{2.5}$$

where, if $|\psi\rangle$ is normalized and the states $|\psi_n\rangle$ are orthonormal, then

$$\sum_n |a_n|^2 = 1. \tag{2.6}$$

This is consistent with the probability interpretation that $|a_n|^2$ is the probability that a suitable measurement will find a system prepared in the state $|\psi\rangle$ to be in the state $|\psi_n\rangle$.

The description of a quantum system is completed by the introduction of operators. An operator $\hat{A}$ acting on any state of this system produces another state $\hat{A}|\psi\rangle$, which, in general, will not be normalized. The Hermitian conjugate $\hat{B}^\dagger$ of an operator $\hat{B}$ is defined by the requirement that for any pair of states $|\psi\rangle$ and $|\phi\rangle$, we have

$$\langle\psi|\hat{B}^\dagger|\phi\rangle = \langle\phi|\hat{B}|\psi\rangle^*. \tag{2.7}$$

It is straightforward to show that the Hermitian conjugate has the following properties:

$$(\hat{B}^\dagger)^\dagger = \hat{B}, \tag{2.8}$$
$$(\hat{B} + \hat{C})^\dagger = \hat{B}^\dagger + \hat{C}^\dagger, \tag{2.9}$$
$$(\hat{B}\hat{C})^\dagger = \hat{C}^\dagger\hat{B}^\dagger, \tag{2.10}$$
$$(\lambda\hat{B})^\dagger = \lambda^*\hat{B}^\dagger, \tag{2.11}$$

Hermitian conjugates   If the operator is written as a matrix then the Hermitian conjugation comprises taking the complex conjugate of the transpose of the matrix. For example, for an operator represented by a two-by-two matrix, we have

$$\hat{B} = \begin{pmatrix} b_{00} & b_{01} \\ b_{10} & b_{11} \end{pmatrix},$$
$$\hat{B}^\dagger = \begin{pmatrix} b_{00}^* & b_{10}^* \\ b_{01}^* & b_{11}^* \end{pmatrix}.$$

where $\hat{C}$ is any other operator and λ is any complex number. An operator $\hat{A}$ that is its own Hermitian conjugate, $\hat{A}^\dagger = \hat{A}$, is said to be a Hermitian operator. These are especially important, as observable quantities, or observables, are associated with Hermitian operators. Strictly speaking, of course, observables are represented by self-adjoint operators, but the distinction between Hermitian and self-adjoint operators will not be important for the subject of this book. The eigenvalues λ_n of a Hermitian operator $\hat{A}$ satisfy the eigenvalue equation

$$\hat{A}|\lambda_n\rangle = \lambda_n|\lambda_n\rangle, \qquad (2.12)$$

where the $|\lambda_n\rangle$ are the eigenstates. The conjugate equation with λ_n replaced by λ_m is

$$\langle\lambda_m|\hat{A}^\dagger = \langle\lambda_m|\hat{A} = \lambda_m^*\langle\lambda_m|, \qquad (2.13)$$

where the Hermiticity of $\hat{A}$ means that $\hat{A}^\dagger = \hat{A}$. If we take the overlap of eqn 2.13 with $|\lambda_n\rangle$, we find

$$\langle\lambda_m|\hat{A}|\lambda_n\rangle = \lambda_m^*\langle\lambda_m|\lambda_n\rangle \qquad (2.14)$$

and, similarly, taking the overlap of eqn 2.12 with $\langle\lambda_m|$ gives

$$\langle\lambda_m|\hat{A}|\lambda_n\rangle = \lambda_n\langle\lambda_m|\lambda_n\rangle. \qquad (2.15)$$

Subtracting eqn 2.15 from eqn 2.14 gives

$$(\lambda_m^* - \lambda_n)\langle\lambda_m|\lambda_n\rangle = 0, \qquad (2.16)$$

so that if $m = n$, we see from eqn 2.3 that $\lambda_n^* - \lambda_n = 0$ and the eigenvalues must be real. If, however, $\lambda_m \neq \lambda_n$ then the states $|\lambda_m\rangle$ and $|\lambda_n\rangle$ are orthogonal and may be chosen to be orthonormal with

$$\langle\lambda_m|\lambda_n\rangle = \delta_{mn}, \qquad (2.17)$$

where δ_{mn} is the Kronecker delta. Hermitian operators therefore have real eigenvalues associated with orthonormal eigenstates.

An ideal measurement of the observable A associated with $\hat{A}$ will yield as its result one of these real eigenvalues of $\hat{A}$. If the normalized state is

$$|\psi\rangle = \sum_n a_n|\lambda_n\rangle, \qquad (2.18)$$

then the probability that a measurement of A gives the result λ_n is $|a_n|^2$. If two orthonormal eigenstates $|\lambda_n\rangle$ and $|\lambda_m\rangle$ have the same eigenvalue λ then the states are said to be degenerate, and the probability of obtaining the result λ is $|a_n|^2 + |a_m|^2$. If all possible states of the system can be expressed in the form of eqn 2.18 then the set $\{|\lambda_n\rangle\}$ is said to be complete. The mean value $\bar{A}$ of A found from measurements on an ensemble of identically prepared systems is the expectation value of $\hat{A}$, given by

$$\bar{A} = \langle\hat{A}\rangle = \sum_n \lambda_n|a_n|^2 = \langle\psi|\hat{A}|\psi\rangle. \qquad (2.19)$$

Self-adjoint operators An operator $\hat{A}$ is said to be self-adjoint if

$$\langle\psi|\hat{A}|\phi\rangle = \langle\phi|\hat{A}|\psi\rangle^*$$

for any pair of states $|\psi\rangle$ and $|\phi\rangle$. The difference between Hermitian and self-adjoint operators is apparent only when dealing with infinite-dimensional state spaces.

The statistical spread of results is often expressed in terms of the variance

$$\Delta A^2 = \langle\psi|(\hat{A} - \langle\hat{A}\rangle)^2|\psi\rangle = \langle\psi|\hat{A}^2|\psi\rangle - \langle\psi|\hat{A}|\psi\rangle^2, \tag{2.20}$$

or the uncertainty $\Delta A = \sqrt{\Delta A^2}$. This uncertainty is zero if and only if $|\psi\rangle$ is an eigenstate of $\hat{A}$.

An important property of operators is that they do not, in general, commute. This means that the state produced by acting with a pair of operators will depend on the order in which they are applied: $\hat{A}\hat{B}|\psi\rangle \neq \hat{B}\hat{A}|\psi\rangle$. The difference is conveniently expressed in terms of the commutator of $\hat{A}$ and $\hat{B}$, which is defined to be

$$[\hat{A}, \hat{B}] = \hat{A}\hat{B} - \hat{B}\hat{A}. \tag{2.21}$$

If this commutator is zero then $\hat{A}$ and $\hat{B}$ are said to commute. If the operators are also Hermitian, then the associated observables A and B are said to be compatible, and the operators $\hat{A}$ and $\hat{B}$ have a common complete set of eigenstates. More generally, the commutator of two Hermitian operators is a skew-Hermitian operator in that $[\hat{A}, \hat{B}]^\dagger = -[\hat{A}, \hat{B}]$, which follows from eqn 2.10. This means that we can write the commutator as $[\hat{A}, \hat{B}] = i\hat{C}$, where $\hat{C}$ is a Hermitian operator. The commutator of an operator $\hat{A}$ and the operator product $\hat{B}\hat{C}$ is easily seen to be

The uncertainty principle   We can derive this from the Cauchy–Schwarz inequality (eqn 2.55) by selecting

$$|\phi_1\rangle = (\hat{A} - \langle\hat{A}\rangle)|\psi\rangle,$$
$$|\phi_2\rangle = (\hat{B} - \langle\hat{B}\rangle)|\psi\rangle,$$

where $|\psi\rangle$ is the state of the system under consideration. It then follows from the Cauchy–Schwarz inequality that

$$\begin{aligned}\Delta A^2 \Delta B^2 &\geq |\langle\psi|\hat{A}'\hat{B}'|\psi\rangle|^2 \\ &= \frac{1}{4}\langle\psi|\{\hat{A}', \hat{B}'\}|\psi\rangle^2 \\ &\quad + \frac{1}{4}|\langle\psi|[\hat{A}', \hat{B}']|\psi\rangle|^2 \\ &\geq \frac{1}{4}|\langle\psi|[\hat{A}, \hat{B}]|\psi\rangle|^2,\end{aligned}$$

where $\hat{A}' = \hat{A} - \langle\hat{A}\rangle$, with a similar definition for $\hat{B}'$. The square root of this gives the uncertainty principle stated in eqn 2.24. States for which the equality holds in eqn 2.24 are sometimes called 'minimum uncertainty states', but a better term is 'intelligent states'. This is because enforcing the equality does not usually minimize the product $\Delta A \, \Delta B$ for any given ΔA. States that achieve this minimum are the minimum-uncertainty-product states.

$$\begin{aligned}[\hat{A}, \hat{B}\hat{C}] &= \hat{B}[\hat{A}, \hat{C}] + [\hat{A}, \hat{B}]\hat{C}, \\ [\hat{B}\hat{C}, \hat{A}] &= \hat{B}[\hat{C}, \hat{A}] + [\hat{B}, \hat{A}]\hat{C}.\end{aligned} \tag{2.22}$$

The anticommutator is defined to be

$$\{\hat{A}, \hat{B}\} = \hat{A}\hat{B} + \hat{B}\hat{A}, \tag{2.23}$$

which is clearly Hermitian if $\hat{A}$ and $\hat{B}$ are Hermitian.

The uncertainties associated with the observables A and B for any given state are bounded by the uncertainty principle:

$$\Delta A \, \Delta B \geq \frac{1}{2}|\langle[\hat{A}, \hat{B}]\rangle|. \tag{2.24}$$

For incompatible observables, the associated operators $\hat{A}$ and $\hat{B}$ will not commute, and the uncertainty principle then places a lower bound on the extent to which it is possible to specify the values of A and B for a given state.

A special class among the Hermitian operators is that of the positive operators. An operator $\hat{A}$ is said to be positive (or, sometimes, positive semidefinite) if for any state $|\psi\rangle$,

$$\langle\psi|\hat{A}|\psi\rangle \geq 0. \tag{2.25}$$

Positive operators are important in the description of quantum states and also in the theory of measurements, as we shall see in Chapter 4.

If we express $|\psi\rangle$ as a superposition of the eigenvectors $|\lambda_n\rangle$ of $\hat{A}$, as in eqn 2.18, then the positivity condition in eqn 2.25 becomes

$$\sum_n \lambda_n |a_n|^2 \geq 0. \tag{2.26}$$

This should hold for all possible states and therefore for all possible probability amplitudes a_n, which tells us that positive operators have non-negative eigenvalues; that is, $\lambda_n \geq 0$. It follows that the equality in eqn 2.25 holds only if $|\psi\rangle$ is a zero-eigenvalue eigenstate of $\hat{A}$, that is, if $\hat{A}|\psi\rangle = 0$.

The outer product of two normalized states $|\phi_1\rangle$ and $|\phi_2\rangle$ is the operator $|\phi_1\rangle\langle\phi_2|$ or its Hermitian conjugate $|\phi_2\rangle\langle\phi_1| = (|\phi_1\rangle\langle\phi_2|)^\dagger$. This outer product is a Hermitian operator if and only if $|\phi_1\rangle = |\phi_2\rangle$. The operator $|\phi_1\rangle\langle\phi_2|$ acting on a state $|\psi\rangle$ produces the state $\langle\phi_2|\psi\rangle|\phi_1\rangle$, that is, the state $|\phi_1\rangle$ multiplied by the complex number $\langle\phi_2|\psi\rangle$. If $\{|\lambda_n\rangle\}$ is a complete orthonormal set of eigenvectors of a Hermitian operator $\hat{A}$ then

$$\hat{A} = \sum_n \lambda_n |\lambda_n\rangle\langle\lambda_n|. \tag{2.27}$$

The identity operator $\hat{\mathrm{I}}$ is the Hermitian operator which when acting on any state $|\psi\rangle$ gives the same state; that is, $\hat{\mathrm{I}}|\psi\rangle = |\psi\rangle$. It follows from eqn 2.27 that

$$\hat{\mathrm{I}} = \sum_n |\lambda_n\rangle\langle\lambda_n| \tag{2.28}$$

as, using eqns 2.17 and 2.18, we find

$$\hat{\mathrm{I}}|\psi\rangle = \sum_m |\lambda_m\rangle\langle\lambda_m| \sum_n a_n |\lambda_n\rangle$$
$$= \sum_n a_n \sum_m \delta_{mn} |\lambda_m\rangle$$
$$= |\psi\rangle. \tag{2.29}$$

The resolution of the identity operator, that is, its expression in terms of the states $|\lambda_n\rangle$ given in eqn 2.28, is, in fact, an alternative statement of the completeness of the set $\{|\lambda_n\rangle\}$.

A function $f(\hat{A})$ of the Hermitian operator $\hat{A}$ in eqn 2.27 is defined to be the operator

$$f(\hat{A}) = \sum_n f(\lambda_n)|\lambda_n\rangle\langle\lambda_n|, \tag{2.30}$$

so that $|\lambda_n\rangle$ is an eigenstate of $f(\hat{A})$ with eigenvalue $f(\lambda_n)$. This means that if we measure the observable $f(A)$ then the possible results will be $\{f(\lambda_n)\}$. It is also sometimes possible to write a function of an operator as a Taylor series

$$f(\hat{A}) = \hat{\mathrm{I}}f(0) + \hat{A}f'(0) + \hat{A}^2 \frac{f''(0)}{2!} + \cdots + \hat{A}^n \frac{f^{(n)}(0)}{n!} + \cdots, \tag{2.31}$$

where $f^{(n)}(0)$ denotes the nth derivative with respect to x of $f(x)$ evaluated at $x = 0$. This expansion is often used for the exponential function, for which

$$\exp(\alpha\hat{A}) = \hat{I} + \alpha\hat{A} + \frac{\alpha^2}{2!}\hat{A}^2 + \cdots + \frac{\alpha^n}{n!}\hat{A}^n + \cdots. \qquad (2.32)$$

This Taylor series can also be used to write functions of non-Hermitian operators for which no eigenstate expansion, in the manner of eqn 2.30, is possible.

The evolution of a state $|\psi(t)\rangle$ is governed by the Schrödinger equation

$$i\hbar\frac{d}{dt}|\psi(t)\rangle = \hat{H}|\psi(t)\rangle, \qquad (2.33)$$

where $\hat{H}$ is the Hamiltonian. This Hermitian operator may, depending on the system and the model being used to study it, be time-dependent or time-independent. The expectation value of a time-independent operator $\hat{A}$ changes with time owing to the evolution of $|\psi(t)\rangle$, so that

$$\frac{d}{dt}\langle\hat{A}\rangle = -\frac{i}{\hbar}\langle\psi(t)|[\hat{A}, \hat{H}]|\psi(t)\rangle. \qquad (2.34)$$

This change in $\langle\hat{A}\rangle$ arises from transitions between the eigenstates $|\lambda_n\rangle$ of $\hat{A}$ induced by $\hat{H}$. If $\hat{A}$ commutes with $\hat{H}$ then $\langle\hat{A}\rangle$, $\langle\hat{A}^2\rangle$, and all higher moments of $\hat{A}$ are constants of the motion, and we say that $\hat{A}$ and its associated observable A are constants of the motion. Substituting the eigenstate expansion of $|\psi\rangle$ given in eqn 2.18 into the Schrödinger equation (eqn 2.33) and taking the overlap with $|\lambda_m\rangle$ gives the amplitude equations

$$\dot{a}_m(t) = -\frac{i}{\hbar}\sum_n\langle\lambda_m|\hat{H}|\lambda_n\rangle a_n(t). \qquad (2.35)$$

This set of coupled differential equations is equivalent to the Schrödinger equation, and from the solution we can construct $|\psi(t)\rangle$.

The formal solution of the Schrödinger equation is

$$|\psi(t)\rangle = \hat{U}(t)|\psi(0)\rangle, \qquad (2.36)$$

where $\hat{U}(t)$ is a unitary operator, for which

$$\hat{U}^\dagger = \hat{U}^{-1}, \qquad (2.37)$$

so that $\hat{U}^\dagger\hat{U} = \hat{I} = \hat{U}\hat{U}^\dagger$. We shall find out more about unitary operators in Section 2.3. The evolution operator $\hat{U}(t)$ itself satisfies the Schrödinger equation

$$i\hbar\frac{d}{dt}\hat{U}(t) = \hat{H}\hat{U}(t), \qquad (2.38)$$

and hence if $\hat{H}$ is time-independent then

$$\hat{U}(t) = \exp\left(-i\frac{\hat{H}t}{\hbar}\right). \qquad (2.39)$$

The Hermitian conjugate operator $\hat{U}^\dagger(t)$ is given by $\hat{U}^\dagger(t) = \exp(i\hat{H}t/\hbar)$, from which, using our definitions of a function of an operator, it is straightforward to show that $\hat{U}^\dagger(t))\hat{U}(t) = \hat{I} = \hat{U}(t)\hat{U}^\dagger(t)$ and hence, as stated in eqn 2.37, that $\hat{U}(t)$ is unitary. The bra vector $\langle\psi(t)|$ is given by $\langle\psi(0)|\hat{U}^\dagger$, where

$$-i\hbar\frac{d}{dt}\hat{U}^\dagger(t) = \hat{U}^\dagger(t)\hat{H}. \tag{2.40}$$

It follows that the overlap of a state vector with itself is conserved: $\langle\psi(t)|\psi(t)\rangle = \langle\psi(0)|\psi(0)\rangle$, so that normalization is retained. Moreover, the overlap of any two state vectors is conserved: $\langle\psi(t)|\phi(t)\rangle = \langle\psi(0)|\phi(0)\rangle$. An alternative to using the Schrödinger equation is to work in the Heisenberg picture, in which operators evolve so that $\hat{A}(t)$ satisfies the Heisenberg equation

$$\frac{d}{dt}\hat{A}(t) = \frac{i}{\hbar}[\hat{H}, \hat{A}] + \frac{\partial}{\partial t}\hat{A}, \tag{2.41}$$

where the partial-derivative term accounts for any explicit time dependence. We should note that the Heisenberg equation in eqn 2.41 is the correct equation of motion for all operators, including those that are not Hermitian.

2.2 Mixed states

If we do not have enough information to specify the state vector but know the probabilities P_n that the system is in a normalized state $|\psi_n\rangle$ then the mean value of A is

$$\bar{A} = \sum_n P_n\langle\psi_n|\hat{A}|\psi_n\rangle. \tag{2.42}$$

We should note that this is different from the expression in eqn 2.19 which holds when we know precisely the normalized state $|\psi\rangle$. It is convenient to introduce the density operator or density matrix $\hat{\rho}$, which is the Hermitian operator

$$\hat{\rho} = \sum_n P_n|\psi_n\rangle\langle\psi_n|. \tag{2.43}$$

If one of the probabilities is unity, that is if $P_n = \delta_{mn}$, then the density operator reduces to the simple form $\hat{\rho} = |\psi_m\rangle\langle\psi_m|$. We refer to density operators of this form, for which the state vector is known, as pure-state density operators or pure states. Density operators of the form in eqn 2.43 represent a statistical mixture of states, or mixed states. It is common practice to use the word *state* for the state of a system, its density operator, and, for pure states, its state vector as well.

The states $|\psi_n\rangle$ need not be orthogonal, but it is always possible to write $\hat{\rho}$ in diagonal form. This follows from the fact that it is Hermitian. The density operator is also positive, as for any state $|\phi\rangle$,

$$\langle\phi|\hat{\rho}|\phi\rangle = \sum_n P_n|\langle\phi|\psi_n\rangle|^2 \geq 0, \tag{2.44}$$

as each term in the sum is positive or zero. It follows that we can write any density operator in the form

$$\hat{\rho} = \sum_m \rho_m |\rho_m\rangle\langle\rho_m|, \tag{2.45}$$

where the ρ_m are the (positive) eigenvalues of the density operator and the $|\rho_m\rangle$ are the corresponding eigenstates.

The mean value of an observable A is given by

$$\bar{A} = \langle\hat{A}\rangle = \text{Tr}(\hat{\rho}\hat{A}), \tag{2.46}$$

where Tr denotes the trace operation, which is carried out by summing the diagonal elements of the operator $\hat{\rho}\hat{A}$ in any basis consisting of a complete orthonormal set of states. Consider, for example, using the basis $\{|\lambda_m\rangle\}$ to calculate the trace as follows:

$$\begin{aligned}
\text{Tr}(\hat{\rho}\hat{A}) &= \sum_m \langle\lambda_m| \left(\sum_n P_n |\psi_n\rangle\langle\psi_n| \right) \hat{A} |\lambda_m\rangle \\
&= \sum_n P_n \langle\psi_n|\hat{A} \sum_m |\lambda_m\rangle\langle\lambda_m|\psi_n\rangle \\
&= \sum_n P_n \langle\psi_n|\hat{A}|\psi_n\rangle, \tag{2.48}
\end{aligned}$$

where we have used eqn 2.28. This has the same form as eqn 2.42, confirming that $\text{Tr}(\hat{\rho}\hat{A})$ is the mean value of A. Three important results follow from eqn 2.48. The first is that the trace is independent of the basis $\{|\lambda_m\rangle\}$ used to calculate $\langle\hat{A}\rangle$. It follows that the trace can be evaluated in any basis. Secondly, if we choose $\hat{A}$ to be the identity operator $\hat{I}$ then we find

$$\text{Tr}(\hat{\rho}) = \sum_n P_n = 1, \tag{2.49}$$

which reflects the fact that the sum of any complete set of probabilities is unity. The trace condition in eqn 2.49 is the analogue of the normalization condition, $\langle\psi|\psi\rangle = 1$, for state vectors. If we evaluate the trace of $\hat{\rho}$ in the basis formed by its eigenvectors then we find

$$\text{Tr}(\hat{\rho}) = \sum_n \rho_n = 1, \tag{2.50}$$

so that the sum of the eigenvalues of $\hat{\rho}$ is unity. Our third result follows on choosing $\hat{A}$ to be $\hat{\rho}$ and writing the density operator in the form of eqn 2.45 so that eqn 2.48 becomes

$$\text{Tr}(\hat{\rho}^2) = \sum_n \rho_n \langle\rho_n| \left(\sum_m \rho_m |\rho_m\rangle\langle\rho_m| \right) |\rho_n\rangle = \sum_n \rho_n^2. \tag{2.51}$$

We can interpret the eigenvalues ρ_n as probabilities, and this means that they are positive. It follows that

$$\text{Tr}(\hat{\rho}^2) = \sum_n \rho_n^2 \le \sum_n \sum_m \rho_n \rho_m = 1, \tag{2.52}$$

The trace operation   The trace of any operator $\hat{B}$ is the sum of the diagonal matrix elements in any orthonormal basis $\{|\lambda_n\rangle\}$,

$$\text{Tr}\hat{B} = \sum_n \langle\lambda_n|\hat{B}|\lambda_n\rangle.$$

An important property of the trace operation is that the trace of a product of operators is invariant under cyclic permutation of these operators, so that

$$\begin{aligned}
\text{Tr}(\hat{\rho}\hat{A}\hat{B}) &= \text{Tr}(\hat{B}\hat{\rho}\hat{A}) \\
&= \text{Tr}(\hat{A}\hat{B}\hat{\rho}). \tag{2.47}
\end{aligned}$$

Other permutations do not, in general, have the same trace, as

$$\text{Tr}(\hat{\rho}\hat{A}\hat{B}) - \text{Tr}(\hat{\rho}\hat{B}\hat{A}) = \text{Tr}(\hat{\rho}[\hat{A},\hat{B}]).$$

The proof of eqn 2.47 follows by evaluating the trace in any basis $\{|\lambda_m\rangle\}$ and using the identity in eqn 2.28 as follows:

$$\begin{aligned}
\text{Tr}(\hat{\rho}\hat{A}\hat{B}) &= \sum_{lmn} \langle\lambda_n|\hat{\rho}|\lambda_m\rangle \\
&\quad \times \langle\lambda_m|\hat{A}|\lambda_l\rangle\langle\lambda_l|\hat{B}|\lambda_n\rangle \\
&= \sum_{lmn} \langle\lambda_l|\hat{B}|\lambda_n\rangle \\
&\quad \times \langle\lambda_n|\hat{\rho}|\lambda_m\rangle\langle\lambda_m|\hat{A}|\lambda_l\rangle \\
&= \text{Tr}(\hat{B}\hat{\rho}\hat{A}).
\end{aligned}$$

with equality holding if and only if one of the ρ_n is unity, with the others being zero. It follows that $\text{Tr}(\hat{\rho}^2) = 1$ for pure states but $\text{Tr}(\hat{\rho}^2) \leq 1$ for mixed states. The quantity $\text{Tr}(\hat{\rho}^2)$ is sometimes called the degree of purity or, more often, simply the purity of the state.

The matrix elements ρ_{nm} of the density operator in any basis $\{|\lambda_n\rangle\}$ are

$$\rho_{nm} = \langle \lambda_n | \hat{\rho} | \lambda_m \rangle = \rho_{mn}^* \tag{2.53}$$

and these are constrained by the inequality

$$\rho_{nm}\rho_{mn} \leq \rho_{nn}\rho_{mm}. \tag{2.54}$$

The proof of this follows from the Cauchy–Schwarz inequality

$$|\langle \phi_1 | \phi_2 \rangle|^2 \leq \langle \phi_1 | \phi_1 \rangle \langle \phi_2 | \phi_2 \rangle, \tag{2.55}$$

where $|\phi_1\rangle$ and $|\phi_2\rangle$ are any two (unnormalized) state vectors. The inequality for our matrix elements in eqn 2.54 follows from the Cauchy–Schwarz inequality on writing $|\phi_1\rangle = \hat{\rho}^{1/2}|\lambda_n\rangle$ and $|\phi_2\rangle = \hat{\rho}^{1/2}|\lambda_m\rangle$, where $\hat{\rho}^{1/2}$ is defined using eqn 2.30. For a pure state, the equality in eqn 2.54 holds for all m and n.

The evolution of the density operator is determined by the Schrödinger equation. If the initial density operator is

$$\hat{\rho}(0) = \sum_n P_n |\psi_n(0)\rangle\langle\psi_n(0)| \tag{2.56}$$

then the solution of the Schrödinger equation (eqn 2.36) and its bra equivalent tell us that the evolved density operator is

$$\hat{\rho}(t) = \sum_n P_n \hat{U}(t)|\psi_n(0)\rangle\langle\psi_n(0)|\hat{U}^\dagger(t)$$
$$= \hat{U}(t)\hat{\rho}(0)\hat{U}^\dagger(t), \tag{2.57}$$

where $\hat{U}(t)$ is the unitary evolution operator. A transformation of this form, with an operator sandwiched between a unitary operator and its inverse, is a unitary transformation. We can obtain an evolution equation for $\hat{\rho}(t)$ by differentiating with respect to time to give

$$\frac{d}{dt}\hat{\rho}(t) = -\frac{i}{\hbar}\left(\hat{H}\hat{U}(t)\hat{\rho}(0)\hat{U}^\dagger(t) - \hat{U}(t)\hat{\rho}(0)\hat{U}^\dagger(t)\hat{H}\right)$$
$$= -\frac{i}{\hbar}[\hat{H}, \hat{\rho}(t)], \tag{2.58}$$

where we have used eqns 2.38 and 2.40. Note the similarity with the Heisenberg equation (eqn 2.41) for the evolution of operators in the Heisenberg picture. The all-important sign difference, however, marks eqn 2.58 as evolution in the Schrödinger picture. In the Schrödinger picture it is the states and density operators that evolve, while in the Heisenberg picture it is the operators representing observables or functions of them that evolve. We should note that the density operator is capable of describing more complicated dynamics, in which the degree

The Cauchy–Schwarz inequality
This important inequality (eqn 2.55) constrains the overlaps of any pair of (un-normalized) states $|\phi_1\rangle$ and $|\phi_2\rangle$. It follows directly from the inequality in eqn 2.3 on writing

$$|\psi\rangle = |\phi_2\rangle - \frac{\langle\phi_1|\phi_2\rangle}{\langle\phi_1|\phi_1\rangle}|\phi_1\rangle.$$

of mixedness can change through evolution of the P_n. Such evolutions occur when the quantum system is coupled to an environment into which information can be lost.

We have introduced the density operator (eqn 2.43) as a description of a statistical ensemble of states, with the interpretation that any single member of the ensemble has been prepared in one of the states $|\psi_n\rangle$ with probability P_n. It is important to note, however, that this interpretation is far from unique, in that a single density operator may represent many distinct ensembles of prepared states and probabilities. For example, the mixed state with density operator

$$\hat{\rho} = \frac{1}{2}|0\rangle\langle 0| + \frac{1}{2}|1\rangle\langle 1|, \tag{2.59}$$

where $|0\rangle$ and $|1\rangle$ are a pair of orthonormal states, can also be written as

$$\hat{\rho} = \frac{1}{3}|0\rangle\langle 0| + \frac{1}{3}\frac{(|0\rangle + \sqrt{3}|1\rangle)}{2}\frac{(\langle 0| + \sqrt{3}\langle 1|)}{2} \\ + \frac{1}{3}\frac{(|0\rangle - \sqrt{3}|1\rangle)}{2}\frac{(\langle 0| - \sqrt{3}\langle 1|)}{2}. \tag{2.60}$$

We might be tempted to interpret eqn 2.59 as an ensemble of systems each of which is prepared in either the state $|0\rangle$ or the state $|1\rangle$ with probability 1/2. Equally, we might interpret eqn 2.60 as an ensemble of systems each of which is prepared in $|0\rangle$, $(|0\rangle + \sqrt{3}|1\rangle)/2$, or $(|0\rangle - \sqrt{3}|1\rangle)/2$ with equal probabilities. The fact that the two density operators are equal, however, means that there is no way to distinguish, even in principle, between these differently prepared but identical mixed states. There is, in fact, an infinite number of ensembles that correspond to the same mixed state. A density operator

$$\hat{\rho} = \sum_l Q_l|\phi_l\rangle\langle\phi_l|, \tag{2.61}$$

where the states $|\phi_l\rangle$ are prepared with the probabilities Q_l, is equal to eqn 2.43 if and only if

$$\sqrt{P_n}|\psi_n\rangle = \sum_l u_{nl}\sqrt{Q_l}|\phi_l\rangle, \tag{2.62}$$

where u_{nl} is the nl-element of a unitary matrix. If the number of states $|\psi_n\rangle$ differs from the number of states $|\phi_l\rangle$, then it is necessary to add rows or columns to the matrix so that it is square and so can be unitary. It is straightforward to demonstrate the sufficiency of this condition by substituting eqn 2.62 into eqn 2.43 to give

$$\sum_n P_n|\psi_n\rangle\langle\psi_n| = \sum_{nlm} u_{nl}u_{nm}^*\sqrt{Q_lQ_m}|\phi_l\rangle\langle\phi_m|$$

$$= \sum_{lm}\left(\sum_n u_{mn}^\dagger u_{nl}\right)\sqrt{Q_lQ_m}|\phi_l\rangle\langle\phi_m|$$

$$= \sum_l Q_l|\phi_l\rangle\langle\phi_l|, \tag{2.63}$$

where the last line follows from the unitarity of u ($\sum_n u^\dagger_{mn} u_{nl} = \delta_{ml}$).

For two independent quantum systems a and b, we can write a composite state as the direct product

$$|\psi\rangle = |\lambda\rangle_a |\phi\rangle_b, \qquad (2.64)$$

where $|\lambda\rangle_a$ and $|\phi\rangle_b$ are states for the a and b systems, respectively. Quantum information problems can involve large numbers of systems and it is convenient to be able to drop the labels. We write

$$|\psi\rangle = |\lambda\rangle \otimes |\phi\rangle, \qquad (2.65)$$

where the symbol $\otimes$ denotes the tensor product of the state spaces. More simply, this symbol separates states (and operators) associated with the two quantum systems and removes the requirement for indices indicating the individual quantum systems. Not all states of the two systems can be written in the form of eqn 2.65, as the superposition principle implies that

$$|\psi\rangle = \sum_n a_n |\lambda_n\rangle \otimes |\phi_n\rangle \qquad (2.66)$$

is also a possible composite state of the two systems. States of this type, which cannot be written as a direct product of the form of eqn 2.65, are called entangled states. These are of great importance in the study of quantum information and we shall describe their remarkable properties in Section 2.5 and in Chapter 5. We choose, as is always possible, the states $\{|\lambda_n\rangle\}$ and $\{|\phi_n\rangle\}$ to be such that each forms a complete, orthonormal basis. The density operator associated with the entangled state in eqn 2.66 is $\hat{\rho} = |\psi\rangle\langle\psi|$. The expectation value of any operator $\hat{A}$ acting only on the space of the first system, spanned by the states $\{|\lambda_n\rangle\}$, is then

$$\begin{aligned}
\langle\psi|\hat{A}|\psi\rangle &= \sum_m \sum_n a^*_m a_n \langle\phi_m| \otimes \langle\lambda_m|\hat{A} \otimes \hat{I}|\lambda_n\rangle \otimes |\phi_n\rangle \\
&= \sum_m \sum_n a^*_m a_n \langle\phi_m|\phi_n\rangle \langle\lambda_m|\hat{A}|\lambda_n\rangle \\
&= \sum_n |a_n|^2 \langle\lambda_n|\hat{A}|\lambda_n\rangle.
\end{aligned} \qquad (2.67)$$

Here $\hat{A} \otimes \hat{I}$ denotes $\hat{A}$ acting on the first system and the identity operator $\hat{I}$ acting on the second. Comparing this with eqns 2.43 and 2.48 shows that the same expectation value is obtained by using the density operator

$$\hat{\rho}_a = \sum_n |a_n|^2 |\lambda_n\rangle\langle\lambda_n| \qquad (2.68)$$

for the first system. This is the reduced operator which provides a complete description of the statistical properties of the single system spanned by the basis $\{|\lambda_n\rangle\}$ but contains no information on the other system. If we restrict our attention to observables associated with only one of the two entangled systems then this takes no account of any

Composite quantum systems
The fact that the state vector for a pair of independent quantum systems is the *product* of those for the individual systems follows from the fact that probabilities for the independent systems are products of those for the individual systems. In particular, if the probability that the properties A and B of the two systems take the values a and b are $P(a)$ and $P(b)$, respectively, then

$$P(a,b) = P(a)P(b).$$

This, together with the quantum rules for calculating probabilities, gives

$$\begin{aligned}
P(a,b) &= |\langle a,b|\psi\rangle|^2 \\
&= |\langle a|\lambda\rangle|^2 |\langle b|\phi\rangle|^2,
\end{aligned}$$

from which we can infer eqn 2.64.

correlations between the two systems. A set of measurements of these observables will at best only enable us to construct the reduced density operator for the system under observation. We can write the expectation value given in eqn 2.67 in the form

$$
\begin{aligned}
\langle \hat{A} \rangle &= \mathrm{Tr}(\hat{A} \otimes \hat{\mathrm{I}} |\psi\rangle\langle\psi|) \\
&= \sum_m \sum_n \langle \phi_m| \otimes \langle \lambda_n|(\hat{A} \otimes \hat{\mathrm{I}} |\psi\rangle\langle\psi|)|\lambda_n\rangle \otimes |\phi_m\rangle \\
&= \mathrm{Tr}_a \left[\hat{A} \, \mathrm{Tr}_b(|\psi\rangle\langle\psi|) \right].
\end{aligned}
\tag{2.69}
$$

Here the subscripts a and b denote traces over the state spaces associated with the two quantum systems spanned by the bases $\{|\lambda_n\rangle\}$ and $\{|\phi_m\rangle\}$ respectively. The operator $\mathrm{Tr}_b(|\psi\rangle\langle\psi|)$ is the reduced density operator for the first system. More generally, if the two systems are described by the density operator $\hat{\rho}_{ab}$ then the reduced density operator for system a is

$$
\begin{aligned}
\hat{\rho}_a &= \sum_m {}_b\langle \phi_m|\hat{\rho}_{ab}|\phi_m\rangle_b \\
&= \mathrm{Tr}_b(\hat{\rho}_{ab}).
\end{aligned}
\tag{2.70}
$$

The reduced density operator for system b is obtained by evaluating the trace of $\hat{\rho}_{ab}$ over the states for system a.

The procedure of obtaining a reduced density operator for one of a pair of entangled systems can be reversed, in the sense that we can represent any mixed state by a state vector for the quantum system of interest entangled with an additional, or ancillary, quantum system. This procedure is the basis of the method of thermofields as developed in finite-temperature quantum field theory. In quantum information theory, it is referred to as purification. The starting point is to write the density operator for the system of interest in diagonal form, that is, in terms of its eigenvalues and eigenvectors as in eqn 2.45. The expectation value of an operator $\hat{A}$ is then

$$
\langle \hat{A} \rangle = \sum_m \rho_m \langle \rho_m|\hat{A}|\rho_m\rangle.
\tag{2.71}
$$

If we introduce a second quantum system, of the same form as the original, then we can prepare the entangled state

$$
|\psi\rangle = \sum_m \sqrt{\rho_m}|\rho_m\rangle \otimes |\rho_m\rangle.
\tag{2.72}
$$

The mean value of A for this entangled state is

$$
\begin{aligned}
\bar{A} &= \langle \psi|\hat{A} \otimes \hat{\mathrm{I}}|\psi\rangle \\
&= \sum_m \sum_n \sqrt{\rho_m \rho_n}\langle \rho_m|\hat{A}|\rho_n\rangle\langle \rho_m|\hat{\mathrm{I}}|\rho_n\rangle \\
&= \sum_m \rho_m \langle \rho_m|\hat{A}|\rho_m\rangle.
\end{aligned}
\tag{2.73}
$$

This has the same value as that obtained from the original density operator in eqn 2.71. It follows that the purified state in eqn 2.72 provides a representation of the system equivalent to that afforded by the original density operator. We should emphasize that the procedure of purification is not unique. The entangled state

$$|\psi\rangle = \sum_m \sqrt{\rho_m} e^{i\theta_m} |\rho_m\rangle \otimes |\lambda_m\rangle, \tag{2.74}$$

where the $\{\theta_m\}$ are any real phases and $\{|\lambda_m\rangle\}$ is any orthonormal basis, gives the same statistical properties for the original system as those for the density operator in eqn 2.45. The introduction of an ancillary system is an important idea in quantum information, which we shall use in our study of generalized measurements in Chapter 4.

2.3 Unitary operators

We have seen that the time evolution of a quantum state is associated with the action of a unitary operator, as in eqn 2.36. This evolution is the information-processing element in quantum computation, with the information extraction associated with measurements. We can view the design of a quantum information processor as the selection of a suitable unitary operator and its implementation by means of the appropriate quantum evolution. Studying the properties of unitary operators, and the state transformations generated by them, helps us to determine what is and what is not possible.

The unitary time evolution operator, $\hat{U}(t) = \exp(-i\hat{H}t/\hbar)$, is generated by a Hermitian operator, the Hamiltonian $\hat{H}$. This relationship with a Hermitian operator is a characteristic of all unitary operators, and we can write any unitary operator in the form

$$\hat{U} = \exp(i\hat{C}), \tag{2.75}$$

where $\hat{C}$ is Hermitian. It then follows that we can realize any unitary operator if we can produce the right Hamiltonian and interaction time, so that $\hat{H}t = -\hat{C}\hbar$. We can prove eqn 2.75 by first noting that any operator can be written in the form $\hat{A}+i\hat{B}$, where $\hat{A}$ and $\hat{B}$ are Hermitian, so that $\hat{A}$ is the Hermitian part of the operator and $i\hat{B}$ is its skew-Hermitian part. If we write our unitary operator and its Hermitian conjugate in this form

$$\hat{U} = \hat{A} + i\hat{B},$$
$$\hat{U}^\dagger = \hat{A} - i\hat{B}, \tag{2.76}$$

then the unitarity of $\hat{U}$ implies that

$$\hat{U}^\dagger\hat{U} = \hat{A}^2 + \hat{B}^2 + i[\hat{A}, \hat{B}] = \hat{I},$$
$$\hat{U}\hat{U}^\dagger = \hat{A}^2 + \hat{B}^2 - i[\hat{A}, \hat{B}] = \hat{I}. \tag{2.77}$$

Unitary operators Unitarity is a property of an operator and its Hermitian conjugate. An operator $\hat{U}$ is unitary if

$$\hat{U}\hat{U}^\dagger = \hat{I} = \hat{U}^\dagger\hat{U}.$$

Note that both these conditions are important. There exist operators for which only one is satisfied. One example is provided by the bare raising and lowering operators for the harmonic oscillator

$$\hat{E} = \sum_{n=0}^{\infty} |n\rangle\langle n+1|,$$
$$\hat{E}^\dagger = \sum_{n=0}^{\infty} |n+1\rangle\langle n|,$$

where the $\{|n\rangle\}$ are the energy eigenstates. These operators are not unitary, as

$$\hat{E}\hat{E}^\dagger = \hat{I},$$
$$\hat{E}^\dagger\hat{E} = \hat{I} - |0\rangle\langle 0|.$$

It follows that $\hat{A}$ and $\hat{B}$ must commute and that $\hat{A}^2 + \hat{B}^2 = \hat{I}$. This allows us to write $\hat{A}$ and $\hat{B}$ as functions of a single Hermitian operator in the form

$$\hat{A} = \cos \hat{C},$$
$$\hat{B} = \sin \hat{C}, \tag{2.78}$$

which implies eqn 2.75. If we can find the eigenstates and eigenvalues of $\hat{C}$ then we can use eqn 2.30 to find the form of the transformed states. Alternatively, we can use the Taylor expansion given in eqn 2.32.

Principal among the properties of a unitary evolution is that the overlaps of the transformed states are the same as for the untransformed ones. If a unitary operator $\hat{U}$ transforms a state $|\psi\rangle$ into a new state $|\psi'\rangle = \hat{U}|\psi\rangle$ then

$$\langle\phi'|\psi'\rangle = \langle\phi|\hat{U}^\dagger\hat{U}|\psi\rangle = \langle\phi|\psi\rangle. \tag{2.79}$$

In particular, the overlap of any transformed normalized state with itself is unity and the overlap of a pair of transformed orthogonal states is zero. It follows that if $\{|\lambda_m\rangle\}$ is a complete basis of orthonormal states then so is $\{\hat{U}|\lambda_m\rangle\}$. We have already commented that the overlap between a pair of state vectors has much in common with the scalar or dot product of two vectors. A rotation about an axis through the origin will preserve the lengths of any set of position vectors and also the angles between them, and hence it will preserve their scalar products. It is sometimes helpful to view a unitary evolution as an analogous rotation in a space spanned by the state vectors.

The expectation value of $\hat{A}$ for a unitarily transformed state $\hat{U}|\psi\rangle$ is the same as that of the operator $\hat{U}^\dagger\hat{A}\hat{U}$ for the untransformed state $|\psi\rangle$:

$$(\langle\psi|\hat{U}^\dagger)\hat{A}(\hat{U}|\psi\rangle) = \langle\psi|(\hat{U}^\dagger\hat{A}\hat{U})|\psi\rangle. \tag{2.80}$$

This means that we can consider a unitary evolution either of the state or of the associated observables. These two possibilities correspond, of course, to analysing the evolution of the system in the Schrödinger and Heisenberg pictures, respectively. If we choose to transform the operators then we can use

$$\exp(-i\hat{C})\hat{A}\exp(i\hat{C}) = \hat{A} - i[\hat{C}, \hat{A}] - \frac{1}{2!}[\hat{C}, [\hat{C}, \hat{A}]] + \frac{i}{3!}[\hat{C}, [\hat{C}, [\hat{C}, \hat{A}]]] + \cdots, \tag{2.81}$$

which follows from the Taylor expansion of the unitary operators as given in eqn 2.32. This can also be applied to calculating the evolution of density operators given in eqn 2.57.

It is often helpful to break a complicated unitary transformation into a sequence of simpler ones. A unitary evolution produced by an operator $\hat{U}$ will be equivalent to a sequence of n unitary operators $\hat{U}_1, \hat{U}_2, \cdots, \hat{U}_n$ if

$$\hat{U} = \hat{U}_n \cdots \hat{U}_2 \hat{U}_1. \tag{2.82}$$

Note that the first operator is the last in this sequence but that this means it acts first on the state to be transformed:

$$\hat{U}|\psi\rangle = \hat{U}_n \cdots \hat{U}_2\hat{U}_1|\psi\rangle. \tag{2.83}$$

The order of the operators is important, of course, as the $\hat{U}_i$ will not necessarily mutually commute. We shall meet a number of examples of this procedure in the coming chapters, notably in Section 2.4 and in Chapter 6, where we shall use the idea to build quantum information-processing elements out of simpler quantum gates.

2.4   Qubits

In information theory, the term 'bit' refers to two related but distinct things. We first encountered it as the unit of information, associated with using logarithms in base 2. The term is also used to describe a physical system with two distinct physical states. These two are connected, of course, by the fact that the physical bit can hold a maximum of one bit of information. A qubit is a quantum system having two distinct, that is, orthogonal, states. We label these states with a zero and a one, $|0\rangle$ and $|1\rangle$. Clearly a qubit can hold one bit of information by virtue of it being possible to prepare it in either of these states. Where a qubit differs from its classical counterpart, however, is that the superposition principle tells us that the qubit can be prepared in any superposition of the states $|0\rangle$ and $|1\rangle$, that is, $a_0|0\rangle + a_1|1\rangle$, where a_0 and a_1 are complex numbers.

Any quantum system with two quantum states can be used to provide a physical implementation of a qubit. Examples that have been realized in the laboratory include the two orthogonal polarization states of a photon, the orientation of a spin-half particle, a pair of electronic energy levels in an atom, ion, or quantum dot, and a pair of paths in an interferometer. We should emphasize, however, that from the perspective of quantum information theory any two-state quantum system can represent a qubit and that the idea of a qubit, like the bit in classical information theory, has a generality that is independent of its physical realization.

It is convenient to treat a qubit as though it were a spin-half particle and to introduce, to describe it, the Pauli operators

$$\begin{aligned}
\hat{I} &= |0\rangle\langle 0| + |1\rangle\langle 1|, \\
\hat{\sigma}_x &= |0\rangle\langle 1| + |1\rangle\langle 0|, \\
\hat{\sigma}_y &= i(|1\rangle\langle 0| - |0\rangle\langle 1|), \\
\hat{\sigma}_z &= |0\rangle\langle 0| - |1\rangle\langle 1|.
\end{aligned} \tag{2.84}$$

These correspond, respectively, to the identity operator and to the x-, y-, and z-components of the angular momentum, in units of $\hbar/2$, for the effective spin-half particle associated with the qubit. It is sometimes convenient to represent the state $a_0|0\rangle + a_1|1\rangle$ as the column vector

$$\psi = \begin{pmatrix} a_0 \\ a_1 \end{pmatrix}. \tag{2.85}$$

In this representation, the four Pauli operators have the forms

$$\hat{I} = \begin{pmatrix} 1 & 0 \\ 0 & 1 \end{pmatrix},$$

$$\hat{\sigma}_x = \begin{pmatrix} 0 & 1 \\ 1 & 0 \end{pmatrix},$$

$$\hat{\sigma}_y = \begin{pmatrix} 0 & -i \\ i & 0 \end{pmatrix},$$

$$\hat{\sigma}_z = \begin{pmatrix} 1 & 0 \\ 0 & -1 \end{pmatrix}. \tag{2.86}$$

The operators associated with the three components of the spin do not mutually commute, but the commutator of any two is proportional to the third:

$$[\hat{\sigma}_x, \hat{\sigma}_y] = 2i\hat{\sigma}_x,$$
$$[\hat{\sigma}_y, \hat{\sigma}_z] = 2i\hat{\sigma}_x,$$
$$[\hat{\sigma}_z, \hat{\sigma}_x] = 2i\hat{\sigma}_y. \tag{2.87}$$

They do, however, mutually anticommute, that is, the anticommutator of any two different spin components is zero:

$$\{\hat{\sigma}_x, \hat{\sigma}_y\} = \{\hat{\sigma}_y, \hat{\sigma}_z\} = \{\hat{\sigma}_z, \hat{\sigma}_x\} = 0. \tag{2.88}$$

The fact that the spin operators do not commute tells us that they have different eigenvectors. The operators $\hat{\sigma}_x$, $\hat{\sigma}_y$, and $\hat{\sigma}_z$ each have the two eigenvalues ± 1, with the corresponding eigenvectors being, respectively, $(|0\rangle \pm |1\rangle)/\sqrt{2}$ for $\hat{\sigma}_x$, $(|0\rangle \pm i|1\rangle)/\sqrt{2}$ for $\hat{\sigma}_y$, and $|0\rangle$ and $|1\rangle$ for $\hat{\sigma}_z$.

It is sometimes helpful to picture the qubit states as points on the surface of a sphere, the Bloch sphere (depicted in Fig. 2.1). This is a sphere of unit radius, with each point on its surface corresponding to a different pure state. Opposite points represent a pair of mutually orthogonal states. The north and south poles correspond to the states $|0\rangle$ and $|1\rangle$, respectively, and the eigenstates of $\hat{\sigma}_x$ and $\hat{\sigma}_y$ are aligned along the x- and y-axes, respectively. More generally, a qubit state

$$|\psi\rangle = \cos\left(\frac{\theta}{2}\right)|0\rangle + e^{i\varphi}\sin\left(\frac{\theta}{2}\right)|1\rangle \tag{2.89}$$

corresponds to a point with spherical polar coordinates θ and φ.

Any single-qubit unitary operator can be written in the form

$$\hat{U} = \exp\left(i\alpha\hat{I} + i\beta\vec{a} \cdot \hat{\vec{\sigma}}\right), \tag{2.90}$$

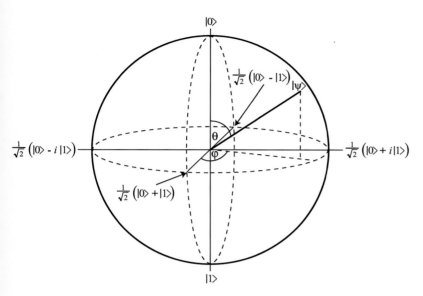

Fig. 2.1 Representation of the pure state $|\psi\rangle$ (eqn 2.89) as a point on the surface of the Bloch sphere.

where α and β are real constants, $\vec{a}$ is a unit vector ($a_x^2 + a_y^2 + a_z^2 = 1$), and $\hat{\vec{\sigma}}$ is the vector operator ($\hat{\sigma}_x, \hat{\sigma}_y, \hat{\sigma}_z$). It is straightforward to show, using either eqn 2.30 or eqn 2.32, that

$$\hat{U} = e^{i\alpha}\left(\cos\beta \hat{I} + i\sin\beta \vec{a}\cdot\hat{\vec{\sigma}}\right). \qquad (2.91)$$

The parameter α simply acts to change the arbitrary phase of the state vector and has no physical consequences. The meaning of β and $\vec{a}$ is most readily appreciated by reference to the Bloch sphere; $\vec{a}$ describes an axis, and 2β is the rotation of the Bloch vector (representing $|\psi\rangle$) about that axis. The action of $\hat{U}$ on the state $|\psi\rangle$ is depicted in Fig. 2.2.

The unitary transformation above can be realized if we can produce a Hamiltonian proportional to $\vec{a}\cdot\hat{\vec{\sigma}}$ and suitably control the interaction time. Methods that have been used include applying radio frequency fields to a nuclear spin, applying laser pulses to atoms or ions, and propagation through birefringent wave plates for polarized photons.

It may not always be convenient to realize a Hamiltonian proportional to $\vec{a}\cdot\hat{\vec{\sigma}}$, but the unitary transformation in eqn 2.90 can also be implemented by a sequence of transformations. A simple example is the product

$$\hat{U} = \exp\left(i\frac{\nu}{2}\hat{\sigma}_z\right)\exp\left(i\frac{\mu}{2}\hat{\sigma}_y\right)\exp\left(i\frac{\lambda}{2}\hat{\sigma}_z\right). \qquad (2.92)$$

This corresponds to Euler's decomposition of a rotation (on the Bloch sphere) into a sequence of three rotations, through the angles λ, μ, and ν about the $z-, y-$, and $z-$ axes, respectively. It is straightforward to show that any transformation of the form of eqn 2.90, with $\alpha = 0$, may be written in this form.

It is clear from eqns 2.86 that any two-by-two matrix can be written as a weighted sum of the four Pauli operators. This means, in turn, that

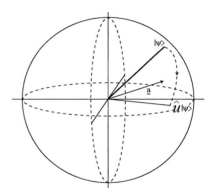

Fig. 2.2 A single-qubit unitary transformation induces a rotation of the Bloch vector.

any operator associated with our qubit can also be expressed in terms of these operators. In particular, we can write the density operator in the form

$$\hat{\rho} = \frac{1}{2}\left(\hat{I} + u\hat{\sigma}_x + v\hat{\sigma}_y + w\hat{\sigma}_z\right), \tag{2.93}$$

where the factor $1/2$ ensures that $\text{Tr}(\hat{\rho}) = 1$. The Hermiticity of the density operator ensures that u, v, and w are real and the positivity of $\hat{\rho}$ requires that $u^2 + v^2 + w^2 \leq 1$. We can associate u, v, and w with the x-, y- and z-components of the Bloch vector. If $u^2 + v^2 + w^2 = 1$ then this vector lies on the surface of the Bloch sphere and corresponds to the $+1$-eigenvalue eigenstate of the operator $u\hat{\sigma}_x + v\hat{\sigma}_y + w\hat{\sigma}_z$. If $u^2 + v^2 + w^2 < 1$ then the Bloch vector describes a point inside the Bloch sphere and corresponds to a mixed state.

The eigenvectors of $\hat{\rho}$, namely $|\rho_+\rangle$ and $|\rho_-\rangle$, are the eigenvectors of $(u\hat{\sigma}_x + v\hat{\sigma}_y + w\hat{\sigma}_z)$ corresponding to the eigenvalues $\pm(u^2 + v^2 + w^2)^{1/2}$. This means that we can write the density operator in eqn 2.93 in diagonalized form as

$$\hat{\rho} = \frac{1}{2}\left[1 + (u^2 + v^2 + w^2)^{1/2}\right]|\rho_+\rangle\langle\rho_+|$$
$$+ \frac{1}{2}\left[1 - (u^2 + v^2 + w^2)^{1/2}\right]|\rho_-\rangle\langle\rho_-|, \tag{2.94}$$

which reduces to the pure state $|\rho_+\rangle\langle\rho_+|$ if $u^2 + v^2 + w^2 = 1$. The action of the unitary operator in eqn 2.91 on $\hat{\rho}$ is to induce a rotation of the Bloch vector about the axis $\vec{a}$. To see this, we write the Bloch vector as $\vec{r} = (u, v, w)$ and split this into components parallel, $(\vec{a} \cdot \vec{r})\vec{a}$, and perpendicular, $\vec{r}_\perp = \vec{r} - (\vec{a} \cdot \vec{r})\vec{a}$, to $\vec{a}$. With this notation the density operator is

$$\hat{\rho} = \frac{1}{2}\left(\hat{I} + \vec{r} \cdot \hat{\vec{\sigma}}\right), \tag{2.95}$$

which transforms to

$$\hat{U}\hat{\rho}\hat{U}^\dagger = \frac{1}{2}\left[\hat{I} + (\vec{a} \cdot \vec{r})\vec{a} \cdot \hat{\vec{\sigma}} + \cos(2\beta)\vec{r}_\perp \cdot \hat{\vec{\sigma}} - \sin(2\beta)(\vec{a} \times \vec{r}_\perp) \cdot \hat{\vec{\sigma}}\right], \tag{2.96}$$

corresponding to a rotation of the Bloch vector through an angle 2β about the axis $\vec{a}$.

Many ideas in quantum information employ multiple qubits. The state of n qubits, each of which is prepared in the state $|0\rangle$, is written

$$|\psi\rangle = \underbrace{|0\rangle \otimes |0\rangle \otimes \cdots \otimes |0\rangle}_{n \text{ terms}}, \tag{2.97}$$

that is, the tensor product of n $|0\rangle$ kets. We can perform single-qubit transformations on all or any of these qubits. If, for example, we wish to apply the unitary operator $\hat{U}$ to the mth qubit then we can write the transformation of the n-qubit state as

$$\underbrace{\hat{I} \otimes \hat{I} \otimes \cdots \otimes}_{m-1 \text{ terms}} \hat{U} \otimes \underbrace{\hat{I} \otimes \cdots \otimes \hat{I}}_{n-m \text{ terms}} |\psi\rangle$$
$$= \underbrace{|0\rangle \otimes |0\rangle \otimes \cdots \otimes}_{m-1 \text{ terms}} (\hat{U}|0\rangle) \underbrace{\otimes \cdots \otimes |0\rangle}_{n-m \text{ terms}}. \tag{2.98}$$

For example the unitary operator $\hat{\sigma}_x \otimes \hat{\sigma}_y \otimes \hat{\sigma}_z$ transforms the three-qubit state $|0\rangle \otimes |0\rangle \otimes |0\rangle$ into

$$\hat{\sigma}_x \otimes \hat{\sigma}_y \otimes \hat{\sigma}_z |0\rangle \otimes |0\rangle \otimes |0\rangle = |1\rangle \otimes i|1\rangle \otimes |0\rangle = i|1\rangle \otimes |1\rangle \otimes |0\rangle. \quad (2.99)$$

It is not always necessary to use the tensor product symbol $\otimes$, and we shall often omit it where there is no danger of misunderstanding. In the above equation, for example, we can write the three-qubit state $|1\rangle \otimes |1\rangle \otimes |0\rangle$ as $|1\rangle|1\rangle|0\rangle$ or $|110\rangle$ without introducing any ambiguity. It would be wrong, however, to write $\hat{\sigma}_x \otimes \hat{\sigma}_y \otimes \hat{\sigma}_z$ as $\hat{\sigma}_x\hat{\sigma}_y\hat{\sigma}_z$. The latter, of course, denotes the product of three Pauli operators acting on the *same qubit* so that $\hat{\sigma}_x\hat{\sigma}_y\hat{\sigma}_z = i\hat{I}$. The set of all possible unitary transformations is not limited to those acting on single qubits alone. More generally, transformations can couple two or more qubits to form entangled states. We shall study such transformations in Chapter 6.

2.5   Entangled states

Entanglement is a property of correlations between two or more quantum systems. These correlations defy classical description and are associated with intrinsically quantum phenomena. For this reason, entanglement has played an important role in the development and testing of quantum theory. It is also a central element in quantum information. It is not easy, however, to give a precise definition of entanglement other than that it is a property of entangled states. The problem is then shifted to defining the entangled states. It is simpler, however, to define states that are not entangled.

We begin our discussion of entangled states by considering a pure state of two quantum systems which we label a and b. We have seen in Section 2.2 that we can write a composite state of these as the direct product

$$|\psi\rangle = |\lambda\rangle_a |\phi\rangle_b. \quad (2.100)$$

For this state we can write down a state $|\lambda\rangle_a$ for system a alone. If the state in eqn 2.100 is prepared then the statistical properties of system a are determined by the state vector $|\lambda\rangle_a$. The superposition principle tells us that any superposition of product states such as 2.100 is also an allowed state of our two systems. Consider, for example, the two-qubit state

$$|\psi\rangle = \frac{1}{\sqrt{2}} \left(|0\rangle_a |0\rangle_b + |1\rangle_a |1\rangle_b \right). \quad (2.101)$$

This comprises a superposition of product states and it is not possible to write this as a product of a state for system a and one for system b, that is in the form of eqn 2.100. States having this property, that they cannot be written as product states, are entangled. It is not sufficient, however, for entanglement that the two-system state is a superposition of product states. Consider, for example, the state

$$\frac{1}{2}(|0\rangle_a|0\rangle_b + |0\rangle_a|1\rangle_b + |1\rangle_a|0\rangle_b + |1\rangle_a|1\rangle_b)$$

$$= \frac{1}{\sqrt{2}}(|0\rangle_a + |1\rangle_a) \otimes \frac{1}{\sqrt{2}}(|0\rangle_b + |1\rangle_b). \qquad (2.102)$$

This state is such a superposition but can be written as a product state and is therefore not entangled.

The impossibility of writing a state as a product may be quite difficult to establish and it is useful to have a more direct test for entanglement. This is provided by forming the density operator associated with the state and taking the trace over one of the systems to obtain the reduced density operator for the other. The density operator for the non-entangled state in eqn 2.100 is

$$\hat{\rho}_{ab} = |\lambda\rangle\langle\lambda| \otimes |\phi\rangle\langle\phi|. \qquad (2.103)$$

Evaluating the trace over the b states gives the reduced density operator for the a system

$$\hat{\rho}_a = |\lambda\rangle\langle\lambda|, \qquad (2.104)$$

which is the density operator for a pure state. Clearly $\hat{\rho}_a^2 = \hat{\rho}_a$ and $\mathrm{Tr}(\hat{\rho}_a^2) = 1$. Any non-entangled pure state can be written in the form of eqn 2.100 and therefore the condition $\mathrm{Tr}(\hat{\rho}_a^2) = 1$ is a signature of a non-entangled state. If we find that $\mathrm{Tr}(\hat{\rho}_a^2) \neq 1$ then the state is entangled.

It is always possible to write an entangled state of our two systems in the form

$$|\psi\rangle = \sum_n a_n|\lambda_n\rangle_a|\phi_n\rangle_b, \qquad (2.105)$$

where the states $\{|\lambda_n\rangle\}$ and $\{|\phi_n\rangle\}$ are orthonormal sets for the a and b systems, respectively. Here, each state $|\lambda_n\rangle$ of the a system is uniquely associated with a state $|\phi_n\rangle$ of the b system. This form of the state is known as the Schmidt decomposition, and the orthonormal states are the eigenstates of the reduced density operators for the a and b systems

$$\hat{\rho}_a = \sum_n |a_n|^2|\lambda_n\rangle\langle\lambda_n|,$$

$$\hat{\rho}_b = \sum_n |a_n|^2|\phi_n\rangle\langle\phi_n|. \qquad (2.106)$$

A proof that this decomposition is always possible is given in Appendix D. Note that the two density operators have the same eigenvalues, and this means that $\mathrm{Tr}(\hat{\rho}_a^2) = \mathrm{Tr}(\hat{\rho}_b^2)$. If the states $\{|\lambda_n\rangle\}$ and $\{|\phi_n\rangle\}$ are the eigenstates of a pair of operators

$$\hat{A} = \sum_n \lambda_n|\lambda_n\rangle_{aa}\langle\lambda_n|,$$

$$\hat{B} = \sum_n \phi_n|\phi_n\rangle_{bb}\langle\phi_n|, \qquad (2.107)$$

and if the eigenvalues are all distinct then it follows that a measurement of $\hat{A}$ uniquely determines the outcome of a measurement of $\hat{B}$ and, of course, vice versa. In this way, we see that the observables A and B are perfectly correlated.

Among the many possible entangled states, the Bell states of two qubits have a special prominence. The reasons for this include their simplicity and the fact that they have been realized in a number of diverse experiments. The four Bell states are conventionally written in the form

$$|\Psi^-\rangle = \frac{1}{\sqrt{2}} \left(|0\rangle \otimes |1\rangle - |1\rangle \otimes |0\rangle \right),$$

$$|\Psi^+\rangle = \frac{1}{\sqrt{2}} \left(|0\rangle \otimes |1\rangle + |1\rangle \otimes |0\rangle \right),$$

$$|\Phi^-\rangle = \frac{1}{\sqrt{2}} \left(|0\rangle \otimes |0\rangle - |1\rangle \otimes |1\rangle \right),$$

$$|\Phi^+\rangle = \frac{1}{\sqrt{2}} \left(|0\rangle \otimes |0\rangle + |1\rangle \otimes |1\rangle \right). \tag{2.108}$$

We see that the first of these is antisymmetric under exchange of the two qubits but that the remaining three are symmetric. In terms of angular momentum, the first is the state of zero total angular momentum and the remaining three span the space of states with unit total angular momentum. The Bell states are the common eigenstates of the three operators $\hat{\sigma}_x \otimes \hat{\sigma}_x$, $\hat{\sigma}_y \otimes \hat{\sigma}_y$, and $\hat{\sigma}_z \otimes \hat{\sigma}_z$, with eigenvalues ± 1, corresponding to the values of the two spin components being equal ($+1$) or opposite (-1). It follows that measuring one of the two qubits immediately reveals the value that would be found if the same spin component were to be measured on the second qubit. The single-qubit operators $\hat{\sigma}_x$, $\hat{\sigma}_y$, and $\hat{\sigma}_z$ do not commute and so are incompatible and have no common eigenstates, which means that it is not possible to simultaneously predetermine the outcome of a measurement of any one of these. A measurement carried out on the first qubit, however, *instantaneously* determines the state of the second, no matter how distant it may be from the first. The apparent instantaneous change exerted by a measurement on a distant quantum system led Einstein, Podolsky, and Rosen to question the validity of quantum mechanics as a fundamental theory, and much attention has been devoted to it. We shall discuss such counter-intuitive *non-local* behaviour more thoroughly in Chapter 5.

Entanglement is not limited to just a pair of quantum systems, and in quantum computing, in particular, very large numbers of qubits may be entangled. For this reason it is useful and interesting to study entangled states of more than two systems. As the number of systems is increased, the variety of possible entangled states grows rapidly. Even with just three qubits we can have states in which two of the qubits are entangled with each other but not with the third, for example

$$\frac{1}{2} \left(|010\rangle - |100\rangle + |011\rangle - |101\rangle \right) = |\Psi^-\rangle \otimes \frac{1}{\sqrt{2}} \left(|0\rangle + |1\rangle \right). \tag{2.109}$$

It is also possible for the three qubits to be fully entangled, with the properties of any one being correlated with both of the others. Two important examples are the GHZ (Greenberger–Horne–Zeilinger) state and the Werner state, which have the forms

$$|\text{GHZ}\rangle = \frac{1}{\sqrt{2}}\left(|000\rangle + |111\rangle\right),$$
$$|\text{W}\rangle = \alpha|001\rangle + \beta|010\rangle + \gamma|100\rangle, \tag{2.110}$$

where α, β, and γ are probability amplitudes, none of which take the value zero. It is clear that for these states all three qubits are mutually entangled. We can demonstrate this in two ways. Firstly, we can form the reduced density operators for the three qubits, $\hat{\rho}_a$, $\hat{\rho}_b$, and $\hat{\rho}_c$, and we find that for each of these $\text{Tr}(\hat{\rho}_i^2) < 1$. For the GHZ state, this gives $\text{Tr}(\hat{\rho}_i^2) = \frac{1}{2}$ for each qubit. A second and rather more direct method is to note that any measurement of one of the three qubits changes the state of the remaining two. For example, observing the third qubit in the Werner state to be in the state $|1\rangle$ changes the state of the other two qubits to $|00\rangle$, but finding the third qubit to be in the state $|0\rangle$ leaves the others in the entangled state $(\beta|01\rangle + \gamma|10\rangle)/(|\beta|^2 + |\gamma|^2)^{1/2}$. For the state in eqn 2.109, however, we find that $\text{Tr}(\hat{\rho}_c^2) = 1$ and that measuring the state of the third qubit does not change the state of the remaining two. For four or more qubits there are more complicated possibilities, including, for example, $|\Psi^-\rangle_{ab}|\Psi^+\rangle_{cd}$, in which qubits a and b are entangled, as are qubits c and d, but the first pair are not entangled with the second. For such states, more subtle methods are required to fully characterize the entanglement.

If two or more systems are entangled then it necessarily follows that at least some of their properties are correlated. For a pair of observables A and B, associated with the two respective systems, this means that the expectation value of the product of the corresponding operators will not, in general, factorize:

$$\langle \hat{A} \otimes \hat{B} \rangle \neq \langle \hat{A} \rangle \langle \hat{B} \rangle. \tag{2.111}$$

The existence of correlated properties is not special to entangled states, however, nor even to quantum physics. We shall find, however, that entangled states exhibit some correlations that cannot be mimicked by classical systems, and so it is useful to reserve the term 'entanglement' for such intrinsically quantum phenomena. All correlated pure states are entangled, but this is not true for mixed states. If a mixed state of two systems a and b has a density operator of the form

$$\hat{\rho} = \hat{\rho}_a \otimes \hat{\rho}_b \tag{2.112}$$

then the two systems are clearly uncorrelated. We refer to a state formed from a mixture of these,

$$\hat{\rho} = \sum_{ij} P(a_i, b_j) \hat{\rho}_a^i \otimes \hat{\rho}_b^j, \tag{2.113}$$

Classical communications and un-entangled states   We can envisage a communication channel characterized by the joint probabilities $\{P(a_i, b_j)\}$. If Alice selects the message a_i then she also prepares her quantum system in the state $\hat{\rho}_a^i$, and if Bob receives the signal b_j then he prepares $\hat{\rho}_b^j$. The a priori density operator for the two quantum systems is then that in eqn 2.113.

as a correlated but not an entangled state. Such states can be prepared by classical communications and do not display the intrinsically quantum correlations associated with entanglement. As with pure states, it is easier to define unentangled mixed states than entangled ones: an unentangled mixed state of two quantum systems is one that can be written in the form of eqn 2.113 with the $\{P(a_i, b_j)\}$ being a set of (positive) joint probabilities. A mixed state is entangled if it cannot be written in this form.

It would be useful if we had a test that could determine whether or not a given mixed state is entangled. At present, however, no such universal test exists and finding one remains an important research problem in quantum information theory. There is, however, a sufficient condition for a mixed state to be entangled and this is based on the transpose of the density operator or, more precisely, its partial transpose. If we choose a basis $\{|n\rangle\}$ in which to represent our density operator then we can write it as a matrix, the density matrix, in which the elements are ρ_{mn}. The corresponding element of the transposed density matrix will be ρ_{nm}. Equivalently, if our density operator is

$$\hat{\rho} = \sum_{mn} \rho_{mn} |m\rangle\langle n| \tag{2.114}$$

then $\hat{\rho}^{\mathrm{T}}$ is obtained by changing $|m\rangle\langle n|$ to $|n\rangle\langle m|$. For a single qubit, the density matrix in the $|0\rangle, |1\rangle$ basis has the form

$$\rho = \begin{pmatrix} \rho_{00} & \rho_{01} \\ \rho_{10} & \rho_{11} \end{pmatrix}, \tag{2.115}$$

where $\rho_{10} = \rho_{01}^*$. The transpose of the density operator $(\hat{\rho}^{\mathrm{T}})$ has the associated density matrix

$$\rho^{\mathrm{T}} = \begin{pmatrix} \rho_{00} & \rho_{10} \\ \rho_{01} & \rho_{11} \end{pmatrix}. \tag{2.116}$$

It is clear that this is an allowed density matrix and, more generally, if $\hat{\rho}$ is a possible density operator for a quantum system then so too is $\hat{\rho}^{\mathrm{T}}$.

Performing the transposition operation on *one* of a pair of quantum systems but not on the second constitutes the *partial transpose*, or PT, operation. If we consider the unentangled state in eqn 2.113 then it is clear that performing the transpose operation on the second system but not on the first will produce a density operator of the form

$$\hat{\rho}^{\mathrm{PT}} = \sum_{ij} P(a_i, b_j) \hat{\rho}_a^i \otimes \hat{\rho}_b^{j\mathrm{T}}. \tag{2.117}$$

That this represents an allowed state of the combined system is clear from the fact that the transpose of any density operator (in this case $\hat{\rho}_b^{j\mathrm{T}}$) is itself a density operator.

Performing the partial transpose for an entangled state, however, does not necessarily lead to an allowed density operator. In particular, for an entangled state we often find that $\hat{\rho}^{\mathrm{PT}}$ has some negative eigenvalues and

Complex conjugate of a density operator It is clear that, unlike the Hermitian conjugate, the transpose of a density operator depends on the basis in which it is expressed. The Hermitian conjugate is a combination of the transpose and the complex conjugate operations and it follows, therefore, that the transpose of a Hermitian operator (such as a density operator) is also its complex conjugate:

$$\hat{\rho}^{\mathrm{T}} = \hat{\rho}^{\dagger*} = \hat{\rho}^*.$$

hence lacks the positivity required by all density operators. A simple example is the Bell state $|\Psi^-\rangle$, which, in the basis $|00\rangle, |01\rangle, |10\rangle, |11\rangle$, has the density matrix

$$\rho = \frac{1}{2} \begin{pmatrix} 0 & 0 & 0 & 0 \\ 0 & 1 & -1 & 0 \\ 0 & -1 & 1 & 0 \\ 0 & 0 & 0 & 0 \end{pmatrix}. \tag{2.118}$$

The partially transposed density matrix is

$$\rho^{\text{PT}} = \frac{1}{2} \begin{pmatrix} 0 & 0 & 0 & -1 \\ 0 & 1 & 0 & 0 \\ 0 & 0 & 1 & 0 \\ -1 & 0 & 0 & 0 \end{pmatrix}, \tag{2.119}$$

which has three positive eigenvalues and one negative one. It follows that ρ^{PT} is not an acceptable density matrix. If $\hat{\rho}$ is an unentangled state then $\hat{\rho}^{\text{PT}}$ is an allowed density operator. It follows that $\hat{\rho}^{\text{PT}}$ not being positive is a sufficient condition for $\hat{\rho}$ to be entangled. For some simple systems, including states of a pair of qubits, it is also a necessary condition. As an example we consider Werner's mixed state, which is a mixture of the randomized two-qubit state $\frac{1}{4}\hat{I} \otimes \hat{I}$ and the first Bell state:

$$\hat{\rho}_{\text{W}} = \frac{p}{4}\hat{I} \otimes \hat{I} + (1-p)|\Psi^-\rangle\langle\Psi^-|. \tag{2.120}$$

The corresponding density operator, in the basis $|00\rangle, |01\rangle, |10\rangle, |11\rangle$, is

$$\rho = \begin{pmatrix} \frac{1}{4}p & 0 & 0 & 0 \\ 0 & \frac{1}{2}\left(1 - \frac{1}{2}p\right) & -\frac{1}{2}(1-p) & 0 \\ 0 & -\frac{1}{2}(1-p) & \frac{1}{2}\left(1 - \frac{1}{2}p\right) & 0 \\ 0 & 0 & 0 & \frac{1}{4}p \end{pmatrix}. \tag{2.121}$$

Performing the partial transpose on this gives the matrix

$$\rho^{\text{PT}} = \begin{pmatrix} \frac{1}{4}p & 0 & 0 & -\frac{1}{2}(1-p) \\ 0 & \frac{1}{2}\left(1 - \frac{1}{2}p\right) & 0 & 0 \\ 0 & 0 & \frac{1}{2}\left(1 - \frac{1}{2}p\right) & 0 \\ -\frac{1}{2}(1-p) & 0 & 0 & \frac{1}{4}p \end{pmatrix}, \tag{2.122}$$

which has a negative eigenvalue if $p < \frac{2}{3}$. It follows that $\hat{\rho}_{\text{W}}$ can be written in the form of eqn 2.113 and hence is not an entangled state if $p \geq \frac{2}{3}$.

Suggestions for further reading

Barnett, S. M. and Radmore, P. M. (1997). *Methods in theoretical quantum optics*. Oxford University Press, Oxford.

Holevo, A. S. (1982). *Probabilistic and statistical aspects of quantum theory*. North Holland, Amsterdam.

Holevo, A. S. (2000). *Statistical structure of quantum theory*. Springer-Verlag, Berlin.

Nielsen, M. A. and Chuang, I. L. (2000). *Quantum computation and quantum information*. Cambridge University Press, Cambridge.

Peres, A. (1995). *Quantum theory: concepts and methods*. Kluwer Academic Publishers, Dordrecht.

Stenholm, S. and Suominen, K.-A. (2005). *Quantum approach to informatics*. Wiley, New Jersey.

Exercises

(2.1) Verify that the Hermitian conjugate operation for two-by-two matrices satisfies all of the requirements in eqns 2.8–2.11.

(2.2) Show that the uncertainty ΔA is zero if and only if the state is an eigenstate of $\hat{A}$.

(2.3) Show that if two Hermitian operators $\hat{A}$ and $\hat{B}$ commute then they have a complete set of common eigenstates.

(2.4) Hermitian operators have real eigenvalues. Is there a similar statement that we can make about skew-Hermitian operators?

(2.5) Simplify the following using commutators and anticommutators:

(a) $\hat{A}\hat{C}[\hat{B},\hat{D}] + \hat{A}[\hat{B},\hat{C}]\hat{D} + [\hat{A},\hat{C}]\hat{D}\hat{B}$ $+\hat{C}[\hat{A},\hat{D}]\hat{B}$;

(b) $\{\hat{A},\hat{B}\}\hat{C} + \hat{B}[\hat{C},\hat{A}]$;

(c) $\hat{B}[\hat{A},\hat{C}\hat{D}] + [\hat{A},\hat{B}]\hat{C}\hat{D} + \hat{B}\hat{C}[\hat{A},\hat{D}]$ $+\hat{B}[\hat{A},\hat{C}]\hat{D}$;

(d) $\hat{A}\hat{B}\hat{C} + \hat{A}\hat{C}\hat{B} + \hat{B}\hat{C}\hat{A} + \hat{C}\hat{B}\hat{A}$.

(2.6) Show that the intelligent states, those that satisfy the equality in the uncertainty relation in eqn 2.24, are eigenstates of the operator $\hat{A} + i\lambda\hat{B}$, where λ is real. What physical properties of the state determine the corresponding eigenvalue and the value of λ?

(2.7) The position and momentum operators, $\hat{x}$ and $\hat{p}$, satisfy the commutation relation $[\hat{x},\hat{p}] = i\hbar$. This leads to Heisenberg's uncertainty principle

$$\Delta x\,\Delta p \geq \hbar/2.$$

The intelligent states and minimum uncertainty-product states are the same because the lower bound is a constant. Find the general form of these states.

[*Hint:* you may find it useful to work in the position representation in which

$$\hat{x}|\psi\rangle \rightarrow x\psi(x)$$
$$\hat{p}|\psi\rangle \rightarrow i\hbar\frac{\partial}{\partial x}\psi(x).]$$

(2.8) Show that the probability that a measurement of the position of a particle gives a value further than $\mu\,\Delta x$ from the expectation value, for any positive μ, is less than or equal to $1/\mu^2$; that is,

$$P\left(|x - \langle x\rangle| \geq \mu\,\Delta x\right) \leq \frac{1}{\mu^2}.$$

(2.9) Show that all positive operators are Hermitian. [*Hint:* you might like to start by noting that any operator $\hat{A}$ can be written in terms of a Hermitian part $(\hat{A} + \hat{A}^\dagger)/2$ and a skew-Hermitian part $(\hat{A} - \hat{A}^\dagger)/2$, and then impose the condition for positivity on $\hat{A}$.]

(2.10) Show that the expansion of a function of a Hermitian operator in terms of eigenstates (eqn 2.30) is equivalent to that based on Taylor's theorem (eqn 2.31). Why might using eqn 2.31 cause problems for some functions?

(2.11) The maximum value of the purity, $\text{Tr}(\hat{\rho}^2)$, is unity. What is the minimum value and what is the corresponding density operator?

(2.12) Evaluate the purity for Werner's mixed state (eqn 2.120).

(2.13) Show that if the equality holds in eqn 2.54 for all n and m then the state is pure.

(2.14) The Cauchy–Schwarz inequality (eqn 2.55) holds as an equality if and only if $|\phi_2\rangle \propto |\phi_1\rangle$. Prove this statement.

[*Hint:* you might like to try writing $|\phi_2\rangle = a|\phi_1\rangle + |\chi\rangle$, where $\langle\chi|\phi_1\rangle = 0$.]

(2.15) Express $\text{Tr}(\hat{\rho}^n(t))$ for the time-evolved density operator (eqn 2.57) in terms of $\hat{\rho}(0)$.

(2.16) Not all the quantities $\mathrm{Tr}(\hat{\rho}^n)$ are independent. To illustrate this idea, write, for a general qubit state, the following quantities as functions of $\mathrm{Tr}(\hat{\rho}) = 1$ and $\mathrm{Tr}(\hat{\rho}^2)$:

(a) $\mathrm{Tr}(\hat{\rho}^3)$;
(b) $\det(\hat{\rho})$, the determinant of the 2×2 matrix associated with $\hat{\rho}$.

As an extra challenge, you might like to prove the statement that 'for a d-dimensional state space only the first d values of n give independent quantities.'

(2.17) We have seen that the condition in eqn 2.62 is a sufficient condition for the density operators in eqns 2.43 and 2.61 to be identical. Show that it is also a necessary condition.

(2.18) If the Hamiltonian is time-dependent then the solution of eqn 2.38 is complicated by the fact that the Hamiltonian operators at two different times will not usually commute. Show that the resulting time evolution operator is

$$\hat{U}(t) = \mathrm{T}\exp\left(-\frac{i}{\hbar}\int_0^t \hat{H}(t')\,dt'\right),$$

where T denotes time ordering. This means that the Taylor series for this operator has the form

$$\hat{U}(t) = \hat{\mathrm{I}} - \frac{i}{\hbar}\int_0^t dt_1 \hat{H}(t_1)$$
$$- \frac{1}{\hbar^2}\int_0^t dt_1 \hat{H}(t_1)\int_0^{t_1} dt_2 \hat{H}(t_2)$$
$$+ \cdots.$$

Verify that this operator is unitary.

(2.19) Show, for any unitary operator $\hat{U}$, that if $\{|\lambda_m\rangle\}$ is a complete basis of orthonormal states then so is $\{\hat{U}|\lambda_m\rangle\}$.

(2.20) Prove eqn 2.81 using the operator expansion in eqn 2.32.

(2.21) If $\hat{A}$ and $\hat{B}$ are non-commuting Hermitian operators then $\exp\left[i\left(\hat{A}+\hat{B}\right)t\right]$ is a unitary operator.

(a) Show that

$$\exp\left[i\left(\hat{A}+\hat{B}\right)t\right]$$
$$= \lim_{n\to\infty}\left[\exp\left(i\frac{\hat{A}t}{n}\right)\exp\left(i\frac{\hat{B}t}{n}\right)\right]^n.$$

(b) We can use this expression for finite n as an approximation. Show that a better approximation is

$$\exp\left[i\left(\hat{A}+\hat{B}\right)t\right]$$
$$\approx \left[\exp\left(i\frac{\hat{B}t}{2n}\right)\exp\left(i\frac{\hat{A}t}{n}\right)\exp\left(i\frac{\hat{B}t}{2n}\right)\right]^n.$$

(2.22) Find the Hermitian conjugate of eqn 2.82 and hence confirm that it is unitary.

(2.23) Show that each of the four Pauli operators is both unitary and Hermitian. It follows, of course, that the square of a Pauli operator is $\hat{\mathrm{I}}$.

(2.24) Evaluate the operator product $\vec{a}\cdot\hat{\vec{\sigma}}\,\vec{b}\cdot\hat{\vec{\sigma}}$.

(2.25) Write down the uncertainty principle for the Pauli operators $\hat{\sigma}_x$ and $\hat{\sigma}_y$. Derive the form of the associated intelligent states.

(2.26) Derive the general form of the minimum uncertainty-product states for $\hat{\sigma}_x$ and $\hat{\sigma}_y$. (These are the states that minimize the value of $\Delta\sigma_x$ for given $\Delta\sigma_y$.)

(2.27) We can write the density operator for a single qubit in the form

$$\hat{\rho} = \int_0^\pi \sin\theta\,d\theta \int_0^{2\pi} d\varphi\, P(\theta,\varphi)|\psi\rangle\langle\psi|,$$

where the state $|\psi\rangle$ is given in eqn 2.89.

(a) Find the diagonal form of the density operator if $P(\theta,\varphi)$ is a constant over the whole Bloch sphere.

(b) Find the diagonal form of the density operator if $P(\theta,\varphi)$ takes a constant value for $0 \leq \theta < \mu$ and is zero for $\mu \leq \theta \leq \pi$, for a given angle μ.

(2.28) (a) Construct a general unitary two-by-two matrix. Hence show that a general single-qubit unitary operator can be written in the forms given in eqns 2.90 and 2.91.

(b) Show that any single-qubit unitary operator given in eqn 2.90, with $\alpha = 0$ can be written in the form given in eqn 2.92 for a suitable choice of the angles λ, μ and ν.

(2.29) Calculate the length of the Bloch vector $(u^2 + v^2 + w^2)^{1/2}$ for the qubit state in eqn 2.93 as a function of $\mathrm{Tr}(\hat{\rho}^2)$.

(2.30) By writing the general qubit state vector

$$|\lambda\rangle_a = \cos(\theta/2)|0\rangle_a + e^{i\varphi}\sin(\theta/2)|1\rangle_a,$$

show that it is impossible to write the entangled state in eqn 2.101 in the form of a product state as in eqn 2.100.

2.31) A simple measure of the degree of entanglement of two quantum systems is the Schmidt number. This is defined for pure states, and for a state written in the form of eqn 2.105 has the form

$$K = \frac{1}{\sum_n |a_n|^4}.$$

(a) Find the minimum and maximum values of the Schmidt number for a given pair of d-state quantum systems and show that the minimum value occurs only if the state is not entangled.

(b) We can use the Schmidt number as a quantitative measure of entanglement with increasing values associated with ever higher degrees of entanglement. Show that, by this measure, the Bell states in eqn 2.108 are maximally entangled.

2.32) Consider the entangled pure state in eqn 2.105 and the operators in eqn 2.107.

(a) Does the perfect correlation between these observables imply that $\Delta(A - B)^2 = 0$?

(b) Calculate $H(A : B)$. What is the relevance of this value?

2.33) Construct the unitary 'swap' operator, the action of which is to exchange the arbitrary states of a pair of qubits: $\hat{U}_{\text{swap}}|\alpha\rangle \otimes |\beta\rangle = |\beta\rangle \otimes |\alpha\rangle$.

[*Hint:* you may find it helpful to consider the symmetry properties of the Bell states.]

2.34) Show that the three operators $\hat{\sigma}_x \otimes \hat{\sigma}_x$, $\hat{\sigma}_y \otimes \hat{\sigma}_y$, and $\hat{\sigma}_z \otimes \hat{\sigma}_z$ all mutually commute, and confirm that the four Bell states are their common eigenstates. Why are the products of the three eigenvalues -1 for all these states?

2.35) Find $\text{Tr}(\hat{\rho}_a^2)$, $\text{Tr}(\hat{\rho}_b^2)$, and $\text{Tr}(\hat{\rho}_c^2)$ for the Werner state in eqn 2.110.

2.36) Consider the four-qubit state $|\Phi^+\rangle_{ab}|\Phi^+\rangle_{cd}$.

(a) Which qubits are entangled?

(b) Show that measuring any one of the qubits changes the state of the remaining three.

(c) Calculate $\text{Tr}(\hat{\rho}_i^2)$ for the reduced density operators of each of the four qubits.

2.37) For each of the following states, determine which of the qubits are mutually entangled:

(a) $\frac{1}{2}(|00\rangle + |01\rangle + |10\rangle - |11\rangle)$;

(b) $\frac{1}{2}(|000\rangle + 2|011\rangle + |110\rangle)$;

(c) $\frac{1}{2}(|000\rangle + i|010\rangle + i|101\rangle - |111\rangle)$;

(d) $\frac{1}{2}(|0000\rangle + |0011\rangle + |1100\rangle + |1111\rangle)$.

2.38) Show that the correlated density operator in eqn 2.113 can be written in the form

$$\hat{\rho} = \sum_{kl} \tilde{P}(k, l)|\psi_k\rangle\langle\psi_k| \otimes |\phi_l\rangle\langle\phi_l|.$$

2.39) Consider a general density matrix for a quantum system with a d-dimensional state space.

(a) Prove that the transpose of this density matrix corresponds to an allowed quantum state.

(b) Show that the eigenvalues of $\hat{\rho}^{\text{T}}$ are the same as those of $\hat{\rho}$.

(c) How are the eigenvectors of $\hat{\rho}^{\text{T}}$ related to the eigenvectors of $\hat{\rho}$?

2.40) Construct the density matrix, in the basis $|00\rangle, |01\rangle, |10\rangle, |11\rangle$, for the two-qubit density operator

$$\hat{\rho} = \frac{p}{4}(|0\rangle + |1\rangle)(\langle 0| + \langle 1|) \otimes (|0\rangle + i|1\rangle)(\langle 0| - i\langle 1|).$$
$$+ (1 - p)|0\rangle\langle 0| \otimes |1\rangle\langle 1|$$

Construct the partial transpose of this density matrix and confirm that it is an acceptable density matrix.

2.41) Write the state $\hat{\rho}_W$ (eqn 2.120) with $p \geq \frac{2}{3}$ in the explicitly unentangled form given in eqn 2.113.

[*Hint:* you might start by considering a mixture of anticorrelated states such as $|0\rangle\langle 0| \otimes |1\rangle\langle 1|$ and $\frac{1}{4}(|0\rangle + |1\rangle)(\langle 0| + \langle 1|) \otimes (|0\rangle - |1\rangle)(\langle 0| - \langle 1|)$.]

Quantum cryptography

The practical implementation of quantum information technologies requires, for the most part, highly advanced and currently experimental procedures. One exception is quantum cryptography, or quantum key distribution, which has been successfully demonstrated in many laboratories and has reached an advanced level of development. It will probably become the first commercial application of quantum information.

In quantum key distribution, Alice and Bob exploit a quantum channel to create a secret shared key comprising a random string of binary digits. This key can then be used to protect a subsequent communication between them. The principal idea is that the secrecy of the key distribution is ensured by the laws of quantum physics. Proving security for practical communication systems is a challenging problem and requires techniques that are beyond the scope of this book. At a fundamental level, however, the ideas are simple and may readily be understood with the knowledge we have already acquired.

Quantum cryptography is the latest idea in the long history of secure (and not so secure) communications and, if it is to develop, it will have to compete with existing technologies. For this reason we begin with a brief survey of the history and current state of the art in secure communications before turning to the possibilities offered by quantum communications.

3.1 Information security

The history of cryptography is a long and fascinating one. As a consequence of the success or, more spectacularly, the failure of ciphers, wars have been fought, battles decided, kingdoms won, and heads lost. In the information age, ciphers and cryptosystems have become part of everyday life; we use them to protect our computers, to shop over the Internet, and to access our money via an ATM (automated teller machine).

One of the oldest and simplest of all ciphers is the transposition or Caesarean cipher (attributed to Julius Caesar), in which the letters are shifted by a known (and secret) number of places in the alphabet. If the shift is 1, for example, then A is enciphered as B, B→C, ···, Y→Z, Z→A. A shift of five places leads us to make the replacements A→F, B→G, ···, Y→D, Z→E. The weakness of this cipher is that there are only 25 possible shifts to try and it is easy to try them all. Doing so will usually lead to only one combination that is intelligible, and this is the

Codes and ciphers The terms 'code' and 'cipher' are often used synonymously but in our subject they have quite distinct meanings. A code is produced by a substitution of the symbols in a message by other symbols; there is not necessarily any attempt at secrecy and the code may be widely published, such as with Morse code and ASCII. A cipher, however, is a message specifically modified so as to protect its meaning.

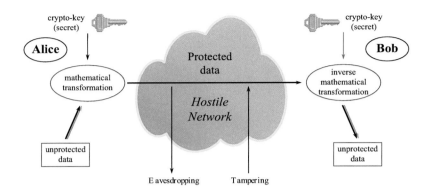

Fig. 3.1 Depiction of the elements of a generic cryptosystem.

deciphered message. Consider, for example, the cipher

YBTXOB QEB FABP LC JXOZE

A test of all possible shifts reveals a sensible message only for a shift of 23 places, which reveals

BEWARE THE IDES OF MARCH

The Caesarean cipher, although very simple to break, illustrates the important elements of a private-key cryptosystem, as depicted in Fig. 3.1. The first of these is the plaintext $\mathcal{P}$, which is the secret message Alice wishes to transmit to Bob. The second is the secret key $\mathcal{K}$, known only to Alice and to Bob, and used by Alice to generate the secure ciphertext $\mathcal{C}$. It is this ciphertext that is transmitted to Bob, who uses $\mathcal{K}$ to transform $\mathcal{C}$ into $\mathcal{P}$. In this way, we can think of the ciphertext as a function of the plaintext and the key, and also the plaintext as a function of the ciphertext and the key:

$$\mathcal{C} = \mathcal{C}(\mathcal{P}, \mathcal{K}),$$
$$\mathcal{P} = \mathcal{P}(\mathcal{C}, \mathcal{K}). \tag{3.1}$$

For the case of the Caesarean cipher, the key is one of the first 25 integers and the mathematical transformation is transposition of the letters. At least in principle, secrecy is ensured by the fact that Eve (an eavesdropper on the network) has access to $\mathcal{C}$ but not to $\mathcal{K}$.

The obvious weakness of the Caesarean cipher is that it is limited to just 25 keys. In a substitution cipher, however, each letter is replaced by another letter or symbol, which results in $26! \approx 4 \times 10^{26}$ possible keys. Clearly an exhaustive key search, corresponding to trying all possible combinations, is impractical. This might lead us to believe that substitution ciphers are highly secure, but this is not the case. The reason is that the letters in a message appear with more or less well-defined frequencies. These will depend slightly on the text and its subject but the overwhelming majority of messages in English will have frequencies close to those given in table 3.1. From this it should be relatively easy to identify the symbols representing the commonest letters: E, T, A, O, and perhaps I. These, together with knowledge of the English language and perhaps the context of the message, can be used to complete the decipherment. The sequence T?E, for example, is very likely to mean that the symbol ? represents the letter H. This technique for breaking substitution ciphers has been understood for centuries, and numerous embellishments upon it have been developed but then defeated. There are some Caesarean and substitution ciphers in the exercises section if you would like to try applying these methods of deciphering.

Given the failures of transposition, substitution, and yet more sophisticated ciphers, it is interesting to ask whether an unbreakable or perfectly secret cipher can exist. In order to assess this question carefully, we require a precise mathematical definition of perfect secrecy. Shannon provided the required definition by considering the task facing Eve, who wishes to reconstruct the plaintext $\mathcal{P}$ given access only to the ciphertext $\mathcal{C}$. The message is perfectly secret if $\mathcal{C}$ gives Eve no information about $\mathcal{P}$; that is, it does not change the probability for any given message. Let the set of possible plaintext messages be $\{\mathcal{P}_i\}$ and let $\{\mathcal{C}_j\}$ be the set of possible ciphertexts. Perfect security then implies that

$$P(\mathcal{P}_i|\mathcal{C}_j) = P(\mathcal{P}_i), \qquad \forall i,j. \tag{3.2}$$

Hence Eve's probability for recovering the original plaintext is the same whether or not she has access to the ciphertext. It is also clear that any possible plaintext should lead to any given ciphertext with equal probability, so that

$$P(\mathcal{C}_j|\mathcal{P}_i) = P(\mathcal{C}_j), \qquad \forall i,j. \tag{3.3}$$

Were this not the case then Eve could associate a given ciphertext with a class or group of messages, and the corresponding information gained by Eve would mean that perfect secrecy was lost.

The condition for perfect secrecy tells us something important about the minimum number of possible keys required. To calculate this minimum number, we can reason as follows. First we note that all possible messages when encrypted with the same key must lead to distinct messages; were this not so, then Bob would be unable to reconstruct the original plaintext. It follows that there must be at least as many possible ciphertexts as plaintexts. Hence perfect secrecy requires the number of possible keys to be at least as great as the number of possible plaintexts. or messages. Were this not the case then there would be some messages that could not be encrypted to a given ciphertext, which violates the perfect-secrecy condition in eqn 3.3.

The first and simplest cipher to achieve perfect secrecy was the Vernam cipher, or one-time pad. The central idea is that the rule for transposing or substituting the letters changes for each symbol and never repeats. In the digital age, all messages are represented by a string of binary digits. Alice's message is encoded in this bit string, using a system such as ASCII, and it is this string that constitutes the plaintext.

Table 3.1 Approximate relative frequencies of the letters in English.

A	8.2%	J	0.1%	S	6.3%		
B	1.5%	K	0.8%	T	9.0%		
C	2.8%	L	4.0%	U	2.8%		
D	4.2%	M	2.4%	V	1.0%		
E	12.7%	N	6.7%	W	2.4%		
F	2.2%	O	7.5%	X	0.1%		
G	2.0%	P	1.9%	Y	2.0%		
H	6.1%	Q	0.1%	Z	0.1%		
I	7.0%	R	6.0%				

The ASCII code   ASCII, or the American Standard Code for Information Interchange, maps 128 distinct symbols onto the 128 different seven-bit strings 0000000 to 1111111. For example, the 26 upper-case letters, A, B, $\cdots$, Z, correspond to the binary strings 1000001, 1000010, $\cdots$, 10011010, and the lower-case letters, a, b, $\cdots$, z, are represented by the numbers 1100001, 1100010, $\cdots$, 1111010. Any message of m characters (including spaces) will correspond to a continuous string of $7m$ bits, with spaces between words encoded as the string 1011100.

Let us suppose that all of the possible messages are encoded as strings of N bits so as to be indistinguishable on the basis of length alone. Shorter messages can be increased to N bits, for example by adding zeros or the ASCII code for the suitable number of spaces. The Vernam cipher uses a key of N randomly chosen bits and its security relies entirely on the secrecy of this, which should ideally be known only to Alice and Bob. The ciphertext is created by modulo 2 addition, which we denote by $\oplus$:

$$0 \oplus 0 = 0, \qquad 0 \oplus 1 = 1,$$
$$1 \oplus 0 = 1, \qquad 1 \oplus 1 = 0. \tag{3.4}$$

Performing this addition bit by bit between the plaintext and the key generates the ciphertext:

$$\begin{aligned} \mathcal{P} & \quad \cdots 001011010 \cdots, \\ \mathcal{K} & \quad \cdots 101110100 \cdots, \\ \mathcal{C} = \mathcal{P} \oplus \mathcal{K} & \quad \cdots 100101110 \cdots. \end{aligned} \tag{3.5}$$

We see that each bit of the plaintext is either left unchanged by this operation or flipped ($0 \to 1$, $1 \to 0$), and if the key is random then each of these possibilities occurs with probability $\frac{1}{2}$. The nett effect is that the ciphertext itself reflects the random nature of the key, in that to anyone without access to the key the ciphertext is random and hence carries no information about the plaintext.

We can demonstrate the perfect secrecy of the Vernam cipher from the properties of the key. The key is a random string of N bits and it follows that it can take any one of 2^N values, with each possible value occurring with probability 2^{-N}. Hence any chosen plaintext will be mapped, by modulo 2 addition of the key, to any of 2^N possible ciphertexts, and each of these is equally likely:

$$P(\mathcal{C}_j|\mathcal{P}_i) = 2^{-N} = P(\mathcal{C}_j), \qquad \forall i, j, \tag{3.6}$$

which is Shannon's criterion for perfect secrecy given in eqn 3.3. It is also clear that an eavesdropper Eve having access only to the ciphertext has no information about the plaintext. This follows from the fact that

$$P(\mathcal{C}_j, \mathcal{P}_i) = P(\mathcal{C}_j|\mathcal{P}_i)P(\mathcal{P}_i) = P(\mathcal{C}_j)P(\mathcal{P}_i), \qquad \forall i, j, \tag{3.7}$$

which, in turn, means that the mutual information shared between the ciphertext and the plaintext is zero:

$$H(\mathcal{P} : \mathcal{C}) = 0. \tag{3.8}$$

Shannon's noisy coding theorem then tells us that no information about $\mathcal{P}$ is carried by $\mathcal{C}$ alone.

Recovering the message from the ciphertext is straightforward for Bob because he has the secret key. All he need do is to perform bit-by-bit modulo 2 addition of the ciphertext and the key:

$$\begin{aligned} \mathcal{C} & \quad \cdots 100101110 \cdots, \\ \mathcal{K} & \quad \cdots 101110100 \cdots, \\ \mathcal{P} = \mathcal{C} \oplus \mathcal{K} & \quad \cdots 001011010 \cdots. \end{aligned} \tag{3.9}$$

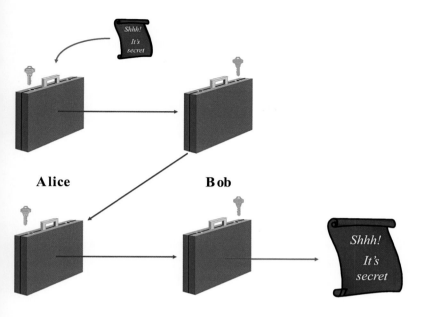

Alice

Bob

Fig. 3.2 The concept underlying Diffie–Hellman key exchange.

This works because the modulo 2 addition of a number with itself gives zero: $\mathcal{K} \oplus \mathcal{K} = \cdots 0000 \cdots$.

The perfect secrecy of the Vernam cipher has an obvious problem, and that is the necessity of communicating the secret key whilst keeping it secret. Alice and Bob could meet up and exchange a number of keys and, indeed, for many years this is what was done. The requirement to meet is, however, a major impediment to the widespread use of secure communications. If, for example, you wished to purchase a copy of this book online, then you would first need to meet with the bookseller in order to agree a key with which to secure your credit card details. If you need to meet up then why not simply purchase the book when you meet? The secure exchange of keys is the problem of key distribution that quantum cryptography, or quantum key distribution, was proposed to address. At the present time, however, the problem is dealt with in a very different way, with the security of the key distribution process being based on the difficulty of performing certain mathematical operations, and we conclude this section with a discussion of two of these ways.

The idea underlying Diffie–Hellman key exchange is depicted in Fig. 3.2. We suppose that Alice prepares a strong case, into which she places a secret message or cryptographic key destined for Bob, and that she then locks the case with a padlock. Alice then dispatches the case to Bob, keeping with her the only key for the padlock. For the purposes of this analogy, we shall assume that Eve is unable to open the locked case. On receipt of the case, Bob is also unable to open the case, but instead he locks it with a second padlock, keeps the only key, and returns the case to Alice. Alice can then unlock her padlock and once again dispatch the case to Bob, who, finally, can open it and retrieve the message. The case has made three journeys but in none of these is the secret

Practical security Perfect secrecy requires the generation, communication, and storage of large numbers of long keys. For all but the most secret messages, this is considered to be too high a price to pay. We usually settle for using shorter keys together with a published scrambling algorithm such as DES (Data Encryption Standard) or AES (Advanced Encryption Standard). Such ciphers are vulnerable to exhaustive key searches (trying each possible key in turn), but their security is based on the observation that this will take a very long time and the belief that no efficient algorithm exists for performing the unscrambling in the absence of the key. It is safest to consider messages encrypted in this way as being secure only for a time much less than that required to perform an exhaustive key search.

message available to Eve. If we could achieve something like this with mathematical transformations of our bit string then Alice and Bob could use it to communicate in secret. Let us see what happens if we try this using the Vernam cipher described above. We suppose that Alice wishes to transmit a plaintext $\mathcal{P}$ to Bob and generates a key $\mathcal{K}_A$, known only to her, to create a ciphertext,

$$\mathcal{C}_A = \mathcal{P} \oplus \mathcal{K}_A, \tag{3.10}$$

which she transmits to Bob. Naturally, neither Bob nor Eve can read this, but Bob can further encrypt the ciphertext using a second key $\mathcal{K}_B$, known only to him, to produce a second ciphertext,

$$\mathcal{C}_B = \mathcal{C}_A \oplus \mathcal{K}_B = \mathcal{P} \oplus \mathcal{K}_A \oplus \mathcal{K}_B, \tag{3.11}$$

which he sends back to Alice. Alice removes her key by a second modulo 2 addition of $\mathcal{K}_A$:

$$\mathcal{C}_{A'} = \mathcal{C}_B \oplus \mathcal{K}_A = \mathcal{P} \oplus \mathcal{K}_B. \tag{3.12}$$

She sends the result to Bob, who can remove $\mathcal{K}_B$ and read $\mathcal{P}$. If the two keys $\mathcal{K}_A$ and $\mathcal{K}_B$ are one-time pads then none of these communications can convey any information to Eve. By combining *all three*, however, Eve can readily recover the original message as follows:

$$\mathcal{C}_A \oplus \mathcal{C}_B \oplus \mathcal{C}_{A'} = (\mathcal{P} \oplus \mathcal{K}_A) \oplus (\mathcal{P} \oplus \mathcal{K}_A \oplus \mathcal{K}_B) \oplus (\mathcal{P} \oplus \mathcal{K}_B)$$
$$= \mathcal{P}. \tag{3.13}$$

The failure of this protocol is a consequence of the simplicity of the mathematical transformation used to create the ciphertext, in this case addition modulo 2. Most more complicated transformations, however, suffer from the problem that the processes do not commute, in that the last encryption operation to be applied needs to be the first removed; Bob needs to remove his padlock before Alice removes hers. As an illustration, we suppose that Alice and Bob each use a 26 letter one-time pad and that the substitution rules for the *first character* in the plaintext are

Note that Alice's key enciphers an initial letter B as B. It is important to include the possibility of such occurrences in our one-time pad, as not doing so would provide some information to Eve: she would know that each letter received must represent one of the other 25.

Alice

plaintext	**A**BCDEFGHIJKLMNOPQ**R**STUVWXYZ
substitution	**X**BJIFWMLQTAZUODEC**S**NKHPGRVY

Bob

plaintext	ABCDEFG**H**IJKLMNOPQRSTUVW**X**YZ
substitution	FPOAZWI**R**JXTCUVEKYGBHNMD**S**LQ .

Suppose that the first letter of the plaintext is A. If we follow the protocol described above then we generate $\mathcal{C}_A = \text{X}$, $\mathcal{C}_B = \text{S}$ and Alice's deciphering step then gives $\mathcal{C}_{A'} = \text{R}$. Finally Bob uses his substitution table to give H rather than A. The problem is that accurate recovery of the plaintext requires Bob to reverse his encryption *before* Alice but, of course, this renders the key distribution protocol insecure.

Diffie and Hellman proposed a solution to the problem by introducing a one-way function: a function that is easy to calculate but with an inverse that is very difficult to evaluate. We start by choosing a large prime number p and a second number, g, that is a primitive root modulo p. This property of g means that for any integer A in the set $\{1, 2, \cdots, p-1\}$ there exists an exponent a in the set $\{0, 1, \cdots, p-2\}$ such that

$$A = g^a \bmod p, \tag{3.14}$$

where $\bmod p$ means modulo p. Calculating A from a, for given p and g is straightforward. The exponent a is the discrete logarithm of A to the base g:

$$a = \mathrm{dlog}_g A. \tag{3.15}$$

Computing discrete logarithms is a difficult problem, in that no efficient algorithm for this is known; finding A from a is easy but finding a from A is hard. Alice and Bob can use this fact to generate a shared secret key by communicating over an insecure channel. First they agree on a large prime p and a primitive root g ($2 \leq g \leq p-2$), and because they need to communicate these we assume that they are also known to Eve. Alice randomly selects an integer a, which she keeps secret, computes A as above, and sends this number to Bob. Bob randomly selects a secret integer b in the set $\{0, 1, \cdots, p-2\}$, computes

$$B = g^b \bmod p, \tag{3.16}$$

and sends the result to Alice. Alice uses a to compute

$$B^a \bmod p = g^{ab} \bmod p \tag{3.17}$$

and Bob uses b to compute

$$A^b \bmod p = g^{ab} \bmod p. \tag{3.18}$$

The common secret key is

$$\mathcal{K} = g^{ab} \bmod p. \tag{3.19}$$

The task of generating $\mathcal{K}$ is easy given either a and B (known to Alice) or b and A (known to Bob) but cannot readily be found from just the publicly known numbers A and B, because of the difficulty in evaluating discrete logarithms.

It would be nice if Alice and Bob did not have to go through the exchange of messages in order to communicate in secret. In public-key cryptography, Bob publishes a pair of numbers with which anyone, say Alice, wishing to send a secret to Bob can encrypt their message. This needs to be done in such a way, however, that only Bob can decrypt the message. The first such scheme is the RSA cryptosystem, named after Rivest, Shamir, and Adleman, who were the first to publish it. Bob

Number theory   Modern cryptography is based on the properties of large integers and so forms a branch of number theory. Some of the most important results for cryptography are described in Appendix E.

starts by generating two large prime numbers p and q and computes the product

$$N = pq. \tag{3.20}$$

Euler's φ-function   The function $\varphi(N)$ is the number of integers a in the set $\{1, 2, \cdots, N\}$ with $\gcd(a, N) = 1$. It is straightforward to show that if p and q are distinct primes then $\varphi(p) = p - 1$, $\varphi(pq) = (p - 1)(q - 1)$ and $\varphi(p^2) = p(p - 1)$.

Bob also selects an integer e with the properties that $1 < e < \varphi(N) = (p - 1)(q - 1)$ and that e is coprime with $(p - 1)(q - 1)$, so that the greatest common divisor of e and $(p - 1)(q - 1)$ is 1:

$$\gcd[e, (p - 1)(q - 1)] = 1. \tag{3.21}$$

Bob's final task is to find a third integer d with the properties $1 < d < \varphi(N)$ and

$$de = 1 \bmod \varphi(N). \tag{3.22}$$

That a suitable number d exists is a consequence of eqn 3.21, and it can be calculated efficiently if p and q are known. Bob's public-key pair is (N, e) and his private key, which he keeps secret, is d. The number N is the RSA modulus, and e and d are the encryption and decryption exponents, respectively. Note that the secrecy of d relies on the secrecy of p and q and, because $N = pq$ is publicly known, this relies on the fact that no efficient and practical factoring algorithm is known.

Quantum factoring   It has been shown that a suitable quantum computer should be able to perform the task of efficient factoring. We shall see how this works in Chapter 7.

Alice can send a plaintext bit string $\mathcal{P} = M < N$ to Bob by raising it to the power e modulo N, using Bob's public key, to generate the ciphertext:

$$\mathcal{C} = M^e \bmod N. \tag{3.23}$$

Bob can decipher the message using his secret decryption exponent:

$$\mathcal{C}^d \bmod N = (M^e)^d \bmod N = M. \tag{3.24}$$

Further details of the mathematics underlying RSA and Diffie–Hellman key exchange can be found in Appendix E.

There are many additional subtleties associated with modern cryptography. Not least amongst these are the need to manage and store keys secretly and the requirement that legitimate parties should be able to identify themselves to each other (all would fail if Eve could convince Alice that she was Bob and Bob that she was Alice). Pursuing these points is interesting in its own right but to do so would take us too far from our main subject of quantum information. Quantum cryptography was devised as a radically different approach to the problem of key distribution, one that relies on the transmission of qubits from Alice to Bob through a quantum channel.

3.2   Quantum communications

A quantum communication channel is one in which the message sent by Alice to Bob is carried by a quantum system. This feature adds three subtleties to the description of a classical communication system as discussed in Section 1.4. These relate to the preparation of the signal by Alice, its measurement by Bob, and the effect on the signal of any activity by an eavesdropper, Eve.

We start by considering the legitimate users of the quantum channel, Alice and Bob. As with the classical channel, we let A represent events in Alice's domain, which lead to the selection of a message from the set $\{a_i\}$ and the preparation of a quantum system in one of the associated states $\{\hat{\rho}_i\}$. The probability that Alice selects the message a_i is $P(a_i)$. We shall assume that both the possible density operators $\{\hat{\rho}_i\}$ and the associated probabilities $P(a_i)$ are known to Bob. This means that Bob can describe the quantum state of the signal, before his measurement, by the a priori density operator

$$\hat{\rho} = \sum_i P(a_i)\hat{\rho}_i. \tag{3.25}$$

Bob's task is to determine, as well as he can, which of the states $\{\hat{\rho}_i\}$ has been prepared and so recover the original message a_i. If the signal states are mutually orthogonal ($\hat{\rho}_i\hat{\rho}_j = 0$, $i \neq j$) then Bob can do this straightforwardly by measuring an observable with these signal states as eigenstates. In general, however, the signal states will not be mutually orthogonal and this means that there is no certain way for Bob to distinguish between them. A formal proof of this must wait until Section 4.4, after we have introduced generalized measurements. We can illustrate the main idea, however, with a simple example. Let us suppose that Alice prepares her quantum system in one of two non-orthogonal pure quantum states, represented by the normalized state vectors $|\psi_1\rangle$ and $|\psi_2\rangle$. We can write the second state in the form

$$|\psi_2\rangle = \alpha|\psi_1\rangle + \beta|\psi_1^\perp\rangle, \tag{3.26}$$

where $\langle\psi_1|\psi_1^\perp\rangle = 0$ and $\alpha \neq 0$. A measurement of the observable corresponding to the Hermitian operator

$$\hat{A} = |\psi_1\rangle\langle\psi_1| - |\psi_1^\perp\rangle\langle\psi_1^\perp| \tag{3.27}$$

is *certain* to give the value $+1$ if the state $|\psi_1\rangle$ was prepared. The non-orthogonality of the two possible signal states, however, means that the same result will occur with probability $|\alpha|^2$ if state $|\psi_2\rangle$ was prepared. It follows that this measurement cannot distinguish with certainty between the two possible signal states. In general, Bob must choose between a range of possible measurements, and this enforced choice is the second intrinsically quantum feature of quantum communications. We shall examine this choice in detail in the following chapter.

In long-distance communications, the inevitable losses on the channel constitute an important source of noise, in that an absorbed quantum in the signal cannot be detected by Bob. In modern fibre-based optical communications, this problem is overcome by the use of repeaters or amplifiers which, respectively, measure and regenerate the signal or amplify its intensity. For quantum channels, however, these process are unsatisfactory as the level of the added noise acts to destroy the quantum information carried. We illustrate the problem by deriving two simple but important results for the measurement or the copying of the state of a single qubit.

If we know the basis in which a qubit has been prepared then there is no problem in measuring the state and then preparing one or more copies of it. For example, if the qubit has been prepared in the state $|0\rangle$ or the state $|1\rangle$ then measuring $\hat{\sigma}_z$ will reveal the identity of the state and provide all the information required to create as many copies as we might desire. If, however, the qubit was prepared in one of a number of non-orthogonal states then no measurement can possibly reveal the state with certainty. We can prove this by means of a simple but general model of the measurement process in which the qubit, in state $|\psi\rangle$, is made to interact with an ancillary quantum system, prepared in the state $|A\rangle$, which represents the measuring device. The interaction is enacted by a unitary transformation, the general form of which is

$$\hat{U} = \hat{\mathbf{I}} \otimes \hat{A}_0 + \hat{\sigma}_x \otimes \hat{A}_x + \hat{\sigma}_y \otimes \hat{A}_y + \hat{\sigma}_z \otimes \hat{A}_z, \qquad (3.28)$$

so that the combined state of our qubit and ancilla becomes

$$\begin{aligned}
|\psi\rangle \otimes |A\rangle &\rightarrow \hat{U}|\psi\rangle \otimes |A\rangle \\
&= |\psi\rangle \otimes \hat{A}_0|A\rangle + \hat{\sigma}_x|\psi\rangle \otimes \hat{A}_x|A\rangle + \hat{\sigma}_y|\psi\rangle \otimes \hat{A}_y|A\rangle \\
&\quad + \hat{\sigma}_z|\psi\rangle \otimes \hat{A}_z|A\rangle.
\end{aligned} \qquad (3.29)$$

The state of the original qubit will be unchanged only if it is an eigenstate of $\hat{U}$. There are, however, no common eigenstates of the Pauli operators $\hat{\sigma}_x$, $\hat{\sigma}_y$, and $\hat{\sigma}_z$. If one of our set of possible states is an eigenstate of one of these Pauli operators then any other state that is not orthogonal to it will not be an eigenstate. If we wish the unknown state of our qubit to be unchanged by the interaction then we require $\hat{A}_x = \hat{A}_y = \hat{A}_z = 0$, but this means that the ancilla is changed in a way that *does not depend* on the state $|\psi\rangle$ and so does not constitute a measurement of the qubit. We can conclude that any measurement of the state of the qubit leads to a change of at least some of the possible states of the qubit.

Amplification to generate copies of the quantum state is a more subtle process, in that it does not include a measurement and so does not reveal any information about the state. It is, nevertheless, impossible to accurately copy the unknown state of a quantum system, and this is elegantly proven in the famous *no-cloning theorem* of Wootters, Zurek, and Dieks. If cloning the unknown state of a qubit, $|\psi\rangle$, were possible then this would mean preparing a second qubit in the blank state $|B\rangle$ and copying the state of the original qubit onto it. This entails performing the transformation

$$|\psi\rangle \otimes |B\rangle \rightarrow |\psi\rangle \otimes |\psi\rangle. \qquad (3.30)$$

For this to be a true cloning transformation it is necessary for the transformation to hold for any possible qubit state $|\psi\rangle$. The no-cloning theorem proves that this is not possible. Suppose that the cloning device works if the qubit is prepared in one of the states $|0\rangle$ and $|1\rangle$ so that

$$\begin{aligned}
|0\rangle \otimes |B\rangle &\rightarrow |0\rangle \otimes |0\rangle, \\
|1\rangle \otimes |B\rangle &\rightarrow |1\rangle \otimes |1\rangle.
\end{aligned} \qquad (3.31)$$

It then follows from the superposition principle (eqn 2.1) that the general qubit state $\alpha|0\rangle + \beta|1\rangle$ will be transformed as

$$(\alpha|0\rangle + \beta|1\rangle) \otimes |B\rangle \rightarrow \alpha|0\rangle \otimes |0\rangle + \beta|1\rangle \otimes |1\rangle, \qquad (3.32)$$

which is not a pair of copies of the original state:

$$(\alpha|0\rangle + \beta|1\rangle) \otimes (\alpha|0\rangle + \beta|1\rangle) = \alpha^2|0\rangle \otimes |0\rangle + \alpha\beta|0\rangle \otimes |1\rangle$$
$$+ \alpha\beta|1\rangle \otimes |0\rangle + \beta^2|1\rangle \otimes |1\rangle. \quad (3.33)$$

It is not hard to see why perfect quantum copying, or cloning, is forbidden. Were this not the case and multiple copies could be made, then it would be possible to identify the state and so make it an observable. For example, if we knew that our qubit had been prepared in an eigenstate of $\hat{\sigma}_z$ or of $\hat{\sigma}_x$, then measuring $\hat{\sigma}_z$ on half of the copies and $\hat{\sigma}_x$ on the other half would allow us to determine the state with a high probability. A more dramatic concern would be the possibility of superluminal communication based on using the Bell state

$$|\Psi^-\rangle = \frac{1}{\sqrt{2}} \left(|0\rangle \otimes |1\rangle - |1\rangle \otimes |0\rangle \right) \qquad (3.34)$$

together with a hypothetical cloning device. If Alice has the first qubit and Bob the second then measuring $\hat{\sigma}_z$ or $\hat{\sigma}_x$ on her qubit would instantaneously transform Bob's qubit into the other eigenstate of the observable chosen by Alice: if Alice chose to measure $\hat{\sigma}_z$ and found the result $+1(-1)$ then Bob's qubit would be left in the state $|1\rangle(|0\rangle)$, but a measurement of $\hat{\sigma}_x$ giving the result $|1\rangle(|0\rangle)$ would leave Bob's qubit in the state $2^{-1/2}(|0\rangle - |1\rangle) \left(2^{-1/2}(|0\rangle + |1\rangle) \right)$. By perfectly cloning multiple copies of his qubit, Bob could determine Alice's choice of observable and hence acquire one bit of information, irrespective of the distance between them. This is in conflict, of course, with the requirements of relativity. It is interesting to note that N. Herbert proposed such an entanglement-based superluminal communication device, and it was in response to this proposal that the no-cloning theorem was first formulated. The no-cloning theorem only forbids *perfect* copying and there exist a number of imperfect cloning schemes, some of which are described in Appendix F.

The problems of measuring or copying the unknown state of a quantum system without changing it are fundamental difficulties facing any eavesdropper. If Eve wishes to share in the information sent from Alice to Bob then she needs to obtain it by interacting with the quantum channel. Any attempt to do so, however, will modify the states of the qubits prepared by Alice and this modification can be used to reveal Eve's presence to Alice and Bob. It is on this ability to detect the activities of an eavesdropper that the security of quantum key distribution is based.

No signalling In fact, Alice's choice of measurement can convey no information to Bob and so there is no conflict with relativity. We shall prove this powerful theorem in Chapter 5.

3.3   Optical polarization

Quantum information, like its classical counterpart, is largely independent of the physical system used to embody it. Any two-state quantum system can represent a qubit and a wide variety have been employed, including electronic energy levels in atoms or ions, nuclear spins, and photon polarization. For this reason, we are presenting our subject in a manner that is largely independent of specific physical implementations. We make one exception, however, and describe in this section the phenomenon of optical polarization and its representation as a qubit. Quantum cryptography relies on the properties of light to realize a quantum channel, and the simplest and most regularly employed property is polarization; this is, of course, the reason for describing it here. Our presentation is based on the more extended treatment given by Fowles (1989).

Light is an electromagnetic phenomenon and we describe it in terms of Maxwell's equations, which, in a linear, isotropic, and homogeneous dielectric medium, take the form

$$\vec{\nabla} \cdot \vec{E} = 0,$$
$$\vec{\nabla} \cdot \vec{H} = 0,$$
$$\vec{\nabla} \times \vec{E} = -\mu_0 \frac{\partial}{\partial t} \vec{H},$$
$$\vec{\nabla} \times \vec{H} = \varepsilon \frac{\partial}{\partial t} \vec{E}, \tag{3.35}$$

where $\vec{E}$ and $\vec{H}$ are the electric and magnetic fields, respectively, and ε is the permittivity of the dielectric. It is convenient to work with *complex fields*, the real parts of which are the observed electric and magnetic fields. A plane wave is characterized by an angular frequency ω and a wavevector $\vec{k}$:

$$\vec{E} = \vec{E}_0 \exp\left[i\left(\vec{k} \cdot \vec{r} - \omega t\right)\right], \qquad \vec{E} = \Re \vec{E},$$
$$\vec{H} = \vec{H}_0 \exp\left[i\left(\vec{k} \cdot \vec{r} - \omega t\right)\right], \qquad \vec{H} = \Re \vec{H}. \tag{3.36}$$

Substituting these into Maxwell's equations produces the four algebraic equations

$$\vec{k} \cdot \vec{E} = 0,$$
$$\vec{k} \cdot \vec{H} = 0,$$
$$\vec{k} \times \vec{E} = \mu_0 \omega \vec{H},$$
$$\vec{k} \times \vec{H} = -\varepsilon \omega \vec{E}. \tag{3.37}$$

These equations tell us that the three vectors $\vec{k}$, $\vec{E}$, and $\vec{H}$ are mutually orthogonal (see Fig. 3.3). The electric and magnetic fields are perpendicular to each other and both lie in the plane that is perpendicular to the direction of propagation. The magnitudes of the fields are related by

$$\frac{B_0}{E_0} = \varepsilon \mu_0 \frac{\omega}{k}, \tag{3.38}$$

Fig. 3.3 The relative orientation of the electric and magnetic fields and the wavevector in a plane wave.

where $B_0 = \mu_0 H_0$ is the magnitude of the magnetic flux density, and the phase velocity of the wave is $\omega/k = c/n = (\varepsilon\mu_0)^{-1/2}$, where n is the refractive index. It follows that $B_0 = nE_0/c$.

The direction and rate of flow of electromagnetic energy are quantified in the Poynting vector

$$\vec{S} = \vec{\mathrm{E}} \times \vec{\mathrm{H}}. \tag{3.39}$$

This vector specifies both the direction and the magnitude of the energy flow; it has the units of watts per square metre. For the plane wave fields in eqn 3.36, with $\vec{E}_0$ and $\vec{H}_0$ real, the Poynting vector is

$$\vec{S} = \vec{E}_0 \times \vec{H}_0 \cos^2\left(\vec{k}\cdot\vec{r} - \omega t\right). \tag{3.40}$$

It is often more natural to work with the cycle-averaged Poynting vector

$$\langle\vec{S}\rangle = \frac{1}{2}\vec{E}_0 \times \vec{H}_0, \tag{3.41}$$

or, more generally, for complex field amplitudes,

$$\langle\vec{S}\rangle = \frac{1}{2}\Re\left(\vec{E}_0 \times \vec{H}_0^*\right). \tag{3.42}$$

This averaging is especially appropriate at optical frequencies, as detectors cannot respond on such short time scales as the optical period. For our plane waves, of course, the Poynting vector is in the same direction as the wavevector $\vec{k}$.

The type of polarization is determined by the orientations of the electric and magnetic fields in the plane perpendicular to the direction of propagation. In order to simplify our discussion we consider plane waves propagating in the positive z-direction, so that the electric and magnetic fields are

$$\begin{aligned}\vec{E} &= \vec{E}_0 \exp\left[i(kz - \omega t)\right],\\ \vec{H} &= \vec{H}_0 \exp\left[i(kz - \omega t)\right].\end{aligned} \tag{3.43}$$

If the amplitudes $\vec{E}_0$ and $\vec{H}_0$ are real, constant vectors (or, more precisely, if the ratio of their x- and y-components is real) then the wave is said to be *linearly polarized*. We have seen that the electric and magnetic fields are orthogonal, and it is conventional in optics to define the direction or plane of polarization by the direction of the electric field. If the x-axis is horizontal and the y-axis vertical, then a field is x-polarized, or horizontally polarized, if the electric field oscillates in the x-direction, and it is vertically polarized if the electric field oscillates in the x-direction (see Fig. 3.4).

Polarization is manipulated and measured by polarization-sensitive optical elements. The simplest of these is the linear polarizer, or polarizing filter. This device absorbs light more strongly if it is, say, x-polarized than if it is y-polarized. Ideally, such a device will fully absorb one direction of linear polarization and leave the other unchanged. If

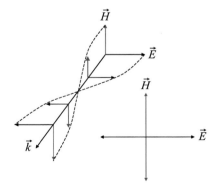

Fig. 3.4 The evolution of the fields for a horizontal linearly polarized field. The right-hand figure depicts the fields as they would appear if the light were propagating out of the page.

the field is polarized at an angle θ to the horizontal then we can write the electric field amplitude in the form

$$\vec{E}_0 = E_1 \cos \theta \vec{\imath} + E_2 \sin \theta \vec{\jmath}, \tag{3.44}$$

where $\vec{\imath}$ and $\vec{\jmath}$ are unit vectors in the x- and y-directions. If the transmission axis of our ideal polarizer is oriented in the x-direction then the transmitted electric field is $E_1 \cos \theta \vec{\imath}$. The intensity of the transmitted light is proportional to the modulus squared of the complex electric field:

$$I_1 = I_0 \cos^2 \theta. \tag{3.45}$$

If the incident light is unpolarized, so that all possible polarizations are present in equal amounts, then one-half of the light will be transmitted and this light will be horizontally polarized.

If the x- and y-components of the field amplitudes are out of phase then the polarization rotates as the field evolves. In the simplest case, the two components have the same amplitude but differ in phase by $\pi/2$ so that

$$\vec{E} = E_0 \left(\vec{\imath} \pm i \vec{\jmath} \right) \exp \left[i(kz - \omega t) \right]. \tag{3.46}$$

A wave of this form is said to be *circularly polarized*. The real part of this complex field is

$$\vec{E} = |E_0| \left[\vec{\imath} \cos(kz - \omega t + \varphi) \mp \vec{\jmath} \sin(kz - \omega t + \varphi) \right], \tag{3.47}$$

where $\varphi = \arg(E_0)$. This field rotates either as time passes or as we change z. The plus (or minus) sign in eqn 3.47 means that the field at a given point in space rotates in a clockwise (or anticlockwise) direction when viewed against the direction of propagation, so that the light is moving towards the observer, and the light is said to be right (or left) circularly polarized. At any given time, the field vectors form a right- (or left-) handed screw when viewed by the same observer (see Fig. 3.5). For the complex field we must associate the amplitude $E_0(\vec{\imath} - \vec{\jmath})$ with right circular polarization and $E_0(\vec{\imath} + \vec{\jmath})$ with left circular polarization. A final possibility is that the x- and y-components of the electric field amplitude can be both of different magnitudes and out of phase so that

$$\vec{E} = (E_0 \vec{\imath} \pm i E_0' \vec{\jmath}) \exp \left[i(kz - \omega t) \right], \tag{3.48}$$

where E_0 and E_0' have the same phase. Fields of this form are said to be *elliptically polarized*. The electric and magnetic field amplitudes both rotate and oscillate in magnitude as we change t or z.

A practical method to change the polarization is to introduce a phase delay between two orthogonal components of the electric field. The simplest of such devices are wave plates, which are formed from a doubly refracting material. This means that the refractive index depends on the direction of polarization. A wave plate will typically have an axis of maximum index n_1 (the slow axis) and an axis of minimum index n_2 (the fast axis) lying at right angles to each other, with both axes

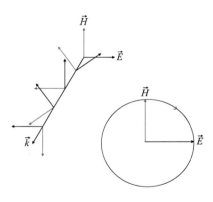

Fig. 3.5 The evolution of the fields for a right-circularly polarized field. The right-hand figure depicts the fields as they would appear if the light were propagating out of the page.

perpendicular to the direction of propagation. If the wave plate has thickness d, then the phase accumulated by the light on propagating through the wave plate will be $n_1 kd$ for light polarized along the slow axis and $n_2 kd$ for the fast axis. Wave plates are usually designed as quarter-wave plates or half-wave plates, corresponding to a relative phase shift between the fast and slow polarizations of $\pi/2$ or π, respectively. For a quarter-wave plate this means choosing the thickness to be

$$d = \frac{\pi}{2k(n_1 - n_2)} = \frac{\lambda_0}{4(n_1 - n_2)}, \tag{3.49}$$

where λ_0 is the wavelength in free space. For a half-wave plate, the thickness will be twice this value. The polarization of light polarized in the direction of the fast or slow axis will be unchanged by a wave plate. A quarter-wave plate will change linearly polarized light at 45° to the fast and slow axes into circularly polarized light, and circularly polarized light into light with linear polarization. A half-wave plate will, in general, induce a rotation of linearly polarized light and change the handedness of circularly polarized light.

The property of polarization depends only on the relative amplitude and phase of the x- and y-components of the electric field, and it is helpful to use a representation of polarization that retains only these properties. A simple and convenient way to do this is to introduce the Jones vector, the elements of which are the x- and y-components of the electric field. If the complex electric field amplitude is

$$\vec{E}_0 = E_{0x}\vec{i} + E_{0y}\vec{j}, \tag{3.50}$$

then the corresponding Jones vector is

$$\begin{bmatrix} E_{0x} \\ E_{0y} \end{bmatrix} = \begin{bmatrix} |E_{0x}| e^{i\phi_x} \\ |E_{0y}| e^{i\phi_y} \end{bmatrix}, \tag{3.51}$$

where $\phi_{x,y}$ is the argument of $E_{0x,y}$. An added advantage for quantum information is that many of the properties of the Jones vector map directly onto those of qubits, when the states of the latter are expressed as column vectors as in eqn 2.85. We do not wish to overemphasize this connection prematurely, however, and for this reason we follow the notation of Fowles and use square brackets for the Jones vector and the associated matrices.

It is sometimes convenient to normalize the Jones vector; we do this by dividing it by $|\vec{E}_0|$. The overall phase of the Jones vector does not affect the polarization, and for this reason we can regard Jones vectors that differ only by a global phase as equivalent. These ideas are reminiscent, of course, of the normalization and arbitrary global phase of quantum state vectors. The Jones vectors for circularly polarized light and for linear polarizations are given in Fig. 3.6. Superpositions of Jones vectors are also Jones vectors and so correspond to allowed states of polarization. It is clear, from their construction, that circular polarization can be

The amplitudes and phases of both components are important for other properties, of course, and appear, for example, in interference phenomena.

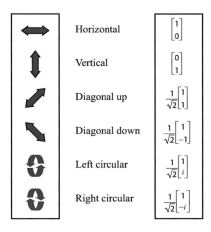

Fig. 3.6 Linear and circular polarizations and their associated Jones vectors.

Device	Orientation	Jones matrix
Linear polarizer	Transmission axis horizontal	$\begin{bmatrix} 1 & 0 \\ 0 & 0 \end{bmatrix}$
	Transmission axis vertical	$\begin{bmatrix} 0 & 0 \\ 0 & 1 \end{bmatrix}$
	Transmission axis at angle θ to horizontal	$\begin{bmatrix} \cos^2\theta & \cos\theta\sin\theta \\ \cos\theta\sin\theta & \sin^2\theta \end{bmatrix}$
Circular polarizer	Right	$\dfrac{1}{2}\begin{bmatrix} 1 & i \\ -i & 1 \end{bmatrix}$
	Left	$\dfrac{1}{2}\begin{bmatrix} 1 & -i \\ i & 1 \end{bmatrix}$
Quarter-wave plate	Fast axis vertical	$\begin{bmatrix} 1 & 0 \\ 0 & -i \end{bmatrix}$
	Fast axis horizontal	$\begin{bmatrix} 1 & 0 \\ 0 & i \end{bmatrix}$
	Fast axis horizontal at angle θ to horizontal	$\begin{bmatrix} \cos^2\theta + i\sin^2\theta & \cos\theta\sin\theta(1-i) \\ \cos\theta\sin\theta(1-i) & \sin^2\theta + i\cos^2\theta \end{bmatrix}$
Half-wave plate	Fast axis vertical or horizontal	$\begin{bmatrix} 1 & 0 \\ 0 & -1 \end{bmatrix}$
	Fast axis horizontal at angle θ to horizontal or vertical	$\begin{bmatrix} \cos^2\theta & \cos\theta\sin\theta \\ \cos\theta\sin\theta & \sin^2\theta \end{bmatrix}$

Fig. 3.7 Polarization-manipulating devices and their associated Jones matrices.

expressed as a superposition of two perpendicular linear polarizations and the converse is also true:

$$\frac{1}{\sqrt{2}}\left(\frac{1}{\sqrt{2}}\begin{bmatrix} 1 \\ -i \end{bmatrix} + \frac{1}{\sqrt{2}}\begin{bmatrix} 1 \\ i \end{bmatrix} \right) = \begin{bmatrix} 1 \\ 0 \end{bmatrix},$$

$$\frac{1}{\sqrt{2}}\left(\frac{1}{\sqrt{2}}\begin{bmatrix} 1 \\ -i \end{bmatrix} - \frac{1}{\sqrt{2}}\begin{bmatrix} 1 \\ i \end{bmatrix} \right) = -i\begin{bmatrix} 0 \\ 1 \end{bmatrix}. \tag{3.52}$$

The Jones formalism is completed by introducing matrices that describe the effect on the polarization of polarization-dependent optical elements. There exists a wide variety of these, but we give the most important in Fig. 3.7. Note that, like the Jones vectors, these are defined only up to an arbitrary global phase. The effect of a single polarization-dependent element is obtained by multiplying the Jones vector $\underline{v}$ by the corresponding Jones matrix $\mathbf{J}$:

$$\underline{v} \to \mathbf{J}\underline{v}. \tag{3.53}$$

A sequence of n such devices will modify the Jones vector as follows:

$$\underline{v} \to \mathbf{J}_n \cdots \mathbf{J}_2 \mathbf{J}_1 \underline{v}, \tag{3.54}$$

where $\mathbf{J}_1, \mathbf{J}_2, \cdots, \mathbf{J}_n$ are the Jones vectors associated with the first, second, $\cdots, n$th optical elements. Once again we see a striking similarity with the rule for applying a sequence of unitary transformations to the state of a qubit (eqn 2.83).

The formal similarity between the Jones vector and the column vector representation of the state of a qubit suggests another representation of polarization. In Section 2.4 we introduced the representation of the state of a qubit as a point on the surface of the Bloch sphere. If we equate the Jones vector with the corresponding qubit states then we arrive at the Poincaré sphere, each point on the surface of which corresponds to a possible polarization.

States on opposite sides of the Bloch sphere are orthogonal, and it is natural to make the same identification for the Poincaré sphere (see Fig. 3.8). This association leads us naturally to define two polarizations as orthogonal if they lie on opposite sides of the Poincaré sphere, and their corresponding complex electric fields $\vec{E}_1$ and $\vec{E}_2$ satisfy the condition

$$\vec{E}_2^{\,*} \cdot \vec{E}_1 = 0. \tag{3.55}$$

When written in terms of the Jones vectors $\underline{v}_1$ and $\underline{v}_2$, this becomes

$$\underline{v}_2^{\dagger} \underline{v}_1 = 0, \tag{3.56}$$

where $\underline{v}_2^{\dagger}$ is the row vector $(v_2^* \ v_1^*)$.

We can move to a quantum treatment of polarization, and thereby introduce a polarization qubit, by working with just one photon. In order to combine the features of the Jones vectors and our qubits we associate a single horizontally polarized photon with the qubit state $|0\rangle$ and a vertically polarized one with the state $|1\rangle$. This choice makes the Bloch and Poincaré spheres equivalent. A selection of some important polarizations and their single-photon qubit states is given in Fig. 3.9. These states form a natural set for use in quantum key distribution.

There is one very important difference between the classical and single-photon Jones vectors, and this relates to the meaning of the entries. For a classical Jones vector, the elements v_x and v_y are proportional to the electric fields present. Passing the light through a linear polarizer with the transmission axis horizontal will reduce the intensity of the light by the factor $|v_x|^2$. If the Jones vector describes only a single photon, however, then v_x is the *probability amplitude* for the photon to pass through the polarizer. A subsequent measurement with an ideal photon counter will detect a photon with probability $|v_x|^2$.

It is useful to be able to superpose fields and their polarizations. A convenient way of achieving this is to use a beam splitter: a partially

Rotation of these devices is described by means of an orthogonal transformation of the associated Jones matrices. If a device is rotated through an angle θ then the corresponding Jones matrix $\mathbf{J}$ transforms as

$$\mathbf{J} \to \mathbf{R}^{\mathrm{T}}(\theta)\,\mathbf{J}\,\mathbf{R}(\theta),$$

where $\mathbf{R}(\theta)$ is the rotation matrix

$$\mathbf{R}(\theta) = \begin{bmatrix} \cos\theta & \sin\theta \\ -\sin\theta & \cos\theta \end{bmatrix}.$$

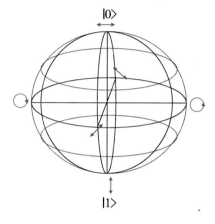

Fig. 3.8 The Poincaré sphere. Note that it is usual to place right circular polarization at the north pole of the sphere and left at the south pole so that all linear polarizations lie on the equator. We have rotated the sphere here so as to match the Bloch sphere.

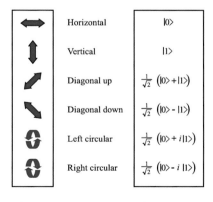

Fig. 3.9 Linear and circular polarizations and the associated single-photon qubit states.

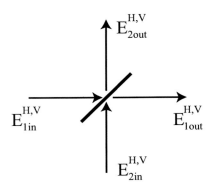

Fig. 3.10 A polarization-independent beam splitter with input and output electric fields.

reflecting and partially transmitting mirror. The reflection and transmission properties are, in general, polarization-dependent but we can consider ideal beam splitters which either are polarization-independent or are specifically designed to reflect one polarization and transmit the perpendicular polarization. A beam splitter combines two input fields to produce two output fields, as depicted in Fig. 3.10. For a polarization-independent beam splitter, which we shall refer to simply as a beam splitter, the output electric fields are related to the inputs by

$$E_{1\,\text{out}}^{H,V} = t_1 E_{1\,\text{in}}^{H,V} + r_1 E_{2\,\text{in}}^{H,V},$$
$$E_{2\,\text{out}}^{H,V} = t_2 E_{2\,\text{in}}^{H,V} + r_2 E_{1\,\text{in}}^{H,V}, \tag{3.57}$$

so that the polarization is preserved. The conservation of energy requires that

$$\left|E_{1\,\text{out}}^{H,V}\right|^2 + \left|E_{2\,\text{out}}^{H,V}\right|^2 = \left|E_{1\,\text{in}}^{H,V}\right|^2 + \left|E_{2\,\text{in}}^{H,V}\right|^2, \tag{3.58}$$

and this condition means that $|t_1|^2 = |t_2|^2$, $|t_i|^2 + |r_i|^2 = 1$, and $t_1^* r_1 + r_2^* t_2 = 0$. It follows that a beam splitter is fully specified by its transmission probability and the phases of any three of its four transmission and reflection coefficients. Two very simple choices are the symmetric beam splitter, for which $t_1 = t_2 = |t|$ and $r_1 = r_2 = i|r|$, and the beam splitter with real coefficients, which has $t_1 = t_2 = |t|$ and $r_1 = -r_2 = |r|$.

A polarizing beam splitter is designed to transmit horizontally polarized light and to reflect vertically polarized light. It follows that the output electric fields are related to the inputs by

$$E_{1\,\text{out}}^{H} = E_{1\,\text{in}}^{H}, \qquad E_{1\,\text{out}}^{V} = E_{2\,\text{in}}^{V},$$
$$E_{2\,\text{out}}^{H} = E_{2\,\text{in}}^{H}, \qquad E_{2\,\text{out}}^{V} = E_{1\,\text{in}}^{V}. \tag{3.59}$$

This device can be used as a linear polarizer, either to prepare a chosen linear polarization or as part of a polarization measurement. It does have an important advantage over the polarizing filter, however, and this is that both polarization components of the beam are preserved and can be measured separately, or recombined at a later stage. We shall see an example of this in Section 4.4.

For a single photon, we can describe the effects of a beam splitter or a polarizing beam splitter using the relationships in eqns 3.57 and 3.59 if we replace the field amplitudes with the probability amplitudes for the photon. If there is more than one photon present, however, then we need to employ the quantum theory of light, in which the electric and magnetic fields are replaced by operators. We present a very brief discussion of this in Appendix G.

3.4　Quantum key distribution

The use of quantum systems to provide information security has its origins in a proposal by S. Wiesner to use it to make unforgeable banknotes. Each of these notes was to include a unique serial number and

20 'light-traps', each of which contains a single polarized photon (or other realization of a qubit) prepared in one of the four states $|0\rangle$, $|1\rangle$, $|0'\rangle = 2^{-1/2} (|0\rangle + |1\rangle)$, and $|1'\rangle = 2^{-1/2} (|0\rangle - |1\rangle)$, corresponding, respectively, to horizontal, vertical, and the two diagonal polarizations (see Fig. 3.9). The sequence of polarization states is known only to the issuing bank and can be identified by reference to the serial number. The bank can check that the banknote is genuine by opening the light-traps and measuring the polarization of each photon in the basis in which it was prepared. If any photon is found to be in the state that is orthogonal to the one prepared then the bank note is counterfeit. A would-be counterfeiter would need to measure the polarization of each of the 20 photons in turn, but would have no way of knowing, even in principle, whether or not he had chosen the correct measurement basis. The best he/she can do is to choose randomly between the $|0\rangle$, $|1\rangle$ and the $|0'\rangle$, $|1'\rangle$ bases and to insert into each of the counterfeit notes a photon prepared with the polarization corresponding to the measurement result. The counterfeiter will choose the basis correctly for a single photon with probability $1/2$, but if the wrong basis is chosen then this will result in an error detected by the bank with probability $1/2$. The possible outcomes are depicted in Fig. 3.11, and it is clear that the probability that any particular photon is identified correctly by the bank as genuine is $3/4$. A counterfeit bank note will only be identified as genuine if all 20 photons pass this test, and this happens with the small probability $(3/4)^{20} = 0.0032$.

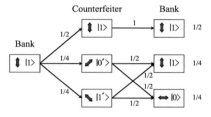

Fig. 3.11 The possible effects of the activities of a counterfeiter of quantum banknotes.

The first protocol for quantum key distribution was proposed by Bennett and Brassard in 1984 and is now known universally as the BB84 protocol. The resources required by Alice and Bob are a quantum channel, over which Alice can send polarized photons to Bob, and a classical channel, over which they can discuss the preparation and measurement events. It is essential, of course, that Eve should not be able to block the transmission over the classical channel and to convince Alice that she is Bob and Bob that she is Alice. Alice and Bob must assume, however, that Eve can intercept and read any message sent on the classical channel, which, for this reason, is sometimes referred to as a public channel.

In BB84 Alice prepares a sequence of single photons, each randomly selected, with equal probability, to be in one of the four polarization states $|0\rangle$, $|1\rangle$, $|0'\rangle$, and $|1'\rangle$. She makes a note of each polarization and associates the bit value 0 with the states $|0\rangle$ and $|0'\rangle$ and the bit value 1 with the states $|1\rangle$ and $|1'\rangle$. The photons are transmitted through a suitable channel, either free space or an optical fibre, to Bob, who randomly selects, with equal probability, to measure each of them in the $|0\rangle$, $|1\rangle$ or $|0'\rangle$, $|1'\rangle$ basis. A sample sequence of 20 photons is given in Fig. 3.12. After the quantum communication, Bob uses the classical channel to tell Alice the basis he used (but not, of course, his measurement result) to measure each photon, and Alice can then tell Bob on which occasions they used the same bases. In the sequence depicted, Alice and Bob have used the same bases in time slots 2, 4, 6, 8, 9, 11, 14, 15,

Phase coding   Polarization is not the only property of photons that has been employed in quantum key distribution. In phase coding, Alice and Bob's communication channel takes the form of an interferometer with two possible paths from Alice to Bob and an output at Bob's end which depends on the phase difference between the paths. If the phase difference is 0 (or an integer multiple of 2π) then the photon leaves through output 0, and if it is an odd-integer multiple of π then it leaves through output 1. Alice can encode the required bits by implementing phase shifts in one of the paths: $|0\rangle \rightarrow 0$, $|1\rangle \rightarrow \pi$, $|0'\rangle \rightarrow 3\pi/2$, $|1'\rangle \rightarrow \pi/2$. Bob can measure in the $|0\rangle$, $|1\rangle$ basis or the $|0'\rangle$, $|1'\rangle$ basis by introducing phase shifts 0 or $\pi/2$ in the same path as that selected by Alice.

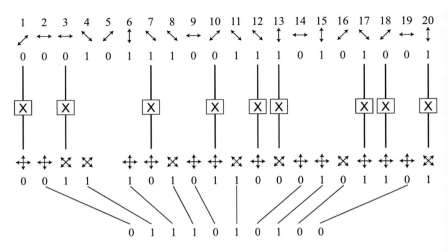

Fig. 3.12 Example of the BB84 protocol in the absence of an eavesdropper.

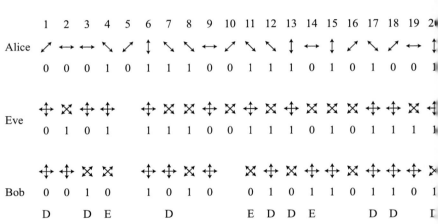

Fig. 3.13 Example of the BB84 protocol in the presence of an eavesdropper.

Eve could, of course, simply disrupt or block the quantum channel and thereby prevent Alice and Bob from establishing a secret key. Her principal objective, however, is to obtain the key and thereby access the intended secret communication.

16 and 19 and should, therefore, share a common bit sequence based on these and can use this as a secret key if there is no eavesdropper present. The bit values in the remaining time slots, where different preparation and measurement bases were used, are uncorrelated and are simply discarded. Naturally, Alice and Bob cannot assume that no eavesdropper is present, and must use their shared bit stream, together with a discussion over their public channel, to test for this possibility.

An eavesdropper can attack this key distribution protocol by measuring the polarization of each photon and then preparing, on the basis of the measurement result, a new photon to send to Bob. The most straightforward way to eavesdrop is for Eve to mimic Bob's behaviour and to measure each photon in the $|0\rangle$, $|1\rangle$ or $|0'\rangle$, $|1'\rangle$ basis. This runs into the same problem as that encountered by a would-be counterfeiter of Wiesner's quantum bank notes. Fig. 3.13 is a representation of a sequence of events as affected by Eve's activities. After a public discussion of the bases used, Alice and Bob discard the bits labelled D and are left with the partially correlated bit strings

Alice   0111010100,

Bob   0011001100.

At this stage they do not know whether or not an eavesdropper has been active, but they can test for this by publicly announcing some of the bits in their respective strings and looking for errors. In principle, the presence of one or more errors reveals that an eavesdropper has been active and the accumulated key should be discarded. Naturally, the bits announced in this public discussion must be discarded and do not form part of the final key. In our unrealistically short example, announcing a set of bits which includes the second, sixth, or seventh bit will reveal the presence of the eavesdropper.

We can and should ask whether Eve has any better measurement strategies and, in particular, whether there is any way in which she can avoid inducing errors in Bob's bit string. For realistic systems, with an intrinsic lower noise level and occasional multiphoton pulses, this is a challenging but important problem. In general, it needs to be addressed for each individual key distribution system. For the ideal system described here, however, the only realistic way for Eve to reduce the number of errors found by Alice and Bob is to measure fewer of the photons. Doing so, however, will reduce her information about the key.

Real communication systems suffer from noise, and so it is inevitable that a bit string of usable length built up using our quantum channel will include some errors, even in the absence of an eavesdropper. If the error rate in this raw key is not too large then Alice and Bob can correct these errors. Caution requires, however, that they should assume that the errors arise as the result of the activities of an eavesdropper. The process of creating a shared secret key from the raw key requires three steps: determining the error rate, removal of the errors, and, finally, privacy amplification.

Alice and Bob can determine the error rate by comparing a proportion of their bits, which are then discarded. If they find that the error rate is too high then it will not be possible to arrive at the desired key. If it is below a critical value (determined by a detailed study of the specific system) then Alice and Bob can proceed to distil a secret key. The remaining errors can be removed by forming sets of bits and comparing their parity over the public channel. If the parities are different then the set contains one (or possibly three) errors, but if they are the same then the set will have no errors (or perhaps two errors). Each parity check leaks information to Eve and so necessitates the discarding of one of the bits from the set. By constructing a suitable sequence of parity checks, all the errors can be removed with high probability. This idea is essentially the same principle as that embodied in Shannon's noisy-channel coding theorem, with the discarded parity-check bits playing the role of the redundant bits.

At the end of the error correction process, Alice and Bob should share a common bit string, but this may be significantly shorter than the raw

Parity of a string   The parity of a string is determined by the number of bits that take the value 1. If this number is even then the parity is 0, and if it is odd then the parity is 1. Learning the parity eliminates half of the possible strings and so corresponds to one bit of information.

key. The remaining task is to estimate, and preferably place an upper bound on, the amount of information that any eavesdropper might have about the remaining key bits. This means determining the maximum probability

$$P_{\text{Eve}} = \frac{1}{2}(1 + \varepsilon) \tag{3.60}$$

that Eve has correctly identified any given bit. Privacy amplification allows Alice and Bob to reduce ε and so reduce Eve's information. This is achieved by selecting groups of m bits and using the parity of each group as a single bit in the final key. This means a reduction in the length of the usable key by a factor of m. Eve will correctly identify this parity if she makes no errors in identifying the m bits or an even number of errors. The probability for this to occur is

$$P_{\text{Eve}} = \left(\frac{1}{2}\right)^m (1 + \varepsilon)^m + \left(\frac{1}{2}\right)^m \frac{m!}{2!(m-2)!}(1 + \varepsilon)^{m-2}(1 - \varepsilon)^2 + \cdots$$
$$= \frac{1}{2}(1 + \varepsilon^m). \tag{3.61}$$

This probability is closer to the zero-information value of $1/2$ than that before privacy amplification given in eqn 3.60. Alice and Bob must decide on the maximum value of P_{Eve} that they can tolerate, and choose m accordingly.

The BB84 protocol is by no means the only scheme for quantum key distribution. Indeed, a wide variety of protocols have been suggested. Here we shall describe briefly just two of these: the two-state protocol of Bennett, and Ekert's entangled-state protocol.

The security of quantum key distribution is based on the fact that Eve cannot discriminate, with certainty, amongst the set of states prepared by Alice. This means, as we have seen, both that she is unable to determine all of the key bits and that her attempts to do so necessarily reveal her presence through the generation of errors in Bob's bit string. The simplest way to realize a set of states that cannot be discriminated is to use just two non-orthogonal quantum states, $|0\rangle$ and

$$|1''\rangle = \cos\left(\frac{\theta}{2}\right)|0\rangle + \sin\left(\frac{\theta}{2}\right)|1\rangle. \tag{3.62}$$

In Bennett's two-state protocol, usually referred to as B92, Alice prepares each of a sequence of photons in the polarization state $|0\rangle$ or $|1''\rangle$ and associates with these the bit values 0 and 1, respectively. Bob selects, randomly, to measure the polarization of each photon in the $|0\rangle$, $|1\rangle$ or the $|0''\rangle$, $|1''\rangle$ basis, where $\langle 0''|1''\rangle = 0$. If Bob measures in the $|0\rangle$, $|1\rangle$ basis and gets the result corresponding to the state $|0\rangle$, then the state prepared by Alice could have been either $|0\rangle$ or $|1''\rangle$. When this occurs, Bob's measurement is inconclusive and he discards the bit. If, however, Bob's measurement gives the result $|1\rangle$, then the state prepared by Alice must have been $|1''\rangle$. In this case, Bob puts a 1 in his bit string and informs Alice that he has successfully identified the bit, although he does not reveal the bit value. The set of possible events is

Alice prepares	Bob's measurement	Bob's result	Action
$\lvert 0\rangle$	$\lvert 0\rangle, \lvert 1\rangle$	$\lvert 0\rangle$	Discard
	$\lvert 0''\rangle, \lvert 1''\rangle$	$\lvert 0''\rangle$	Record bit value 0
		$\lvert 1''\rangle$	Discard
$\lvert 1''\rangle$	$\lvert 0\rangle, \lvert 1\rangle$	$\lvert 0\rangle$	Discard
		$\lvert 1\rangle$	Record bit value 1
	$\lvert 0''\rangle, \lvert 1''\rangle$	$\lvert 1''\rangle$	Discard

Fig. **3.14** The possible events in the B92 protocol.

listed in Fig. 3.14. The protocol is completed by determining the error rate, eliminating the errors, and then performing privacy amplification as described in our discussion of BB84.

Each new protocol brings with it its own subtleties and B92, in particular, is vulnerable if the losses in the quantum channel are too high. To see this, we note that Eve can perform the same measurements as Bob and arrive at the same conclusions. If her measurement result is conclusive then she knows what state to send to Bob, and if it is inconclusive then she can choose to send nothing. In principle, Eve can escape detection by substituting a zero-loss channel between her and Bob and so hide her activities. Eve actually has a better strategy, which is unambiguous state discrimination. This requires a generalized measurement; a single measurement will correctly identify the state as $\lvert 0\rangle$ or $\lvert 1''\rangle$ with probability $1 - |\langle 0 | 1''\rangle|$ and give an ambiguous result with probability $\langle 0 | 1''\rangle|$. We shall describe measurements of this type in Section 4.4.

In both the BB84 and the B92 protocol, Alice selects a bit value and then prepares a state associated with it. Ekert's entangled-state protocol is different in that Alice prepares a sequence of pairs of qubits, with each pair prepared in the *same* entangled state,

$$\lvert \Psi^-\rangle = \frac{1}{\sqrt{2}} \left(\lvert 0\rangle \otimes \lvert 1\rangle - \lvert 1\rangle \otimes \lvert 0\rangle \right). \qquad (3.63)$$

She keeps the first qubit for herself and sends the second to Bob. If Alice and Bob both measure their photon in the $\lvert 0\rangle, \lvert 1\rangle$ basis or the $\lvert 0'\rangle, \lvert 1'\rangle$ basis, then their results should be perfectly *anticorrelated*. Subsequent public discussion can establish those occasions on which Alice and Bob chose the same measurement basis, and perfectly correlated bit strings can be achieved if Bob simply flips the value of each of his bits ($0 \leftrightarrow 1$). The legitimate users of the channel can establish a secret key from their measurement results in the $\lvert 0\rangle, \lvert 1\rangle$ and $\lvert 0'\rangle, \lvert 1'\rangle$ bases in precisely the same way as for BB84. The use of entangled states, however, suggests a more subtle approach to eavesdropper detection. We shall show, in Section 5.1, that entangled states exhibit correlations which are very different from those found for any non-entangled state. A measurement by Eve on one of an entangled pair, followed by preparation of a new photon to send to Bob, necessarily leaves Alice's photon and the photon prepared by Eve in an unentangled state. If Alice and Bob perform occasional measurements in the Breidbart basis,

$$|0_\text{B}\rangle = \cos\left(\frac{\pi}{8}\right)|0\rangle + \sin\left(\frac{\pi}{8}\right)|1\rangle,$$
$$|1_\text{B}\rangle = \sin\left(\frac{\pi}{8}\right)|0\rangle - \cos\left(\frac{\pi}{8}\right)|1\rangle, \tag{3.64}$$

then they can compare the results of these measurements with those of their partner performed in either the $|0\rangle$, $|1\rangle$ or the $|0'\rangle$, $|1'\rangle$ basis to test Bell's inequality. If Bell's inequality is violated then the states that Alice and Bob share are entangled, and there has been no eavesdropper activity. If, however, Bell's inequality is satisfied then their two-photon states are not entangled and they can infer that Eve has been listening in. In practice, of course, Alice and Bob will need to use the observed level of violation of Bell's inequality to place a bound on the possible information available to Eve. They will need to measure the error rate for their raw key, eliminate the errors, and perform privacy amplification in order to reach a final key. We shall discuss Bell's inequality and other tests for entanglement in Section 5.1.

We conclude on a cautionary note with a discussion of a suggestion to use quantum key distribution as a means to achieve the cryptographic task of bit commitment. The challenge is to find a way in which Alice can commit a bit value, 0 or 1, in such a way that she cannot later change her mind. The subtlety which complicates this task is that we also require that Bob should not be able to determine the bit until Alice reveals it to him. A sequence of such committed bits might represent, for example, a sealed bid in a financial transaction or the sealed move in an adjourned game of chess. They can do this via the services of a third party, who is trusted by both Alice and Bob, but our challenge is to achieve bit commitment without relying on anyone else. One idea would be for Alice to give Bob a qubit that she has prepared in one of the four states used in the BB84 protocol, $|0\rangle$, $|1\rangle$, $|0'\rangle$, and $|1'\rangle$. If she wishes to commit the bit value 0 then she prepares her qubit, with equal probabilities, in one of the states $|0\rangle$ or $|1\rangle$. In order to commit the value 1, she prepares the qubit in the state $|0'\rangle$ or $|1'\rangle$. It is clear that Bob cannot determine the bit as, for him, the qubit states corresponding to the two possible committed bit values, $\hat{\rho}_0$ and $\hat{\rho}_1$, are identical:

$$\begin{aligned}
\hat{\rho}_0 &= \frac{1}{2}\left(|0\rangle\langle 0| + |1\rangle\langle 1|\right) \\
&= \frac{1}{2}\left(|0'\rangle\langle 0'| + |1'\rangle\langle 1'|\right) \\
&= \hat{\rho}_1.
\end{aligned} \tag{3.65}$$

Alice can reveal the bit to Bob and then prove that she committed it earlier by telling Bob the state that she prepared, so that he can check this by measuring his qubit. A naive Bob might think that, because he has the qubit, Alice is unable to change the committed bit value. Alice can cheat, however, by preparing an entangled state $|\Psi^-\rangle$ of two qubits and send one of the pair to Bob as her committed bit. She can measure the qubit she has kept in either the $|0\rangle$, $|1\rangle$ or the $|0'\rangle$, $|1'\rangle$ basis at any stage and so reveal *either* 'committed' bit value to Bob for him to check. It is now known that it is impossible to use quantum communications to provide unconditionally secure bit commitment. As always with cryptography, just because a protocol looks secure, it does not follow that it is!

Suggestions for further reading

Barnett, S. M. and Radmore, P. M. (1997). *Methods in theoretical quantum optics*. Oxford University Press, Oxford.

Born, M. and Wolf, E. (1980). *Principles of optics* (6th edn). Pergamon, Oxford.

Bouwmeester, D., Ekert, A., and Zeilinger, A. (eds) (2000). *The physics of quantum information*. Springer-Verlag, Berlin.

Buchmann, J. A. (2001). *Introduction to cryptography*. Springer, New York.

Fowles, G. R. (1989). *Introduction to modern optics*. Dover, New York.

Gisin, N., Ribordy, G., Tittel, W., and Zbinden, H. (2002). Quantum cryptography. *Reviews of Modern Physics* **74**, 145.

Hardy, G. H. and Wright, E. M. (2008). *An introduction to the theory of numbers* (6th edn). Oxford University Press, Oxford.

Hunter, J. (1964). *Number theory*. Oliver and Boyd, Edinburgh.

Jaeger, G. (2007). *Quantum information: an overview*. Springer, New York.

Lo, H.-K., Popescu, S., and Spiller, T. (eds) (1998). *Introduction to quantum computation and information*. World Scientific, Singapore.

Loepp, S. and Wootters, W. K. (2006). *Protecting information: from classical error correction to quantum cryptography*. Cambridge University Press, Cambridge.

Loudon, R. (2000). *The quantum theory of light* (3rd edn). Oxford University Press, Oxford.

Macchiavello, C., Palma, G. M., and Zeilinger, A. (eds) (2000). *Quantum computation and quantum information theory*. World Scientific, Singapore.

Phoenix, S. J. D. and Townsend, P. D. (1995). Quantum cryptography: how to beat the code breakers using quantum mechanics. *Contemporary Physics* **36**, 165.

Piper, F. and Murphy, S. (2002). *Cryptography: a very short introduction*. Oxford University Press, Oxford.

Scarani, V., Iblisdir, S., Gisin, N., and Acín, A. (2005). Quantum cloning. *Reviews of Modern Physics* **77**, 1225.

Singh, S. (1999). *The code book*. Fourth Estate, London.

Singh, S. (2000). *The science of secrecy*. Fourth Estate, London.

Van Assche, G. (2006). *Quantum cryptography and secret-key distillation*. Cambridge University Press, Cambridge.

Exercises

(3.1) Decipher the following text produced using a Caesarean cipher:

QA BPQA I LIDOMZ Q AMM JMNWZM UM

(3.2) Breaking a cipher can be complicated by removing the helpful word breaks, but this does not greatly complicate the task of breaking a Caesarean cipher. Decipher the following Caesarean ciphertext:

XALWP EAJPB KMPDA SKNHZ EOXNK WZWJZ SEZA

(3.3) Why does a transposition cipher have only 25 keys?

(3.4) A substitution cipher is broken by using the letter frequencies chart. A good way to start is to identify the three most frequently occurring letters and try to associate these with the letters E, T and A. Decipher the following substitution ciphertext:

HR LHKAQEK HU BUN LBN
IHGKEXYHUV B GNGKCW LC SBU FXCIHSK
LHKA SCXKBHUKN HC LHKA B
FXQYBYHZHKN CMEBZ KQ EUHKN KAC
TBZEC QR B FANGHSBZ MEBUKHKN
KACU KACXC CJHGKG BU CZCWCUK QR
FANGHSBZ XCBZHKN SQXXCGFQUIUV KQ
KAHG FANGHSBZ MEBUKHKN

(3.5) It will not always be the case, of course, that the most common symbol occurring represents E, T, or A, and symbols other than Roman letters can be used. Try deciphering the following substitution ciphertext:

$\xi\mu \, \xi\iota \, \kappa\pi\delta\epsilon\tau\eta \, \mu\theta\kappa\mu \, \psi\gamma\rho\sigma\epsilon\mu\xi\beta\delta \, \rho\kappa\psi\theta\xi\beta\tau\iota$
$\xi\beta\tau A\xi\mu\kappa\lambda\phi\chi \, \xi\beta A\gamma\phi A\tau \, \eta\tau A\xi\psi\tau\iota \, \nu\theta\xi\psi\theta \, \sigma\tau\pi\zeta\gamma\pi\rho$
$\phi\gamma\delta\xi\psi\kappa\phi \, \zeta\epsilon\beta\psi\mu\xi\gamma\beta\iota \, \mu\theta\kappa\mu \, \eta\gamma \, \beta\gamma\mu \, \theta\kappa A\tau \, \kappa \, \iota\xi\beta\delta\phi\tau$
$A\kappa\phi\epsilon\tau\eta \, \xi\beta A\tau\pi\iota\tau \, \mu\theta\xi\iota \, \phi\gamma\delta\xi\psi\kappa\phi$
$\xi\pi\pi\tau A\tau\pi\iota\xi\lambda\xi\phi\xi\mu\chi \, \xi\iota \, \kappa\iota\iota\gamma\psi\xi\kappa\mu\tau\eta \, \nu\xi\mu\theta \, \sigma\theta\chi\iota\xi\psi\kappa\phi$
$\xi\pi\pi\tau A\tau\pi\iota\xi\lambda\xi\phi\xi\mu\chi \, \kappa\beta\eta \, \pi\tau\alpha\epsilon\xi\pi\tau\iota \, \kappa \, \rho\xi\beta\xi\rho\kappa\phi$
$\theta\tau\kappa\mu \, \delta\tau\beta\tau\pi\kappa\mu\xi\gamma\beta \, \sigma\tau\pi \, \rho\kappa\psi\theta\xi\beta\tau \, \psi\chi\psi\phi\tau$

$\mu\pi\sigma\xi\psi\kappa\phi\phi\chi \, \gamma\zeta \, \mu\theta\tau \, \gamma\pi\eta\tau\pi \, \gamma\zeta \, B\mu \, \zeta\gamma\pi \, \tau\kappa\psi\theta$
$\xi\pi\pi\tau A\tau\pi\iota\xi\lambda\phi\tau \, \zeta\epsilon\beta\psi\mu\xi\gamma\beta$

(3.6) Show that the two conditions for perfect secrecy given in eqns 3.2 and 3.3 are equivalent.

(3.7) Why is it necessary, in the Vernam cipher, to use the key only once if perfect secrecy is to be ensured?

(3.8) Show that 2 is a primitive root modulo 13 but that 4 is not.

(3.9) Calculate a Diffie–Hellman key for $p = 17$ and $g = 3$. What happens if a or b takes the value 0 or 1?

(3.10) Prove the following properties of the Euler φ-function:

(a) $\varphi(p) = p - 1$,
(b) $\varphi(pq) = (p - 1)(q - 1)$,
(c) $\varphi(p^2) = p(p - 1)$,

where p and q are distinct prime numbers.

(3.11) An impractically simple RSA system has $N = 247$ and $e = 5$.

(a) Choose a suitable three-decimal-digit plaintext $\mathcal{P}$ and calculate the corresponding ciphertext $\mathcal{C}$.
(b) By factoring N, show that $d = 173$.
(c) Use the private key d to recover $\mathcal{P}$ from $\mathcal{C}$.

(3.12) Alice prepares a quantum system in one of two possible mixed states with density operators $\hat{\rho}_1$ and $\hat{\rho}_2$. Under what conditions will it be possible for Bob to distinguish between these two possibilities with certainty?

(3.13) We discussed measuring the observable associated with the operator in eqn 3.27 as a means by which to discriminate between the non-orthogonal states $|\psi_1\rangle$ and $|\psi_2\rangle$. Suppose instead that we measured the quantity

$$\hat{B} = |\psi_1\rangle\langle\psi_1| - |\psi_2\rangle\langle\psi_2|. \qquad (3.66)$$

(a) What are the possible values resulting from such a measurement?

(b) Calculate the probabilities for each possible result for each of the two possible signal states $|\psi_1\rangle$ and $|\psi_2\rangle$.

3.14) A source produces single-photon pulses at a rate of 1 MHz and launches these into an ultralow-loss fibre with an absorption of $0.1\,\mathrm{dB\,km^{-1}}$. If each photon carries 1 bit of information then calculate the maximum transmission rate for each of the following channels:

(a) a local link of 10 km;
(b) a link between London and Glasgow (672 km);
(c) a link between London and New York (5585 km).

Recall that the loss in dB is defined in terms of the power at two points P_1 and P_2, in this case 1 km apart:

$$\text{Loss} = 10 \log_{10} \frac{P_1}{P_2}.$$

3.15) Show that a general unitary operator acting on a qubit and an ancilla has the form of eqn 3.28. What constraints does unitarity impose on the operators $\hat{A}_i$?

3.16) Let $|\mu\rangle$ denote one of the two eigenstates of $\hat{\sigma}_z$ or one of the eigenstates of $\hat{\sigma}_x$, with each possibility being equally likely. Suppose that we have $2N$ copies of a qubit, so that the combined state is $|\mu\rangle \otimes |\mu\rangle \otimes \cdots \otimes |\mu\rangle$. We might try to identify the state by measuring $\hat{\sigma}_z$ on N qubits and $\hat{\sigma}_x$ on the remaining N. What is the probability that this process will correctly identify the state?

3.17) Show that even if perfect cloning were limited to making a single copy then it would still be possible to use it and entangled states in order to communicate information superluminally.

3.18) Plane waves are a theoretical abstraction, and real laboratory light beams are all of finite spatial extent. This inevitably leads to small field components in the direction of propagation. The electric field near the focus of a monochromatic laser beam propagating in the z-direction has the form

$$\vec{E} = (\alpha \hat{\imath} + \beta \hat{\jmath}) E_0 \exp\left(-\frac{x^2 + y^2}{2w^2}\right) e^{i(k_z z - \omega t)}$$
$$+ E_z \hat{k},$$

where $\hat{\imath}$, $\hat{\jmath}$, and $\hat{k}$ are the unit vectors in the x-, y- and z- directions, respectively, and $k_z \approx k$.

(a) Calculate E_z and show that it is typically much smaller in magnitude than the other components.

(b) Find the form of $\vec{H}$ in this region and show that it too has a z-component.

(3.19) The Poynting vector is the rate of flow of electromagnetic energy per unit area. Verify this interpretation by proving that

$$\frac{\partial}{\partial t} \int_V w\, dV = -\int_A \vec{S} \cdot d\vec{A},$$

where w is the electromagnetic energy density

$$w = \frac{1}{2}\left(\varepsilon E^2 + \mu_0 H^2\right).$$

(3.20) Light with partial linear polarization is a mixture of linearly polarized light, with intensity I_{pol}, and unpolarized light, with intensity I_{unpol}. The degree of polarization is defined to be

$$P = \frac{I_{\mathrm{pol}}}{I_{\mathrm{pol}} + I_{\mathrm{unpol}}}.$$

If this light is passed through a polarizing filter then the intensity of the transmitted light will vary if the polarizer is rotated. If $I_{\max}$ and $I_{\min}$ are the maximum and minimum values of the transmitted light then show that the degree of polarization is also given by

$$P = \frac{I_{\max} - I_{\min}}{I_{\max} + I_{\min}}.$$

(3.21) Calculate the magnetic field associated with the circularly polarized electric field in eqn 3.46.

(3.22) Show that the general electric field amplitude

$$\vec{E}_0 = E_x \hat{\imath} + E_y \hat{\jmath},$$

for arbitrary complex E_x and E_y, corresponds to an elliptical polarization.

(3.23) (a) Write the Jones matrices in Fig. 3.7 as superpositions of the Pauli matrices.

(b) Which of the Jones matrices are unitary? What physical significance can you attach to this?

(3.24) Show that imposing the conservation of energy on the beam splitter relations in eqn 3.57 leads to the conditions $|t_1|^2 = |t_2|^2$, $|t_i|^2 + |r_i|^2 = 1$, and $t_1^* r_1 + r_2^* t_2 = 0$. Hence, invert the beam splitter relations to express the input fields in terms of the outputs.

(3.25) In Appendix G, there is a fully quantum treatment of beam splitters. Calculate the effect of a symmetric beam splitter, with $t = 1/\sqrt{2}$ and $r = i/\sqrt{2}$, on the following two-photon states:

  (a) $\hat{a}_{1\,\text{in}}^{H\dagger} \hat{a}_{1\,\text{in}}^{V\dagger} |\text{vac}\rangle$;

  (b) $\hat{a}_{1\,\text{in}}^{H\dagger} \hat{a}_{2\,\text{in}}^{V\dagger} |\text{vac}\rangle$;

  (c) $2^{-1/2} \left(\hat{a}_{1\,\text{in}}^{H\dagger} \hat{a}_{2\,\text{in}}^{V\dagger} + \hat{a}_{1\,\text{in}}^{V\dagger} \hat{a}_{2\,\text{in}}^{H\dagger} \right) |\text{vac}\rangle$;

  (d) $2^{-1/2} \left(\hat{a}_{1\,\text{in}}^{H\dagger} \hat{a}_{2\,\text{in}}^{V\dagger} - \hat{a}_{1\,\text{in}}^{V\dagger} \hat{a}_{2\,\text{in}}^{H\dagger} \right) |\text{vac}\rangle$.

  Explain your results in each case.

(3.26) A counterfeiter decides to forge Wiesner banknotes without measuring the photons in a genuine note, but rather by simply guessing the polarization of each photon. What is the probability that one of his notes will be accepted by the bank as genuine?

(3.27) A counterfeiter tries to beat the Wiesner scheme by measuring in the basis

$$|+, \theta\rangle = \cos\left(\frac{\theta}{2}\right) |0\rangle + \sin\left(\frac{\theta}{2}\right) |1\rangle,$$

$$|-, \theta\rangle = -\sin\left(\frac{\theta}{2}\right) |0\rangle + \cos\left(\frac{\theta}{2}\right) |1\rangle,$$

and then preparing a photon after each measurement in the state $|+, \theta\rangle$ or $|-, \theta\rangle$ corresponding to the measurement outcome. What is the probability that his forgery will go undetected?

(3.28) In our description of the BB84 protocol, it is Bob who tells Alice which basis he used for each photon and Alice who then told Bob which results to keep and which to discard. What difference would it make, if any, if it was Alice who tells Bob which basis she used and Bob who told Alice which bits to keep and which to discard?

(3.29) Preparing single photons is technically challenging and highly attenuated laser pulses are sometimes used. The number of photons in such a pulse is Poisson-distributed, with the probability for n photons being present in a single pulse being

$$P(n) = e^{-\bar{n}} \frac{\bar{n}^n}{n!}.$$

  (a) Show that the mean number of photons is $\bar{n}$ and that the uncertainty is $\Delta n = \sqrt{\bar{n}}$.

  (b) If a pulse contains more than one photon then a technologically advanced Eve could take and store a photon from each of the pulses containing more than one photon, leaving the remainder for Bob. For what fraction of the pulses would Eve then have access to Alice and Bob's key bit?

(c) Current laser-based quantum key distribution systems use pulses with a mean photon number of 0.1. What probability should Alice and Bob assign, for each key bit, that Eve has correctly identified that bit? (You should assume that no errors are detected, so that Eve's activity is limited to the taking and storing of photons for future measurement.)

(3.30) In a BB84 system, Eve measures every photon, selecting for each one between the two bases employed by Alice and Bob.

  (a) In their public discussion, Alice and Bob announce the bit values for M bits. What is the probability that no errors will be found?

  (b) If Eve could escape detection, what fraction of her bits would be correct?

  (c) Would Alice and Bob's subsequent public discussion help her at all?

(3.31) It was suggested at an early stage that Eve might measure in a basis that is intermediate between the $|0\rangle$, $|1\rangle$ and $|0'\rangle$, $1'\rangle$ bases. The Breidbart basis has as its elements the states

$$|0_B\rangle = \cos\left(\frac{\pi}{8}\right) |0\rangle + \sin\left(\frac{\pi}{8}\right) |1\rangle,$$

$$|1_B\rangle = \sin\left(\frac{\pi}{8}\right) |0\rangle - \cos\left(\frac{\pi}{8}\right) |1\rangle.$$

If she gets the result corresponding to the state $|0_B\rangle$ (or $|1_B\rangle$) then she assigns the bit value 0 (or 1).

  (a) What is the probability that any one of Eve's measurements will give the correct bit value?

  (b) Assuming that she prepares and transmits a photon to Bob in the Breidbart basis, then what is the probability that this will lead to an error in Bob's bit string?

(3.32) Why is it not possible to use redundancy to combat errors on the quantum channel in quantum key distribution?

(3.33) Alice and Bob find that the error rate is q per bit, and use parity checks to correct the errors in their raw key of N bits. Estimate the length of the resulting corrected key, assuming that the error correction is performed efficiently.

3.34) Before privacy amplification, the probability that Eve correctly assigns the value of any given bit is $P_{\text{Eve}} = \frac{1}{2}(1 + \varepsilon)$.

    (a) Calculate the corresponding mutual information per bit between the key agreed by Alice and Bob, $\mathcal{K}$, and the corresponding string obtained by Eve, $\mathcal{E}$.

    (b) Show that privacy amplification leads to a reduction in $H(\mathcal{E} : \mathcal{K})$ that is approximately exponential in m.

3.35) If the error rate was q per bit and all of the errors have been removed, what fraction of the remaining bits should Alice and Bob assume are known to Eve? (You may assume a lossless quantum channel in which each pulse of light contains precisely one photon and that Eve has measured a fraction of the bits in the Breidbart basis, leaving the remainder unchanged.)

3.36) For an error rate q per bit, show, for the system assumed in the previous question, that the final key must be shorter than the corrected key by at least the factor

$$m > \frac{\log \mu}{\log[3(2 + \sqrt{2})q] - 2},$$

where the probability that Eve knows any one bit is required to be

$$P_{\text{Eve}} \leq \frac{1}{2}(1 + \mu).$$

(3.37) This problem relates to the B92 protocol, based on the two states $|0\rangle$ and $|1''\rangle$ as defined in eqn 3.62.

    (a) Calculate the probability that any given photon will result in a bit of the raw key. (You may assume that each pulse contains precisely one photon and that there are no losses to worry about.)

    (b) For $\theta = \pi/2$, estimate the losses that can be tolerated if a technologically advanced Eve is to be denied access to the key.

(3.38) Alice prepares the entangled state given in eqn 3.63 and sends one of the qubits to Bob. If Eve measures this and prepares a new qubit for Bob, selected on the basis of her measurement outcome, show that the resulting state of Alice and Bob's qubits will be of the form of eqn 2.113 and so be correlated but not entangled.

Generalized measurements

4

Extracting information from a quantum system inevitably requires the performance of a measurement, and it is no surprise that the theory of measurement plays a central role in our subject. The physical nature of the measurement process remains one of the great philosophical problems in the formulation of quantum theory. Fortunately, however, it is sufficient for us to take a pragmatic view by asking what measurements are possible and how the theory describes them, without addressing the physical mechanism of the measurement process. This is the approach we shall adopt. We shall find that it leads us to a powerful and general description of both the probabilities associated with measurement outcomes and the manner in which the observation transforms the quantum state of the measured system.

4.1 Ideal von Neumann measurements

The simplest form of measurement was given a mathematical formulation by von Neumann, and we shall refer to measurements of this type as *von Neumann measurements* or projective measurements. It is this description of measurements that is usually introduced in elementary quantum theory courses. We start with an observable quantity A represented by a Hermitian operator $\hat{A}$, the eigenvalues of which are the possible results of the measurement of A. The relationship between the operator, its eigenstates $\{|\lambda_n\rangle\}$, and its (real) eigenvalues $\{\lambda_n\}$ is expressed by the eigenvalue equation

$$\hat{A}|\lambda_n\rangle = \lambda_n|\lambda_n\rangle. \tag{4.1}$$

The eigenstates form a complete orthonormal set, and this allows us to express the operator in terms of its eigenstates and eigenvalues:

$$\hat{A} = \sum_n \lambda_n|\lambda_n\rangle\langle\lambda_n|. \tag{4.2}$$

The probability that a measurement of A will give the result λ_n is

$$P(\lambda_n) = \langle\lambda_n|\hat{\rho}|\lambda_n\rangle = \text{Tr}\left(\hat{\rho}|\lambda_n\rangle\langle\lambda_n|\right), \tag{4.3}$$

where $\hat{\rho}$ is the density operator representing the state of the system under observation immediately prior to the measurement. This probability has the simple physical meaning that if the measurement were repeated on a very large ensemble of identically prepared systems, then $P(\lambda_n)$ is the fraction of measurements that would give the result λ_n.

It is helpful to write the probability in eqn 4.3 in a different form. We introduce the projector $\hat{P}_n = |\lambda_n\rangle\langle\lambda_n|$ so that

$$P(\lambda_n) = \text{Tr}(\hat{\rho}\hat{P}_n). \tag{4.4}$$

This allows us to deal in a straightforward manner with the possibility that the eigenstates of $\hat{A}$ may be degenerate, so that a number of eigenstates share a common eigenvalue. Suppose, for example, that there are three orthonormal eigenstates, $|\lambda_n^1\rangle, |\lambda_n^2\rangle$, and $|\lambda_n^3\rangle$, corresponding to the single eigenvalue λ_n. The probability that a measurement of A will give the result λ_n is then simply

$$P(\lambda_n) = \sum_{j=1}^{3} \text{Tr}\left(\hat{\rho}|\lambda_n^j\rangle\langle\lambda_n^j|\right). \tag{4.5}$$

We can write this in the more compact form of eqn 4.4, where $\hat{P}_n$ is the projector onto the (three-dimensional) space of eigenstates of $\hat{A}$ with eigenvalues λ_n:

$$\hat{P}_n = |\lambda_n^1\rangle\langle\lambda_n^1| + |\lambda_n^2\rangle\langle\lambda_n^2| + |\lambda_n^3\rangle\langle\lambda_n^3|. \tag{4.6}$$

A von Neumann measurement is one in which the probability for a given outcome is given by eqn 4.4, where $\hat{P}_n$ is a projector onto one or more orthonormal states. The properties of these projectors are summarized in Table 4.1.

Table 4.1 Properties of projectors

I. They are Hermitian operators	$\hat{P}_n^\dagger = \hat{P}_n$
II. They are positive operators	$\hat{P}_n \geq 0$
III. They are complete	$\sum_n \hat{P}_n = \hat{I}$
IV. They are orthonormal	$\hat{P}_i\hat{P}_j = \hat{P}_i\delta_{ij}$

Positivity and Hermiticity Recall that these conditions are not strictly independent in that condition I Hermiticity, is implied by condition II positivity.

The first three properties of the projectors have natural physical meanings. The projectors are Hermitian because they represent observable quantities. They are positive because their expectation values are probabilities and so must be positive (or zero) for all possible states. The third condition ensures that the sum of the probabilities for all possible measurement results will be unity for all possible states. The fourth property of the projectors does not have a convincing measurement interpretation, and we shall find that generalized measurements do not respect it.

The description of a von Neumann measurement is completed by a rule for determining the state of the system immediately following a measurement. The rule is that if a measurement of A gives the result λ_n, associated with a unique (non-degenerate) eigenstate $|\lambda_n\rangle$, then the post-measurement state is $|\lambda_n\rangle$. This means that a second measurement of A, carried out immediately after the first, will give the same result, λ_n. If the measurement result is associated with more than one eigenstate then the new state is obtained from the pre-measurement density operator $\hat{\rho}$ by acting on it with the projector associated with the measurement outcome. Hence our measurement of A giving the result λ_n will, in this case, be accompanied by a change in the density operator of the form

$$\hat{\rho} \rightarrow \hat{\rho}'_n = \frac{\hat{P}_n \hat{\rho} \hat{P}_n}{\text{Tr}(\hat{P}_n \hat{\rho} \hat{P}_n)}, \tag{4.7}$$

where the numerator ensures the normalization of the new density operator. We can use the cyclic property of the trace (eqn 2.47) together with property IV of our projectors to write this denominator as $\text{Tr}(\hat{P}_n \hat{\rho} \hat{P}_n) = (\hat{P}_n \hat{\rho}) = P(\lambda_n)$. Hence the density operator representing our post-measurement state can be written as

$$\hat{\rho}'_n = \frac{\hat{P}_n \hat{\rho} \hat{P}_n}{P(\lambda_n)}, \tag{4.8}$$

that is, the projection of $\hat{\rho}$ onto the space of eigenstates associated with the measurement result divided by the prior probability for the observed measurement outcome. Clearly, a second measurement of A carried out immediately will give the same result as the first.

The above description of the post-measurement state tacitly assumes that we have knowledge of the measurement outcome. Clearly, this need not be the case. Consider, for example, the situation in quantum cryptography in which an eavesdropper is active. Alice can prepare a qubit in a pure state and send it to Bob. Eve, the eavesdropper, may perform a measurement but, obviously, the result will remain unknown to Alice and Bob. It is useful, therefore, to be able to describe the post-measurement state of a system without knowledge of the measurement result. We can do this using the prescription given at the beginning of Section 2.2 that our density operator is the sum of all possible operators, weighted by their associated probabilities. If we know the pre-measurement density operator $\hat{\rho}$ and that the measurement is associated with the projectors $\{\hat{P}_n\}$ but we do not know the measurement outcome, then we describe the post-measurement state by the density operator

$$\hat{\rho}' = \sum_n P(\lambda_n)\hat{\rho}'_n$$

$$= \sum_n \hat{P}_n \hat{\rho} \hat{P}_n. \tag{4.9}$$

Clearly, the fact that the measurement has been performed changes the state even if we do not know the measurement result. This, of course, is one of the reasons why quantum key distribution works. The difference between the density operators in eqns 4.8 and 4.9 highlights the significance of information in quantum theory. The two density operators are different because, for the former, we know something extra (the measurement outcome). The state we assign to the post-measurement system depends on the amount of information available to us.

4.2 Non-ideal measurements

The von Neumann description of a measurement is insufficiently general for the simple reason that most observations that we can perform are not of this type. The real world is noisy, and this ensures that our observations will include errors. It is also often far from true that real measurements leave a quantum system in anything like an eigenstate. More importantly, it is often advantageous to deliberately design a measurement that deviates from the von Neumann form. We shall consider examples of such specifically designed measurements in Section 4.4. In this section we consider the effects of noise-induced errors on ideal von Neumann measurements.

Consider a device for determining whether a qubit is in the state $|0\rangle$ or the state $|1\rangle$. An ideal von Neumann measurement would be described by the pair of projectors

$$\hat{P}_0 = |0\rangle\langle 0|,$$
$$\hat{P}_1 = |1\rangle\langle 1|. \tag{4.10}$$

Suppose, however, that a source of noise means that the measuring device records the wrong state with probability p. This means that if the system is prepared in the state $|0\rangle$ then the measurement will give the result 0 with probability $1-p$ and the result 1 with probability p. This is the quantum analogue of the symmetric noisy channel discussed in Section 1.4. For a state described by a density operator $\hat{\rho}$, the probabilities for each of the two measurement outcomes are

$$P(0) = (1-p)\mathrm{Tr}(\hat{\rho}\hat{P}_0) + p\mathrm{Tr}(\hat{\rho}\hat{P}_1),$$
$$P(1) = (1-p)\mathrm{Tr}(\hat{\rho}\hat{P}_1) + p\mathrm{Tr}(\hat{\rho}\hat{P}_0). \tag{4.11}$$

We can write these in a form similar to eqn 4.4, that is, as

$$P(0) = \mathrm{Tr}(\hat{\rho}\hat{\pi}_0),$$
$$P(1) = \mathrm{Tr}(\hat{\rho}\hat{\pi}_1), \tag{4.12}$$

by introducing the operators

$$\hat{\pi}_0 = (1-p)\hat{P}_0 + p\hat{P}_1 = (1-p)|0\rangle\langle 0| + p|1\rangle\langle 1|,$$
$$\hat{\pi}_1 = (1-p)\hat{P}_1 + p\hat{P}_0 = (1-p)|1\rangle\langle 1| + p|0\rangle\langle 0|. \tag{4.13}$$

These are not projectors, as they obey the first three of the properties of projectors but not the last one: $\hat{\pi}_0\hat{\pi}_1 = p(1-p)\hat{I}$. They do, nevertheless, represent measurement probabilities.

We can extend the above example to describe any noise-affected, that is, non-ideal, von Neumann measurement by using the conditional probabilities introduced in Section 1.2. In order to proceed it is convenient to introduce two variables: we let i denote the outcome of a (hypothetical) ideal von Neumann measurement and r denote the outcome of the real measurement. An ideal von Neumann measurement of an observable A will give one of the results $\{\lambda_n\}$ with probabilities calculated using eqn 4.4, that is,

$$P(i = \lambda_n) = \text{Tr}(\hat{\rho}\hat{P}_n). \tag{4.14}$$

The statistical errors associated with the operation of the measuring device are described by the set of conditional probabilities $P(r = \lambda_m | i = \lambda_n)$. This is the probability that the measurement gives the result λ_m given that an ideal measurement would have given λ_n. Bayes' rule then gives the probability that the measured result is λ_m:

$$P(r = \lambda_m) = \sum_n P(r = \lambda_m | i = \lambda_n) P(i = \lambda_n)$$

$$= \sum_n P(r = \lambda_m | i = \lambda_n) \text{Tr}(\hat{\rho}\hat{P}_n). \tag{4.15}$$

Again we can write these probabilities in the form

$$P(r = \lambda_m) = \text{Tr}(\hat{\rho}\hat{\pi}_m) \tag{4.16}$$

by introducing the operators

$$\hat{\pi}_m = \sum_n P(r = \lambda_m | i = \lambda_n) \hat{P}_n. \tag{4.17}$$

These operators are clearly Hermitian, as $\hat{P}_n^\dagger = \hat{P}_n$. They are also positive, as the conditional probabilities are positive or zero and the projectors are positive operators. That the sum of the operators is the identity operator follows from the fact that the conditional probabilities are indeed probabilities, so that $\sum_m P(r = \lambda_m | i = \lambda_n) = 1$, together with the fact that the projectors sum to the identity operator. They are not projectors, however, as they are not orthonormal.

The change in the state associated with the outcome of a non-ideal measurement can be quite dramatic. For example, a photodetector detects the presence of a photon by absorbing it. After the measurement, there is no photon left. Even if the state transforms in accord with the hypothetical ideal von Neumann measurement, as in eqn 4.8, it will not usually be the case that a second non-ideal measurement will give the same result as the first. The most elegant way to obtain the post-measurement state is to use the theory of operations, which we describe in the final section of this chapter.

4.3 Probability operator measures

We have seen that the first three properties of the projectors have a natural interpretation in terms of the measurement process. The fact that the expectation values of the operators are the probabilities for the corresponding outcomes of a measurement enforces these conditions.

The fourth property, orthonormality of the projectors, does not have a similar significance and, as we saw in the preceding section, the operators describing non-ideal measurements do not respect it.

In quantum information, we are often interested in determining the best possible measurement to perform in any given situation. This means that it is useful to have a simple mathematical formulation which is sufficiently general to describe any possible measurement. We develop the theory by introducing a set of probability operators $\hat{\pi}_m$, such that the probability that a measurement on a system described by a density operator $\hat{\rho}$ gives the result m is

$$P_m = \text{Tr}(\hat{\rho}\hat{\pi}_m). \tag{4.18}$$

The set of operators forms a probability operator measure (POM), also known as a positive operator-valued measure (POVM). We also refer to the probability operators $\hat{\pi}_m$ as the elements of the probability operator measure, or as POM elements. The POM elements are *defined* by their properties, which we summarize in Table 4.2.

POMs or POVMs A measure is a function which assigns a number, in this case a probability, to the subsets of a given set. The elements of the measure are the probability operators and this is the reason for calling it a probability operator measure. The often-used expression 'positive operator-valued measure' expresses the fact that the elements of the measure, the probability operators, are positive operators. Calling the set of operators a POM reminds us of their physical significance, while the term POVM recalls their mathematical properties.

Table 4.2 Properties of probability operators

I. They are Hermitian operators	$\hat{\pi}_n^\dagger = \hat{\pi}_n$
II. They are positive operators	$\hat{\pi}_n \geq 0$
III. They are complete	$\sum_n \hat{\pi}_n = \hat{I}$

Note that there is no restriction on the number of elements in a POM and that this can be greater or less than the dimension of the state space of the system being monitored. This contrasts, of course, with the projectors, which, being orthogonal, cannot exceed the dimension of the state space. Any set of operators satisfying all the properties of a POM represents a possible measurement and, moreover, the outcomes of any measurement can be described in terms of a POM. These statements, which we shall justify below, mean that we can optimize our measurement strategy by considering all possible POMs and only then consider how the optimal measurement can be realized. We shall discuss optimal measurements and the associated POMs in the next section.

That the probabilities associated with any measurement can be described by a POM follows from considering the most general way in which we can carry out a measurement. We start by preparing an ancillary quantum system, or *ancilla*, in a known quantum state $|A\rangle_a$ and then cause a controlled interaction to occur between the system to be measured and our ancilla. The result of this is to create the state $\hat{U}|\psi\rangle \otimes |A\rangle_a$, where both $|A\rangle_a$ and $\hat{U}$ are selected by the observer and $|\psi\rangle$ is the state of the system being observed. A von Neumann measurement is then performed on both the system and the ancilla. This corresponds to projecting the now entangled state onto a complete set of system–ancilla states $\{|m\rangle \otimes |l\rangle_a\}$. The probability that any given result appears is

$$P(m, l) = |_a\langle l| \otimes \langle m|\hat{U}|\psi\rangle \otimes |A\rangle_a|^2 = \langle\psi|\hat{\pi}_{ml}|\psi\rangle, \qquad (4.19)$$

where $\hat{\pi}_{ml}$ is the probability operator

$$\hat{\pi}_{ml} = {}_a\langle A|\hat{U}^\dagger|m\rangle \otimes |l\rangle_{aa}\langle l| \otimes \langle m|\hat{U}|A\rangle_a. \qquad (4.20)$$

It is clear that these operators are Hermitian and they are also positive, in that for any state $|\phi\rangle$ we find $\langle\phi|\hat{\pi}_{ml}|\phi\rangle = |_a\langle A|\otimes\langle\phi|\hat{U}^\dagger|m\rangle\otimes|l\rangle_a|^2 \geq 0$. The third property follows from the completeness of the measurement states $\{|m\rangle \otimes |l\rangle_a\}$:

$$\sum_{m,l} \hat{\pi}_{ml} = {}_a\langle A|\hat{U}^\dagger \sum_m |m\rangle\langle m| \sum_l |l\rangle_{aa}\langle l|\hat{U}|A\rangle_a$$

$$= {}_a\langle A|\hat{I} \otimes \hat{I}_a|A\rangle_a = \hat{I}. \qquad (4.21)$$

We conclude that any measurement that we might devise can be described by a POM. We can rewrite the probability in eqn 4.19 in a suggestive form,

$$P(m, l) = {}_a\langle A| \otimes \langle\psi|\hat{P}_{ml}|\psi\rangle \otimes |A\rangle_a, \qquad (4.22)$$

where $\hat{P}_{ml}$ is a projector onto the entangled state $\hat{U}^\dagger|m\rangle \otimes |l\rangle_a$:

$$\hat{P}_{ml} = \hat{U}^\dagger|m\rangle \otimes |l\rangle_{aa}\langle l| \otimes \langle m|\hat{U}. \qquad (4.23)$$

The probability in eqn 4.22 suggests a general projective measurement on the state $|\psi\rangle\otimes|A\rangle_a$ and so we can picture a generalized measurement as a *comparison* between the system to be measured and an ancillary quantum system prepared in a state of our choosing. One important example of such a comparison is the measurement of a two-qubit state $|\psi\rangle\otimes|A\rangle_a$ in the Bell-state basis (eqn 2.108). The Bell states are simultaneous eigenstates of the three operators $\hat{\sigma}_x\otimes\hat{\sigma}_x$, $\hat{\sigma}_y\otimes\hat{\sigma}_y$, and $\hat{\sigma}_z\otimes\hat{\sigma}_z$. We can view this measurement, therefore, as determining whether the values of these spin components are the same (with the measured value being $+1$) or different (corresponding to the value -1) for the qubit under scrutiny and our specially prepared ancilla. Such a measurement provides information about the three incompatible observables corresponding to the operators $\hat{\sigma}_x$, $\hat{\sigma}_y$, and $\hat{\sigma}_z$ and can be viewed as a simultaneous, imperfect measurement of them.

We should also demonstrate that every POM is realizable, at least in principle, as a generalized measurement. The proof of this is a consequence of Naimark's theorem. We shall not attempt to prove the statement in full generality but, rather, shall demonstrate it for the special case of a POM describing a measurement on a qubit. We consider a POM with N elements of the form $\hat{\pi}_j = |\Psi_j\rangle\langle\Psi_j|$ with $j = 1, \cdots, N$, where the

$$|\Psi_j\rangle = \psi_{j0}|0\rangle + \psi_{j1}|1\rangle \qquad (4.24)$$

are, in general, unnormalized state vectors. The third POM property constrains the coefficients ψ_{j0} and ψ_{j1} to obey

$$\sum_{j=1}^{N} |\psi_{j0}|^2 = 1 = \sum_{j=1}^{N} |\psi_{j1}|^2,$$

$$\sum_{j=1}^{N} \psi_{j0}\psi_{j1}^* = 0 = \sum_{j=1}^{N} \psi_{j0}^*\psi_{j1}. \tag{4.25}$$

Our task is to represent the vectors $|\Psi_j\rangle$ as projections onto the qubit state space of a set of N orthonormal states $|\Psi_j'\rangle$ in an extended state space. Our POM would then describe a von Neumann measurement in this space. In order to show that this is always possible, we consider our two-state qubit to be a subsystem of an N-dimensional state space spanned by the orthonormal states $|0\rangle, |1\rangle, |2\rangle, \cdots, |N-1\rangle$. Within this space we introduce the vectors

$$|\Phi_i\rangle = \sum_{j=0}^{N-1} \psi_{ji}^*|j\rangle, \tag{4.26}$$

Extending the state space   The extra states can be other states of the quantum system not used to represent the qubit, for example extra energy levels in an atom or ion. They can also be formed by introducing an ancilla so that the required N orthonormal states are, for example, $|0\rangle \otimes |0\rangle_a$, $|1\rangle \otimes |0\rangle_a, |0\rangle \otimes |1\rangle_a, |0\rangle \otimes |2\rangle_a, \cdots, |0\rangle \otimes |N-2\rangle_a$.

where $i = 0, \cdots, N-1$. It follows from the conditions in eqn 4.25 that $|\Phi_0\rangle$ and $|\Phi_1\rangle$ are orthonormal. It is straightforward to choose the $N-2$ remaining vectors $|\Phi_2\rangle, \cdots, |\Phi_{N-1}\rangle$ so that the $\{|\Phi_i\rangle\}$ form an orthonormal basis spanning the N-dimensional state space. The orthonormality of these states can be expressed simply in the form

$$\langle \Phi_k|\Phi_i\rangle = \sum_{j=0}^{N-1} \psi_{ji}^*\psi_{jk} = \delta_{ik}. \tag{4.27}$$

It is helpful to express the relationship between our two bases $\{|i\rangle\}$ and $\{|\Phi_i\rangle\}$ by means of a unitary operator $\hat{U}$:

$$|\Phi_i\rangle = \hat{U}|i\rangle. \tag{4.28}$$

Because $\hat{U}$ is unitary, we can construct another orthonormal basis using $\hat{U}^\dagger$:

$$|\Psi_j'\rangle = \hat{U}^\dagger|j\rangle = \sum_{i=0}^{N-1} \psi_{ji}|i\rangle$$

$$= |\Psi_j\rangle + \sum_{i=2}^{N-1} \psi_{ji}|i\rangle. \tag{4.29}$$

The orthonormality of these states means that we can perform a von Neumann measurement in this basis, the results of which are associated with the probability operators:

$$P(j) = \langle \Psi_j'|\hat{\rho}|\Psi_j'\rangle$$
$$= \langle \Psi_j|\hat{\rho}|\Psi_j\rangle$$
$$= \text{Tr}(\hat{\rho}\hat{\pi}_j). \tag{4.30}$$

Hence our general POM elements can be realized as a von Neumann measurement in our extended N-dimensional state space. Equivalently, we can act on our system with the operator $\hat{U}$ and then perform a measurement in the $\{|i\rangle\}$ basis.

A simple example may help to illustrate the construction of a POM as a von Neumann measurement in an extended state space. Consider the three-element POM with elements $\hat{\pi}_j = |\Psi_j\rangle\langle\Psi_j|$, where

$$|\Psi_1\rangle = \frac{1}{\sqrt{2}}(\tan\theta|0\rangle + |1\rangle),$$

$$|\Psi_2\rangle = \frac{1}{\sqrt{2}}(\tan\theta|0\rangle - |1\rangle),$$

$$|\Psi_3\rangle = \sqrt{1 - \tan^2\theta}|0\rangle, \tag{4.31}$$

for some angle $0 \leq \theta \leq \pi/4$. Our first task is to extend the state space, and this can be achieved by introducing a second, ancillary qubit prepared in the state $|0\rangle$. We can construct from the non-orthogonal states in eqn 4.31 a complete set of orthonormal states for the two qubits in the form

$$|\Phi_1\rangle = \frac{1}{\sqrt{2}}(|1\rangle \otimes |0\rangle + \tan\theta|1\rangle \otimes |0\rangle + \sqrt{1 - \tan^2\theta}|1\rangle \otimes |1\rangle),$$

$$|\Phi_2\rangle = \frac{1}{\sqrt{2}}(|1\rangle \otimes |0\rangle - \tan\theta|1\rangle \otimes |0\rangle - \sqrt{1 - \tan^2\theta}|1\rangle \otimes |1\rangle),$$

$$|\Phi_3\rangle = \sqrt{1 - \tan^2\theta}|1\rangle \otimes |0\rangle - \tan\theta|1\rangle \otimes |1\rangle,$$

$$|\Phi_4\rangle = |0\rangle \otimes |1\rangle. \tag{4.32}$$

If our qubit was prepared in the state $|\psi\rangle$ then the probability that our von Neumann measurement gives any one of the four possible results associated with the basis states in eqn 4.32 is

$$P(j) = |\langle\Phi_j||\psi\rangle \otimes |0\rangle|^2. \tag{4.33}$$

For $j = 1, 2, 3$, these are simply $\langle\psi|\hat{\pi}_j|\psi\rangle$, so that the measurement probabilities are precisely those associated with the required three-element POM. For $j = 4$, the probability is zero as $|\Psi_4\rangle$ is orthogonal to the initially prepared state.

We conclude this section with a brief description, within the language of POMs, of the classic problem of measuring simultaneously the position and momentum of a quantum particle. The observables are incompatible as they do not possess a common set of eigenstates. Position and momentum are continuous-valued observables and this leads us to seek continuous-valued POM elements $\hat{\pi}(x_m, p_m)$, where x_m and p_m are the values of the position and momentum given by the measurement. The probability *density* for the joint measurement is

$$\mathcal{P}(x_m, p_m) = \text{Tr}\left[\hat{\rho}\hat{\pi}(x_m, p_m)\right], \tag{4.34}$$

the normalization of which requires the POM elements to satisfy

$$\int dx_m \int dp_m \, \hat{\pi}(x_m, p_m) = \hat{\mathbf{I}}. \tag{4.35}$$

Mixed-state POMs Our probability operators can be a weighted sum of projectors; that is, they can be proportional to mixed-state density operators. We can realize these as von Neumann measurements in an enlarged state space with the von Neumann measurement comprising projectors onto more than one orthonormal state.

It is desirable that the probability density in eqn 4.34 should be as close as possible to the product of the probability densities for x and p associated with the state being measured. This leads us to consider the minimum-uncertainty-product states for x and p,

$$|x_m, p_m\rangle = (2\pi\sigma^2)^{-1/4} \int dx \exp\left[-\frac{(x-x_m)^2}{4\sigma^2} + ip_m x\right] |x\rangle, \quad (4.36)$$

where $|x\rangle$ is the eigenstate of $\hat{x}$ with eigenvalue x. The properties of these eigenstates are described in Appendix H. The minimum-uncertainty-product states given by eqn 4.36 have the smallest uncertainties in x and p and so are as close as it is possible to get to a simultaneous eigenstate of $\hat{x}$ and $\hat{p}$. They are not mutually orthogonal, but they are complete in the sense that

$$\frac{1}{2\pi\hbar} \int dx_m \int dp_m |x_m, p_m\rangle\langle x_m, p_m| = \hat{I}. \quad (4.37)$$

Comparing this with the POM condition in eqn 4.35 leads us to identify

$$\hat{\pi}(x_m, p_m) = \frac{1}{2\pi\hbar} |x_m, p_m\rangle\langle x_m, p_m| \quad (4.38)$$

as the POM elements which optimize the accuracy of the joint measurement. The probability density for the position measurement is

$$\mathcal{P}(x_m) = \int dx \, \langle x|\hat{\rho}|x\rangle \exp\left[-\frac{(x-x_m)^2}{2\sigma^2}\right], \quad (4.39)$$

which is a convolution of the true position probability density, $\langle x|\hat{\rho}|x\rangle$, and that for the states $|x_m, p_m\rangle$. It follows that the variance for the results of the position measurement is

$$\mathrm{Var}(x_m) = \Delta x^2 + \sigma^2, \quad (4.40)$$

while that for the momentum is

$$\mathrm{Var}(p_m) = \Delta p^2 + \frac{\hbar^2}{4\sigma^2}. \quad (4.41)$$

We see that the act of measuring *both* x and p has introduced an additional spread in the measurement results.

4.4   Optimized measurements

The great utility of the POM formalism becomes clear when we seek the optimal measurement in any given situation. Any measurement can be described by a POM and, also, any POM corresponds to a realizable measurement. This means that we can optimize the measurement by considering the mathematical problem of finding the optimal POM. Having found the optimal POM, we can then seek an experimentally feasible way of realizing it. In this section we shall follow this prescription: we first construct optimal POMs for a range of measurement problems and then describe how they can be realized as measurements on optical-polarization qubits.

The form of the optimal measurement depends crucially on the *a priori* information that we have. This is clearly the case in the general quantum communication problem introduced in Section 3.2. We suppose that Alice (the transmitting party) prepares an individual quantum system in one of a set of N possible quantum states with density operators $\{\hat{\rho}_i\}$ and selects the state $\hat{\rho}_j$ with probability p_j. Bob (the receiving party) knows both the set of possible states and the preparation probabilities p_i. His problem is to determine as best as he can the value of i selected by Alice and encoded in her choice of the state $\hat{\rho}_i$. In general, the states will be non-orthogonal and it follows that no measurement can discriminate with certainty between the possible states. To show this, let us suppose that a measurement does exist which can discriminate with certainty between a pair of *non-orthogonal* states $|\psi_1\rangle$ and $|\psi_2\rangle$. This would mean that there exists a pair of probability operators $\hat{\pi}_1$ and $\hat{\pi}_2$ with the properties

$$\begin{aligned} \langle\psi_1|\hat{\pi}_1|\psi_1\rangle = 1, && \langle\psi_1|\hat{\pi}_2|\psi_1\rangle = 0, \\ \langle\psi_2|\hat{\pi}_1|\psi_2\rangle = 0, && \langle\psi_2|\hat{\pi}_2|\psi_2\rangle = 1. \end{aligned} \quad (4.42)$$

These are inconsistent, however, with the positivity and completeness of the probability operators. The completeness and positivity, together with the first condition of eqn 4.42, tell us that

$$\hat{\pi}_1 = |\psi_1\rangle\langle\psi_1| + \hat{A}, \quad (4.43)$$

where $\hat{A}$ is Hermitian and positive and $\hat{A}|\psi_1\rangle = 0$. If we insert this form into the third condition of eqn 4.42, we find

$$\langle\psi_2|\hat{\pi}_1|\psi_2\rangle = |\langle\psi_1|\psi_2\rangle|^2 + \langle\psi_2|\hat{A}|\psi_2\rangle. \quad (4.44)$$

The positivity of $\hat{A}$ means that the smallest possible value of this probability is $|\langle\psi_1|\psi_2\rangle|^2$, which is not zero, as $|\psi_1\rangle$ and $|\psi_2\rangle$ are not orthogonal.

We have to accept either the possibility of errors in Bob's determination of Alice's choice i or of some indeterminate or inconclusive measurement outcomes. The optimal, or best, measurement strategy for Bob to adopt depends on what we mean by 'best'. We shall describe state discrimination with minimum error and also the possibility of unambiguous state discrimination.

In seeking to minimize the probability of error in discriminating between the N states $\{\hat{\rho}_i\}$, we need to find N probability operators $\hat{\pi}_i$, one for each of the N possible signal states $\hat{\rho}_i$. If our measurement gives the result i, associated with the probability operator $\hat{\pi}_i$, then we decide that the signal state was $\hat{\rho}_i$. The probability that this procedure will correctly identify the state is then

$$P_{\text{corr}} = \sum_{j=1}^{N} p_j \text{Tr}(\hat{\pi}_j \hat{\rho}_j). \quad (4.45)$$

Our task is to maximize this, or equivalently, to minimize the error probability

$$P_{\text{err}} = 1 - P_{\text{corr}} = 1 - \sum_{j=1}^{N} p_j \text{Tr}(\hat{\pi}_j \hat{\rho}_j). \qquad (4.46)$$

The conditions for minimizing the error have been described by Helstrom and Holevo. We shall first state these conditions and then prove them. The first condition on the optimal probability operators is that they satisfy the operator equation

$$\hat{\pi}_j (p_j \hat{\rho}_j - p_k \hat{\rho}_k) \hat{\pi}_k = 0, \qquad \forall j, k. \qquad (4.47)$$

The second condition is

$$\hat{\Gamma} - p_j \hat{\rho}_j \geq 0, \qquad \forall j, \qquad (4.48)$$

where $\hat{\Gamma}$ is the Hermitian operator

$$\hat{\Gamma} = \sum_i p_i \hat{\rho}_i \hat{\pi}_i. \qquad (4.49)$$

The inequality in eqn 4.48 means that $\hat{\Gamma} - p_j \hat{\rho}_j$ is a positive operator. If we sum eqn 4.47 over k, using the completeness of the probability operators, then we get

$$\hat{\pi}_j (p_j \hat{\rho}_j - \hat{\Gamma}) = 0, \qquad \forall j, \qquad (4.50)$$

and if we sum it over j then we get (after relabelling)

$$(p_j \hat{\rho}_j - \hat{\Gamma}) \hat{\pi}_j = 0, \qquad \forall j. \qquad (4.51)$$

In order to demonstrate the optimality of the probability operators satisfying these conditions, we consider another measurement, associated with the probability operators $\{\hat{\pi}_j'\}$. The difference between the error probability obtained with the primed probability operators and the minimum is

The positivity of the trace of the product of two positive operators $\hat{A}$ and $\hat{B}$ follows on writing $\hat{B} = \hat{B}^{1/2}\hat{B}^{1/2}$ and using the cyclic property of the trace:

$$\text{Tr}(\hat{A}\hat{B}) = \text{Tr}(\hat{B}^{1/2}\hat{A}\hat{B}^{1/2})$$
$$= \sum_n \langle n|\hat{B}^{1/2}\hat{A}\hat{B}^{1/2}|n\rangle$$
$$= \sum_n \langle \tilde{n}|\hat{A}|\tilde{n}\rangle \geq 0,$$

where $|\tilde{n}\rangle = \hat{B}^{1/2}|n\rangle$. Each term in the summation is positive or zero, as $\hat{A}$ is a positive operator.

$$
\begin{aligned}
P'_{\text{err}} - P_{\text{err}}^{\min} &= -\sum_{j=1}^{N} p_j \text{Tr}(\hat{\pi}_j' \hat{\rho}_j) + \text{Tr}(\hat{\Gamma}) \\
&= \text{Tr}\left(\hat{\Gamma} - \sum_{j=1}^{N} p_j \hat{\pi}_j' \hat{\rho}_j\right) \\
&= \text{Tr}\left[\sum_{j=1}^{N} (\hat{\Gamma} - p_j \hat{\rho}_j) \hat{\pi}_j'\right],
\end{aligned} \qquad (4.52)
$$

where we have used the completeness of the probability operators $\hat{\pi}_j'$. The operators $\hat{\pi}_j'$ are positive, because they are probability operators, and so too are the $\hat{\Gamma} - p_j \hat{\rho}_j$. The trace of a product of two positive operators is greater than or equal to zero and hence $P'_{\text{err}} - P_{\text{err}}^{\min} \geq 0$.

If we can find a POM satisfying the conditions given in eqns 4.47 and 4.48 then we shall have the minimum possible error probability. It is not generally true, however, that the minimum-error POM will be unique. Note that if $P'_{\text{err}} - P^{\text{min}}_{\text{err}} = 0$ then it must follow that

$$(\hat{\Gamma} - p_j\hat{\rho}_j)\hat{\pi}'_j = 0, \qquad \forall j, \qquad (4.53)$$

which should be compared with eqn 4.51. A POM satisfying these conditions is a second measurement strategy for minimizing the error probability. Note that the $\hat{\pi}'_j$ will also satisfy eqn 4.51 with $\hat{\Gamma}$ replaced by $\hat{\Gamma}'$. It follows that $\hat{\Gamma} = \hat{\Gamma}'$ and that all minimum-error measurements will have the same $\hat{\Gamma}$ operator. The analysis presented here demonstrates the sufficiency of the conditions in eqns 4.47 and 4.48 for a minimum-error POM. That they are also necessary is proven in Appendix I.

If there are only two possible states $\hat{\rho}_1$ and $\hat{\rho}_2$, with prior probabilities p_1 and $p_2 = 1 - p_1$, then the minimum-error measurement will be a von Neumann measurement. The required probability operators, $\hat{\pi}_1$ and $\hat{\pi}_2 = \hat{I} - \hat{\pi}_1$, are projectors onto the eigenstates of $p_1\hat{\rho}_1 - p_2\hat{\rho}_2$ with positive and negative eigenvalues, respectively. The condition in eqn 4.47 is satisfied automatically because $\hat{\pi}_1$ and $\hat{\pi}_2$ are projectors onto orthogonal subspaces of the Hermitian operator $p_1\hat{\rho}_1 - p_2\hat{\rho}_2$. The conditions in eqn 4.48 are also satisfied:

$$\hat{\Gamma} - p_2\hat{\rho}_2 = p_1\hat{\rho}_1\hat{\pi}_1 + p_2\hat{\rho}_2\hat{\pi}_2 - p_2\hat{\rho}_2$$
$$= (p_1\hat{\rho}_1 - p_2\hat{\rho}_2)\hat{\pi}_1 \geq 0, \qquad (4.54)$$

because $\hat{\pi}_1$ is a projector onto the positive-eigenvalue state space of $p_1\hat{\rho}_1 - p_2\hat{\rho}_2$. A similar argument shows that eqn 4.48 is also satisfied for $j = 1$. The minimum error probability is

$$P^{\text{min}}_{\text{err}} = 1 - \text{Tr}(p_1\hat{\rho}_1\hat{\pi}_1 + p_2\hat{\rho}_2\hat{\pi}_2)$$
$$= \frac{1}{2}[1 - \text{Tr}(|p_i\hat{\rho}_1 - p_2\hat{\rho}_2|)], \qquad (4.55)$$

where $|p_i\hat{\rho}_1 - p_2\hat{\rho}_2| = (p_i\hat{\rho}_1 - p_2\hat{\rho}_2)(\hat{\pi}_1 - \hat{\pi}_2)$ is the magnitude of $p_i\hat{\rho}_1 - p_2\hat{\rho}_2$. As an example, consider a qubit prepared, with equal probability ($p_1 = \frac{1}{2} = p_2$), in one of the two pure states

$$|\psi_1\rangle = \cos\theta|0\rangle + \sin\theta|1\rangle,$$
$$|\psi_2\rangle = \cos\theta|0\rangle - \sin\theta|1\rangle, \qquad (4.56)$$

where $0 \leq \theta \leq \pi/4$. These states are not orthogonal, as $\langle\psi_1|\psi_2\rangle = \cos(2\theta)$, unless $\theta = \pi/4$. Solving for the orthonormal eigenvectors of $p_1\hat{\rho}_1 - p_2\hat{\rho}_2 = (|\psi_1\rangle\langle\psi_1| - |\psi_2\rangle\langle\psi_2|)/2$ leads us to the probability operators

$$\hat{\pi}_1 = \frac{1}{2}(|0\rangle + |1\rangle)(\langle 0| + \langle 1|) = |\hat{\pi}_1\rangle\langle\hat{\pi}_1|,$$
$$\hat{\pi}_2 = \frac{1}{2}(|0\rangle - |1\rangle)(\langle 0| - \langle 1|) = |\hat{\pi}_2\rangle\langle\hat{\pi}_2|, \qquad (4.57)$$

Minimum-error discrimination between two states For two states, we can also derive the minimum-error measurement without appealing to the conditions in eqns 4.47 and 4.48. The error probability is

$$P_{\text{err}} = p_1\text{Tr}(\hat{\rho}_1\hat{\pi}_2) + p_2\text{Tr}(\hat{\rho}_2\hat{\pi}_1)$$
$$= p_1 + \text{Tr}[(p_1\hat{\rho}_1 - p_2\hat{\rho}_2)\hat{\pi}_2],$$

where we have used the fact that $\hat{\pi}_1 = \hat{I} - \hat{\pi}_2$. This error probability is clearly minimized if $\hat{\pi}_2$ is the projector onto the negative-eigenvalue eigenstates of $p_1\hat{\rho}_1 - p_2\hat{\rho}_2$.

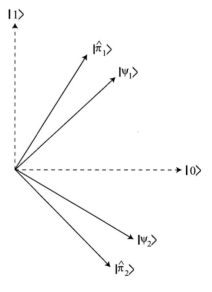

Fig. 4.1 The states $|\psi_1\rangle$ and $|\psi_2\rangle$ can be distinguished with minimum error by means of a von Neumann measurement with the two projectors corresponding to the orthogonal states denoted $|\hat{\pi}_1\rangle$ and $|\hat{\pi}_2\rangle$.

which are clearly projectors. The arrangement of the states $|\psi_1\rangle$ and $|\psi_2\rangle$, and the measurement states $|\hat{\pi}_1\rangle$ and $|\hat{\pi}_2\rangle$ is depicted in Fig. 4.1. The corresponding minimum error probability is

$$P_{\text{err}}^{\min} = \frac{1}{2}[1 - \sin(2\theta)], \tag{4.58}$$

which has a minimum value of zero for $\theta = \pi/4$, when the states are orthogonal and perfect discrimination is possible. It has a maximum value of one-half for $\theta = 0$, when the two states are identical and there is no better strategy than guessing.

Minimum-error state discrimination becomes more interesting when there are more than two possible states. In these situations the general form of the optimal POM is not known, although a variety of special cases have been derived. Often we can be guided by intuition and symmetry in constructing probability operators, the optimality of which can be tested using eqns 4.47 and 4.48. A good starting point in this process is the square-root measurement or 'pretty good' measurement. This has the probability operators

$$\hat{\pi}_i = p_i \hat{\rho}^{-1/2} \hat{\rho}_i \hat{\rho}^{-1/2}, \tag{4.59}$$

where $\hat{\rho} = \sum_{j=1}^{N} p_j \hat{\rho}_j$ is the a priori density operator. A simple example is the so-called trine ensemble, which is three equiprobable ($p_i = \frac{1}{3}$) qubit states of the form

$$|\psi_1\rangle = \frac{1}{2}(|0\rangle + \sqrt{3}|1\rangle),$$
$$|\psi_2\rangle = \frac{1}{2}(|0\rangle - \sqrt{3}|1\rangle),$$
$$|\psi_3\rangle = |0\rangle. \tag{4.60}$$

These states are equispaced on a great circle of the Bloch sphere, as depicted in Fig. 4.2 with overlap $|\langle\psi_i|\psi_j\rangle| = \frac{1}{2}$ for $i \neq j$. The a priori density operator is $\hat{\rho} = \frac{1}{2}\hat{I}$, so the probability operators for the square-root measurement are

$$\hat{\pi}_j = \frac{2}{3}|\psi_j\rangle\langle\psi_j|. \tag{4.61}$$

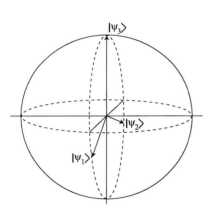

Fig. 4.2 The Bloch vectors for the three trine states.

It is straightforward to show that these satisfy the minimum-error conditions of eqns 4.47 and 4.48 and hence that, for the trine ensemble, the square-root measurement gives the minimum probability of error:

$$P_{\text{err}}^{\min} - 1 - \sum_{i=1}^{3} \frac{1}{3}\left(\frac{2}{3}|\langle\psi_i|\psi_i\rangle|^2\right) = \frac{1}{3}. \tag{4.62}$$

We might suppose that the existence of a minimum probability of error precludes the possibility of error-free or unambiguous state discrimination. This would not be correct, however, and unambiguous discrimination is possible for a set of linearly independent states. There is no contradiction here, as unambiguous state discrimination relies on the possibility of an inconclusive measurement outcome, which does not

assist in identifying the state. In order to appreciate the main ideas,
it suffices to consider a qubit prepared in one of the two pure states
given in eqn 4.56. The qubit state space is spanned by the orthogonal
states $|\psi_1\rangle$ and $|\psi_1^\perp\rangle = \sin\theta|0\rangle - \cos\theta|1\rangle$. If we perform a von Neumann
measurement in this basis then the projectors

$$\hat{P}_1 = |\psi_1\rangle\langle\psi_1|,$$
$$\hat{P}_1^\perp = |\psi_1^\perp\rangle\langle\psi_1^\perp| \tag{4.63}$$

correspond to observing the qubit to be in the state $|\psi_1\rangle$ or not to be
in this state. The latter of these cannot occur if $|\psi_1\rangle$ was prepared, and
so allows us to conclude *unambiguously* that the qubit was prepared in
$|\psi_2\rangle$. If the measurement result corresponds to $|\psi_1\rangle$, however, then either
state could have been prepared and the measurement is inconclusive.
A measurement of this type will only minimize the probability for an
inconclusive result if p_1 is sufficiently small.

A more interesting situation occurs if p_1 and p_2 are comparable in size.
In this case, the probability for an inconclusive measurement outcome
is minimized by a generalized measurement in which three outcomes are
possible. These correspond to error-free identification of $|\psi_1\rangle$ and of $|\psi_2\rangle$
and to the inconclusive result. The requirement that the conclusive re-
sults are error-free leads us to write the probability operators associated
with these in the form

$$\hat{\pi}_1 = A|\psi_2^\perp\rangle\langle\psi_2^\perp|,$$
$$\hat{\pi}_2 = A|\psi_1^\perp\rangle\langle\psi_1^\perp|,$$
$$\hat{\pi}_? = \hat{I} - \hat{\pi}_1 - \hat{\pi}_2, \tag{4.64}$$

where A is a positive number. Clearly, if the system was prepared in the
state $|\psi_1\rangle$ then the measurement outcome corresponding to identification
of $|\psi_2\rangle$ is impossible, as $\langle\psi_1|\hat{\pi}_2|\psi_1\rangle = A|\langle\psi_1|\psi_1^\perp\rangle|^2 = 0$. Identifying the
state $|\psi_2\rangle$ as $|\psi_1\rangle$ is similarly impossible. minimizing the probability
of an inconclusive result corresponds to making A as large as possible
without violating the positivity of $\hat{\pi}_?$. This procedure gives $A = (1 + \langle\psi_1|\psi_2\rangle|)^{-1}$, so that the probability for the inconclusive result to occur
is

$$P_? = |\langle\psi_1|\psi_2\rangle|. \tag{4.65}$$

This result is noteworthy in that it is a probability given by the modulus of the overlap of two states rather than the more familiar modulus squared. We note that the probability operators in eqn 4.64 with
the largest possible value of A correspond to the probability operators
$\hat{\tau}_j = |\Psi_j\rangle\langle\Psi_j|$ with the $|\Psi_j\rangle$ given in eqn 4.31 and $\hat{\pi}_3 = \hat{\pi}_?$. A geometri-
cal interpretation of the unambiguous state-discrimination measurement
is represented in Fig. 4.3.

In a sense, unambiguous state discrimination is more like unambiguous state *elimination* in that we have explicitly constructed probability
operators which tell us that the system was not prepared in one particular state. If there are more than two states and these are linearly

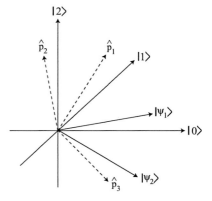

Fig. 4.3 We can realize our unambiguous discrimination measurement by
three projectors in a three-dimensional
state space. These correspond to three
orthogonal vectors, with $\hat{p}_{1,2}$ orthogo-
nal to $|\psi_{2,1}\rangle$ and $\hat{p}_3$ corresponding to
the ambiguous result.

dependent then unambiguous discrimination will not be possible. The reason for this is that there will not exist a state which is orthogonal to all but one of the possible states. Unambiguous state elimination, however, is always possible, subject to the possibility that there might also be an inconclusive result. This process, if successful, allows us to determine one of the states which was not prepared. As an example, consider the trine ensemble of qubit states in eqn 4.60. The probability operators

$$\hat{\pi}_i^\perp = \frac{2}{3}|\psi_i^\perp\rangle\langle\psi_i^\perp| \tag{4.66}$$

form a POM and the probability operator $\hat{\pi}_i^\perp$ corresponds to determining that the state $|\psi_i\rangle$ was not the one that was prepared.

Our aim has been to emphasize the universal nature of the ideas comprising quantum information. Reference to specific physical systems, however, can help to illustrate ideas and clarify concepts. We conclude this section with a brief description of some optimal measurements on states of photon polarization. Our reasons for selecting optical polarization are that we have already discussed it in the preceding chapter and that the measurements described here have been demonstrated in the laboratory. Our qubit in this case is a single photon with horizontal and vertical polarization states $|H\rangle$ and $|V\rangle$, which we associate with the qubit states $|0\rangle$ and $|1\rangle$, respectively (see Section 3.3). The key component in our description is the polarizing beam splitter, which is designed to reflect one of the linear polarizations and to reflect the orthogonal polarization. The orientations of the reflected and transmitted polarizations can be selected either by rotating the beam splitter or by rotating the polarization of the light by means of a suitable half-wave plate.

The two states in eqn 4.56 correspond to distinct linear polarizations at an angle θ to the horizontal so that the amplitudes associated with the horizontal and vertical polarizations are $\cos\theta$ and $\pm\sin\theta$, respectively. We have seen that the minimum-error measurement is simply a von Neumann measurement which, in this case, corresponds to measuring orthogonal linear polarizations. For equiprobable states this means measuring in the polarization basis $\{2^{-1/2}(|H\rangle+|V\rangle), 2^{-1/2}(|H\rangle-|V\rangle)\}$, which corresponds to measuring the polarization at an angle of $\pi/4$ to the horizontal. This measurement is readily achieved by means of a suitably oriented polarizing beam splitter together with a pair of photodetectors, one for each of the reflected and transmitted beams.

The minimum-error discrimination for the trine ensemble and unambiguous state discrimination described above can both be achieved using the experimental configuration depicted in Fig. 4.4. The device resembles, in its layout, a Mach–Zehnder interferometer with lower and upper paths U and L. The polarizing beam splitters transmit horizontally polarized light and reflect vertically polarized light. The input beam splitter, therefore, acts to entangle the polarization of an arbitrarily polarized photon with the path through the interferometer:

$$a_H|H\rangle + a_V|V\rangle \rightarrow a_H|H\rangle \otimes |L\rangle + a_V|V\rangle \otimes |U\rangle. \tag{4.67}$$

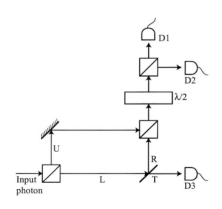

Fig. 4.4 Experimental configuration for minimum-error trine-state discrimination and for unambiguous state discrimination.

This is, of course, the extension of the state space required to realize a generalized measurement. Let us assume, for the sake of brevity, that the path lengths in the upper and lower arms are so chosen that the optical phase shifts due to propagation are equal and we do not need to consider them further. The partially reflecting mirror, or ordinary beam splitter, in the lower arm is designed to transmit the photon with probability T and to reflect it with probability $R = 1 - T$. We shall assume, for the purposes of this example, that the probability amplitudes for transmission and reflection are simply $T^{1/2}$ and $R^{1/2}$, respectively. If the photon is transmitted then it will be detected in photodetector 3, and if it is reflected then it will proceed to the output polarizing beam splitter. Hence the effect of the partially reflecting mirror is to transform our state in eqn 4.67 as

$$a_H|H\rangle \otimes |L\rangle + a_V|V\rangle \otimes |U\rangle \quad \rightarrow$$
$$a_H R^{1/2}|H\rangle \otimes |L\rangle + a_V|V\rangle \otimes |U\rangle + a_H T^{1/2}|D3\rangle, \quad (4.68)$$

where the state $|D3\rangle$ denotes recording the photon in detector 3. The output beam splitter coherently recombines the two beams, sending them both in the same direction, so that our state transforms as

$$a_H R^{1/2}|H\rangle \otimes |L\rangle + a_V|V\rangle \otimes |U\rangle + a_H T^{1/2}|D3\rangle \rightarrow$$
$$(a_H R^{1/2}|H\rangle + a_V|V\rangle) \otimes |U\rangle + a_H T^{1/2} |D3\rangle. \quad (4.69)$$

The half-wave plate is oriented so as to rotate the polarization states as $|H\rangle \rightarrow 2^{-1/2}(|H\rangle + |V\rangle)$ and $|V\rangle \rightarrow 2^{-1/2}(|H\rangle - |V\rangle)$. The final polarizing beam splitter then sends horizontally polarized light to detector 1 and vertically polarize light to detector 2, associated with the states $|D1\rangle$ and $|D2\rangle$ respectively. Hence the nett effect of the device depicted in Figure 4.4 is to transform a general input state $a_H|H\rangle + a_V|V\rangle$, in the fashion

$$a_H|H\rangle + a_V|V\rangle \rightarrow 2^{-1/2}(a_H R^{1/2} + a_V)|D1\rangle + 2^{-1/2}(a_H R^{1/2} - a_V)|D2\rangle$$
$$+ a_H T^{1/2}|D3\rangle. \quad (4.70)$$

The probabilities for detecting the photon in detectors 1, 2, and 3 are $|a_H R^{1/2} + a_V|^2$, $|a_H R^{1/2} - a_V|^2$, and $|a_H|^2 T$, respectively and it follows, therefore, that the associated probability operators are

$$\hat{\pi}_1 = \frac{1}{2}(R^{1/2}|H\rangle + |V\rangle)(R^{1/2}\langle H| + \langle V|)$$
$$= \frac{1}{2}(R^{1/2}|0\rangle + |1\rangle)(R^{1/2}\langle 0| + \langle 1|),$$
$$\hat{\pi}_2 = \frac{1}{2}(R^{1/2}|H\rangle - |V\rangle)(R^{1/2}\langle H| - \langle V|)$$
$$= \frac{1}{2}(R^{1/2}|0\rangle - |1\rangle)(R^{1/2}\langle 0| - \langle 1|),$$
$$\hat{\pi}_3 = T|H\rangle\langle H| = T|0\rangle\langle 0|. \quad (4.71)$$

If we choose $T = 2/3$ (and therefore $R = 1/3$), we recover the probability operators of eqn 4.61 needed to discriminate with minimum error between the states of the trine ensemble. Selecting $T = 1 - \tan^2 \theta$ (for $0 \leq \theta \leq \pi/4$) gives the probability operators, associated with eqn 4.31, which are needed for unambiguous discrimination between the two states in eqn 4.56. It is helpful to note that this value of T ensures that if the photon is not recorded in detector 3, then the output polarizing beam splitter transforms the two initially non-orthogonal polarization states into orthogonal polarizations and these can be distinguished perfectly by detectors 1 and 2.

4.5   Operations

It remains for us to address the state of the quantum system after a measurement is performed. This is important, as we shall sometimes wish to carry out a second observation. The von Neumann description, that the state is left in an eigenstate of the measured observable corresponding to the measurement outcome, is insufficiently general on two grounds. Firstly, many, even most, real measurements are very destructive to the quantum state: in photodetection, for example, the light is necessarily absorbed so that after the measurement there are no photons left. Secondly, it does not tell us how to describe the post-measurement state after a generalized measurement is performed. Providing a simple but sufficiently general treatment of these problems leads us to the language of operations and effects.

In seeking a description of the post-measurement state, we can be guided by the fact that it will always be described by a density operator and that this will constrain the form of the possible changes. A suitable point to start, therefore, is to ask the question 'what is the most general way in which we can change a density operator?' Quantum theory is linear in the density operator and this means that the transformed density operator will be a linear operator-function of the original density operator. It follows that we can transform the density operator by pre-multiplying and post-multiplying $\hat{\rho}$ by an arbitrary pair of operators:

$$\hat{\rho} \rightarrow \hat{A}\hat{\rho}\hat{B}. \tag{4.72}$$

Or, more generally, we might have a sum of such terms:

$$\hat{\rho} \rightarrow \sum_i \hat{A}_i \hat{\rho} \hat{B}_i. \tag{4.73}$$

The properties of the density operator, that it is a Hermitian and positive operator of unit trace, together with the fact that the transformation in eqn 4.73 should be the same for any initial $\hat{\rho}$, constrain the forms of $\hat{A}_i$ and $\hat{B}_i$. The Hermiticity property, in particular, suggests that we choose $\hat{B}_i = \pm \hat{A}_i^\dagger$. If we choose the plus sign then this also ensures the positivity of the transformed density operator, as $\langle \psi | \hat{A}_i \hat{\rho} \hat{A}_i^\dagger | \psi \rangle = \langle \phi_i | \hat{\rho} | \phi_i \rangle \geq 0$, where $|\phi_i\rangle$ is the unnormalized state $\hat{A}_i^\dagger | \psi \rangle$. It turns out that the resulting transformation

$$\hat{\rho} \rightarrow \sum_i \hat{A}_i \hat{\rho} \hat{A}_i^\dagger \tag{4.74}$$

s of the most general form. We refer to the operators $\hat{A}_i$ and $\hat{A}_i^\dagger$ as an effect, or a pair of effect operators, and to the transformation in eqn 4.74 as an operation. The requirement that the trace of the density operator s unity leads, on using the cyclic property of the trace, to the further condition that

$$\sum_i \hat{A}_i^\dagger \hat{A}_i = \hat{I}. \tag{4.75}$$

The operator combination $\hat{A}_i^\dagger \hat{A}_i$ is positive and this, together with the completeness property stated in eqn 4.74, leads us to associate this with he probability operators:

$$\hat{\pi}_i = \hat{A}_i^\dagger \hat{A}_i. \tag{4.76}$$

If the result of a generalized measurement associated with the POM $\{\hat{\pi}_i\}$ is j then the density operator changes as

$$\hat{\rho} \to \frac{\hat{A}_j \hat{\rho} \hat{A}_j^\dagger}{\mathrm{Tr}(\hat{A}_j \hat{\rho} \hat{A}_j^\dagger)}, \tag{4.77}$$

where the denominator is the a priori probability for obtaining the re-sult j and ensures the unit trace of the transformed density operator. This transformation is the required generalization of the von Neumann lescription in eqn 4.7; these are equivalent if the $\hat{A}_j$ are projectors. If he measurement result is not known then the density operator changes according to eqn 4.76, with each of the outcomes j weighted by its as-sociated probability:

$$\hat{\rho} \to \sum_j \mathrm{Tr}(\hat{A}_j \hat{\rho} \hat{A}_j^\dagger) \frac{\hat{A}_j \hat{\rho} \hat{A}_j^\dagger}{\mathrm{Tr}(\hat{A}_j \hat{\rho} \hat{A}_j^\dagger)} = \sum_j \hat{A}_j \hat{\rho} \hat{A}_j^\dagger. \tag{4.78}$$

This is the general form given in eqn 4.74 required by linearity and the properties of density operators.

Equation 4.76 does not allow us, of course, to determine the effect operators from the probability operators. In particular, we can satisfy t by writing

$$\hat{A}_i = \hat{U}_i \hat{\pi}_i^{1/2}, \tag{4.79}$$

where the $\hat{U}_i$ are any unitary operators. Hence, at best, knowing only he probability operators can determine the post-measurement state up to an arbitrary unitary transformation. If the post-measurement state s important then we need to know the associated effects.

Now that we have the means to describe the post-measurement state, we can deal with sequences of measurements. Suppose that we perform irst a measurement with outcomes i, associated with effect operators $\hat{A}_i$, and then a second with outcomes j, associated with effect operators $\hat{B}_j$. f the state prior to the first measurement is described by the density operator $\hat{\rho}$ then the probability that the first measurement gives the esult i is

$$P(i) = \mathrm{Tr}(\hat{A}_i^\dagger \hat{A}_i \hat{\rho}), \tag{4.80}$$

Complete positivity   The proof that eqn 4.74 is the most general form relies on the property of complete pos-itivity: that the transformation is al-lowed even if the system is entangled with another. A derivation of this re-sult is given in Appendix J.

Effects and probability opera-tors   The positivity of the operator combination $\hat{A}_i^\dagger \hat{A}_i$ follows on writing $\langle \psi | \hat{A}_i^\dagger \hat{A}_i | \psi \rangle = \langle \varpi_i | \varpi_i \rangle$, where $|\varpi_i\rangle = \hat{A}_i |\psi\rangle$. It is sometimes necessary to generalize the relationship in eqn 4.76 by writing $\hat{\pi}_i = \sum_k \hat{A}_{ik}^\dagger \hat{A}_{ik}$. This is equivalent, of course, to combining probability operators by writing $\hat{\pi}_i = \sum_k \hat{\pi}_{ik}$.

and this result is accompanied by the transformation in eqn 4.77. The probability that the second measurement then gives the result j is

$$P(j|i) = \frac{\text{Tr}(\hat{B}_j^\dagger \hat{B}_j \hat{A}_i \hat{\rho} \hat{A}_i^\dagger)}{\text{Tr}(\hat{A}_i^\dagger \hat{A}_i \hat{\rho})}, \tag{4.81}$$

so that the probability that the two measurements give the results i and j is

$$P(i,j) = P(j|i)P(i) = \text{Tr}(\hat{A}_i^\dagger \hat{B}_j^\dagger \hat{B}_j \hat{A}_i \hat{\rho}). \tag{4.82}$$

Hence the combined probability operator for the two measurements is

$$\hat{\pi}_{ij} = \hat{A}_i^\dagger \hat{B}_j^\dagger \hat{B}_j \hat{A}_i. \tag{4.83}$$

If the results i and j are known then the density operator following the second measurement is transformed as

$$\hat{\rho} \to \frac{\hat{B}_j \hat{A}_i \hat{\rho} \hat{A}_i^\dagger \hat{B}_j^\dagger}{P(i,j)}. \tag{4.84}$$

If, however, the outcome of neither measurement is known then we must weight the terms given by eqn 4.84 by their probability of occurrence, and the density operator transforms as

$$\hat{\rho} \to \sum_{i,j} \hat{B}_j \hat{A}_i \hat{\rho} \hat{A}_i^\dagger \hat{B}_j^\dagger. \tag{4.85}$$

These expressions can be seen as the natural extension of those obtained for a single measurement by treating $\hat{B}_j \hat{A}_i$ as a single effect operator associated with the set of probability operators $\hat{\pi}_{ij}$.

The formalism of operations is sufficiently general that we can use it to describe the action on $\hat{\rho}$ of any process. One simple but important example is the evolution associated with the solution of the Schrödinger equation,

$$\hat{\rho}(0) \to \hat{\rho}(t) = \exp\left(-i\frac{\hat{H}t}{\hbar}\right) \hat{\rho}(0) \exp\left(i\frac{\hat{H}t}{\hbar}\right), \tag{4.86}$$

in which we recognize the effect as the unitary time evolution operator $\hat{A} = \exp(-i\hat{H}t/\hbar)$. It is perhaps less obvious that non-unitary evolution, associated with dissipative or decohering dynamics, can also be described in terms of effects. As a simple example, we consider a qubit embodied in a pair of electronic energy levels, $|g\rangle$ and $|e\rangle$, of a single atom. The excited state $|e\rangle$ will decay to the ground state $|g\rangle$, owing to spontaneous emission of a photon, at a rate 2Γ. This dissipative process is fully described by the evolution of the matrix elements of $\hat{\rho}$:

$$\begin{aligned}
\langle e|\hat{\rho}(t)|e\rangle &= \langle e|\hat{\rho}(0)|e\rangle \exp(-2\Gamma t), \\
\langle g|\hat{\rho}(t)|g\rangle &= \langle g|\hat{\rho}(0)|g\rangle + \langle e|\hat{\rho}(0)|e\rangle[1 - \exp(-2\Gamma t)], \\
\langle g|\hat{\rho}(t)|e\rangle &= \langle g|\hat{\rho}(0)|e\rangle \exp(-\Gamma t), \\
\langle e|\hat{\rho}(t)|g\rangle &= \langle e|\hat{\rho}(0)|g\rangle \exp(-\Gamma t).
\end{aligned} \tag{4.87}$$

We can write the evolved density operator in terms of two effects:

$$\hat{\rho}(t) = \hat{A}_N(t)\hat{\rho}(0)\hat{A}_N^\dagger(t) + \hat{A}_Y(t)\hat{\rho}(0)\hat{A}_Y^\dagger(t), \tag{4.88}$$

where the effect operators are

$$\hat{A}_N(t) = \exp(-\Gamma t)|e\rangle\langle e| + |g\rangle\langle g|,$$
$$\hat{A}_Y(t) = [1 - \exp(-2\Gamma t)]^{1/2}|g\rangle\langle e|. \tag{4.89}$$

The previous discussion suggests that there might be a measurement interpretation for these effect operators, and this is indeed the case. In order to determine the nature of the observation, we introduce a pair of probability operators associated with the effects,

$$\hat{\pi}_N(t) = \hat{A}_N^\dagger(t)\hat{A}_N(t) = \exp(-2\Gamma t)|e\rangle\langle e| + |g\rangle\langle g|,$$
$$\hat{\pi}_Y(t) = \hat{A}_Y^\dagger(t)\hat{A}_Y(t) = [1 - \exp(-2\Gamma t)]|e\rangle\langle e|. \tag{4.90}$$

These clearly satisfy the requirements for forming a POM and so must describe a possible measurement. The probability that the atom will decay by spontaneous emission between times 0 and t is $[1 - \exp(-2\Gamma t)]$ if the atom was initially prepared in its excited state and is zero if it was in its ground state. This means that we can associate the two effects with the detection or absence of a detection (by an ideal detector) of a photon emitted by spontaneous emission. The effect operators $\hat{A}_N$ and $\hat{A}_Y$ describe a measurement of whether or not the atom has decayed to its ground state, and the subscripts are associated with the answers 'No' and 'Yes', respectively. If we are looking for the emitted photon then the presence or absence of it leads us to write the evolved density operator in the form of eqn 4.77 with $j = Y$ or N, respectively. Detecting a photon, not surprisingly, leaves the atom in its ground state. Failure to detect a photon is a measurement result and so changes the state: it is associated, in particular, with the decay of the off-diagonal matrix elements $\langle g|\hat{\rho}|e\rangle$ and $\langle e|\hat{\rho}|g\rangle$. The description of dissipation in terms of measurement, carried out on the environment, is the key idea in quantum trajectory methods. These have been applied widely to study the effects of dissipation on quantum systems.

It is by no means necessary for there to exist a natural measurement interpretation for any given operation, although the above analysis strongly suggests that we should always be able to contrive one. The method of operations remains a useful description even in these cases. As an example, we consider a communication channel in which a qubit is subjected to a spin flip about the x, y, or z direction with probability p_x, p_y, or p_z, respectively, or is left unchanged with probability $1 - p_x - p_y - p_z$. The spin flips are associated with the unitary operators $\hat{\sigma}_x$, $\hat{\sigma}_y$, and $\hat{\sigma}_z$, respectively, so that the density operator transforms as

$$\hat{\rho} \rightarrow (1 - p_x - p_y - p_z)\hat{\rho} + p_x\hat{\sigma}_x\hat{\rho}\hat{\sigma}_x + p_y\hat{\sigma}_y\hat{\rho}\hat{\sigma}_y + p_z\hat{\sigma}_z\hat{\rho}\hat{\sigma}_z. \tag{4.91}$$

This describes a randomization of the qubit state and can be used to model the effects of noise on the qubit. In particular, if $p_x = p_y = p_z =$

1/4 then the resulting density operator is $\hat{I}/2$ irrespective of the initial state. In this limit the communication is completely dominated by noise and no communication of information is possible.

The operations formalism provides us with a way of describing general changes to the state of a quantum system. This generality means that we can use it to determine which processes are possible within quantum theory and which are not. If a process we would like to perform cannot be described in this way then it cannot be done; we can use this to determine powerful bounds on what may be achieved. The idea is best illustrated by means of an example, and we shall consider Chefles's process of *state separation*. Suppose that we have a quantum system which we know to have been prepared in one of two non-orthogonal states $|\psi_1\rangle$ and $|\psi_2\rangle$. Our task in state separation is to transform the system in such a way that these states become $|\psi'_1\rangle$ and $|\psi'_2\rangle$, respectively, with

$$|\langle\psi'_1|\psi'_2\rangle| < |\langle\psi_1|\psi_2\rangle| \tag{4.92}$$

so that the transformed states have a smaller overlap, or are nearer to being orthogonal, than the originals. Clearly, this process cannot be guaranteed to succeed; were it otherwise, then repeated application of the process would render the states orthogonal and therefore fully distinguishable. It may be possible, however, to achieve state separation with a finite probability of success, P_S, and we would like to know how large this can be.

We introduce an effect operator $\hat{A}_S$ associated with successful state separation (and another, $\hat{A}_F$, associated with failure). The operator $\hat{A}_S$ acts on the states $|\psi_j\rangle$ to produce

$$\hat{A}_S|\psi_j\rangle = \mu|\psi'_j\rangle \tag{4.93}$$

so that $P_S = \langle\hat{A}_S^\dagger\hat{A}_S\rangle = |\mu|^2$. (It is not necessary to assume that μ is the same for both of the states, but doing so simplifies and shortens the derivation of P_S.) In order to find a bound on P_S, it suffices to consider the effect of $\hat{A}_S$ on a normalized superposition of $|\psi_1\rangle$ and $|\psi_2\rangle$:

$$\hat{A}_S\frac{|\psi_1\rangle + e^{i\phi}|\psi_2\rangle}{2^{1/2}[1 + \text{Re}(\langle\psi_1|\psi_2\rangle e^{i\phi})]^{1/2}} = \mu\frac{|\psi'_1\rangle + e^{i\phi}|\psi'_2\rangle}{2^{1/2}[1 + \text{Re}(\langle\psi_1|\psi_2\rangle e^{i\phi})]^{1/2}}. \tag{4.94}$$

The success probability for this operation cannot exceed unity, and this means that the length of the state vector must be less than unity for all values of ϕ:

$$|\mu|^2\frac{1 + \text{Re}(\langle\psi'_1|\psi'_2\rangle e^{i\phi})}{1 + \text{Re}(\langle\psi_1|\psi_2\rangle e^{i\phi})} \leq 1, \qquad \forall\phi. \tag{4.95}$$

This leads us to the bound

$$P_S = |\mu|^2 \leq \frac{1 - |\langle\psi_1|\psi_2\rangle|}{1 - |\langle\psi'_1|\psi'_2\rangle|}, \tag{4.96}$$

which has a natural interpretation in terms of unambiguous discrimination between the pair of states $|\psi_1\rangle$ and $|\psi_2\rangle$. We have seen that, for equiprobable states, the maximum probability for obtaining a conclusive result is

$$P_{\text{Conc}} = 1 - P_? = 1 - |\langle \psi_1 | \psi_2 \rangle|. \tag{4.97}$$

It follows, because this is the maximum value, that state separation followed by unambiguous state discrimination cannot increase this so that $P_S P'_{\text{Conc}} \leq P_{\text{Conc}}$, or

$$P_S \leq \frac{P_{\text{Conc}}}{P'_{\text{Conc}}} = \frac{1 - |\langle \psi_1 \psi_2 \rangle|}{1 - |\langle \psi'_1 \psi'_2 \rangle|}, \tag{4.98}$$

which is eqn 4.96.

The no-cloning theorem tells us that we cannot make a perfect copy of an unknown quantum state. It is possible, with a given probability, to create a copy if it is known to have been prepared in one of the two states $|\psi_1\rangle$ and $|\psi_2\rangle$. This means performing the transformation

$$|\psi_i\rangle \otimes |B\rangle \rightarrow |\psi_i\rangle \otimes |\psi_i\rangle, \tag{4.99}$$

where $|B\rangle$ is the initial 'blank' state of a suitable ancilla. We can view this as an example of state separation with $|\psi'_i\rangle = |\psi_i\rangle \otimes |\psi_i\rangle$, as

$$|\langle \psi'_1 | \psi'_2 \rangle| = |\langle \psi_1 | \psi_2 \rangle|^2 < |\langle \psi_1 | \psi_2 \rangle|. \tag{4.100}$$

The state separation bound in eqn 4.98 then gives a bound on the probability that the cloning will be successful:

$$P_{\text{Clone}} \leq P_S = \frac{1 - |\langle \psi_1 | \psi_2 \rangle|}{1 - |\langle \psi_1 | \psi_2 \rangle|^2} = \frac{1}{1 + |\langle \psi_1 | \psi_2 \rangle|}, \tag{4.101}$$

which is the Duan–Guo bound for perfect cloning, as described in Appendix F.

Suggestions for further reading

Bergou, J. A. (2007). Quantum state discrimination and selected applications. *Journal of Physics: Conference Series* **84**, 012001.

Braginsky, V. B. and Khalili, F. A. (1992). *Quantum measurement.* Cambridge University Press, Cambridge.

Busch, P., Grabowski, M., and Lahti, P. (1995). *Operational quantum physics.* Springer, Berlin.

Chefles, A. (2000). Quantum state discrimination. *Contemporary Physics* **41**, 401.

Hayashi, H. (2006). *Quantum information: an introduction.* Springer-Verlag, Berlin.

Helstrom, C. W. (1976). *Quantum detection and estimation theory.* Academic Press, New York.

Holevo, A. S. (1982). *Probabilistic and statistical aspects of quantum theory.* North Holland, Amsterdam.

Holevo, A. S. (2001). *Statistical structure of quantum theory.* Springer-Verlag, Berlin.

Kraus, K. (1983). *States, effects and operations.* Springer-Verlag, Berlin.

Paris, M. and Řeháček, J. (eds) (2004). *Quantum state estimation.* Lecture Notes in Physics, No. 649. Springer, Berlin.

Peres, A. (1993). *Quantum theory: concepts and methods.* Kluwer Academic, Dordrecht.

von Neumann, J. (1983). *Mathematical foundations of quantum mechanics.* Princeton University Press, Princeton, NJ.

Wheeler, J. A. and Zurek, W. H. (1983). *Quantum theory and measurement.* Princeton University Press, Princeton, NJ.

Exercises

(4.1) Show that the tabulated four properties of projectors are satisfied by:

(a) the qubit projectors

$$\hat{P}_0 = |0\rangle\langle 0|,$$
$$\hat{P}_1 = |1\rangle\langle 1|;$$

(b) the qutrit (three-state system) projectors

$$\hat{P}_0 = |0\rangle\langle 0|,$$
$$\hat{P}_{1,2} = |1\rangle\langle 1| + |2\rangle\langle 2|.$$

(4.2) Show that the requirement that $\sum_n P(\lambda_n) = 1$ for all possible states requires that the projectors are complete.

(4.3) Is the identity operator $\hat{I}$ a complete set of projectors?

(4.4) What would be the post-measurement state for each of the measurement outcomes in Exercise (4.1), part (b), on the qutrit states

(a) $|0\rangle$;
(b) $3^{-1/2}(|0\rangle + |1\rangle + |2\rangle)$;

and the mixed state with density operator

(c) $\frac{1}{4}(|0\rangle + |1\rangle)(\langle 0| + \langle 1|) + \frac{1}{4}(|1\rangle - |2\rangle)(\langle 1| - \langle 2|)$?

(4.5) Under what conditions will the density be unchanged by a von Neumann measurement for which we do not know the measurement outcome? (That is, $\hat{\rho}' = \hat{\rho}$ in eqn 4.9.)

(4.6) Show, using eqn 4.9, that $\text{Tr}(\hat{\rho}'^2) \leq \sum_n P(\lambda_n)\text{Tr}(\hat{\rho}_n'^2)$. How do you interpret this result?

(4.7) Under what conditions are the operators defined in eqn 4.17 projectors?

(4.8) A photon counter detects photons with efficiency η. This means that each photon is registered with probability η but fails to be registered with probability $1 - \eta$. Obtain the operator $\hat{\pi}_m$ corresponding to the probability of registering m counts in terms of the projectors $\hat{P}_n$ onto the field states with n photons. Check that the $\hat{\pi}_m$ satisfy the first three properties of projectors but not the fourth.

(4.9) Assume that the ideal von Neumann measurements associated with the projectors $\hat{P}_n$ transform the density operator according to eqn 4.8. Find the probability that two non-ideal measurements performed in quick succession and described by the operators in eqn 4.17 will give the same result.

(4.10) A qubit is to be compared with an ancillary qubit by simultaneous measurement of the three observables corresponding to the mutually commuting operators $\hat{\sigma}_x \otimes \hat{\sigma}_x$, $\hat{\sigma}_y \otimes \hat{\sigma}_y$, and $\hat{\sigma}_x \otimes \hat{\sigma}_z$. We can view the results of these as simultaneous *unsharp* values of the three incompatible spin components $\hat{\sigma}_x$, $\hat{\sigma}_y$, and $\hat{\sigma}_z$. Show that the average values found, s_i, are

$$s_i = a_i\langle\hat{\sigma}_i\rangle, \quad i = x, y, z,$$

and that the real constants a_i are constrained by the inequality $\sum_{i=x,y,z} a_i^2 \leq 1$.

(4.11) Construct the unitary operator in eqn 4.28 and confirm that it is indeed unitary. Hence prove the orthonormality of the states $\{|\Psi_j'\rangle\}$.

(4.12) Show that the states given in eqn 4.36 are position–momentum minimum-uncertainty-product states and calculate Δx.

4.13) The states in eqn 4.36 are said to be overcomplete, in that they are complete but not orthogonal.

(a) Evaluate the overlap $\langle x_m, p_m | x'_m, p'_m \rangle$.

(b) Prove the completeness relation in eqn 4.37.

4.14) Confirm the forms of the variances for the results of joint measurements of x and p given in eqns 4.40 and 4.41. Show that the product of these variances is bounded by the inequality

$$\text{Var}(x_m)\text{Var}(p_m) \geq \hbar^2.$$

Under what conditions does the equality hold?

4.15) Confirm that if the conditions in eqns 4.47 and 4.48 hold then the operator $\hat{\Gamma}$ in eqn 4.49 is Hermitian.

4.16) Verify that the operators given in eqn I.6, in Appendix I, form a POM.

4.17) In the BB84 protocol, Alice selects from four equally probable states: $|0\rangle, |1\rangle, |0'\rangle = 2^{-1/2}(|0\rangle + |1\rangle)$, and $|1'\rangle = 2^{-1/2}(|1\rangle - |0\rangle)$. For what values of μ does the POM $\{\mu|0\rangle\langle 0|, \mu|1\rangle\langle 1|, (1 - \mu)|0'\rangle\langle 0'|, (1-\mu)|1'\rangle\langle 1'|\}$ minimize the error probability for discriminating between these states? Confirm that the $\hat{\Gamma}$ operators are the same in each case.

4.18) Find the minimum-error POM for the two pure states in eqn 4.56 with arbitrary prior probabilities p_1 and p_2.

4.19) Confirm that the operators associated with the square-root measurement in eqn 4.59 form a POM.

4.20) Is the minimum-error measurement for the two states in eqn 4.56, with prior probabilities p_1 and p_2, also a square-root measurement?

4.21) The symmetric states are a set of N equiprobable states of the form

$$|\psi_j\rangle = \hat{V}^{j-1}|\psi_1\rangle, \qquad j = 1, \cdots, N,$$

where $\hat{V}$ is a unitary operator obeying the condition $\hat{V}^N = \hat{I}$. Show that the square-root measurement gives the minimum possible error probability for discriminating between these states.

4.22) It may be the case that the error probability is minimized by simply choosing a single state $\hat{\rho}_k$ and that no measurement can do better than this. In such cases we have $\hat{\pi}_i = \hat{I}\delta_{ik}$. Under what conditions does this no-measurement POM give the minimum probability of error in discriminating between a set of states?

4.23) The mirror-symmetric qubit states are

$$|\psi_1\rangle = \cos\theta|0\rangle + \sin\theta|1\rangle,$$
$$|\psi_2\rangle = \cos\theta|0\rangle - \sin\theta|1\rangle,$$
$$|\psi_3\rangle = |0\rangle,$$

with prior probabilities $p_{1,2} = p$ and $p_3 = 1 - 2p$. Find the minimum-error POM.

[*Hint:* the ensemble of states is unchanged by the transformation $|0\rangle \rightarrow |0\rangle, |1\rangle \rightarrow -|1\rangle$. Try probability operators with the same symmetry:

$$\hat{\pi}_1 = \frac{1}{2}(a|0\rangle + |1\rangle)(a\langle 0| + \langle 1|),$$
$$\hat{\pi}_2 = \frac{1}{2}(a|0\rangle - |1\rangle)(a\langle 0| - \langle 1|),$$
$$\hat{\pi}_3 = (1 - a^2)|0\rangle\langle 0|.]$$

4.24) The Bayes cost $C_{ij} \geq 0$ is the penalty we pay if we identify the state $\hat{\rho}_j$ as $\hat{\rho}_i$. If there are M states with density operators $\hat{\rho}_j$ with prior probabilities p_j then the average Bayes cost is

$$\bar{C} = \text{Tr}\sum_{i=1}^{M} \hat{w}_i\hat{\pi}_i,$$

where

$$\hat{w}_i = \sum_{j=1}^{M} p_j C_{ij}\hat{\rho}_j.$$

Show that the probability operators which minimize the Bayes cost satisfy the conditions

$$\hat{\pi}_j(\hat{w}_j - \hat{w}_i)\hat{\pi}_i = 0, \qquad \forall i, j,$$
$$\hat{w}_i - \hat{\Upsilon} \geq 0, \qquad \forall i,$$

where $\hat{\Upsilon}$ is the Hermitian operator $\sum_{j=1}^{M} \hat{\pi}_j\hat{\rho}_j$.

4.25) Confirm that in the BB84 protocol Eve, an eavesdropper, has her best chance of identifying the transmitted bit value ('0' or '1') by measuring in the Breidbart basis. This means associating these bit values with projectors onto the states $\cos(\pi/8)|0\rangle + \sin(\pi/8)|1\rangle$ and $\cos(\pi/8)|1\rangle - \sin(\pi/8)|0\rangle$, respectively.

4.26) Confirm that choosing $A = (1 + |\langle\psi_1|\psi_2\rangle|)^{-1}$ minimizes the probability for an inconclusive measurement outcome in unambiguous discrimination between the equiprobable states in eqn 4.56.

4.27) By replacing the POM elements in eqn 4.64 by

$$\hat{\pi}_1 = A|\psi_2^\perp\rangle\langle\psi_2^\perp|,$$
$$\hat{\pi}_2 = B|\psi_1^\perp\rangle\langle\psi_1^\perp|,$$
$$\hat{\pi}_? = \hat{I} - \hat{\pi}_1 - \hat{\pi}_2,$$

find the minimum value of $P_?$ for general prior probabilities p_1 and p_2.

(4.28) How would the B92 protocol change if Eve and/or Bob used unambiguous state discrimination?

(4.29) Write down a set of probability operators for unambiguous state elimination for the mirror-symmetric qubit states given in problem Exercise (4.23). Under what conditions is this possible without including a probability operator corresponding to an inconclusive outcome?

(4.30) Confirm that if the operators $\hat{A}_i$ and $\hat{B}_j$ are effects then the operators $\hat{\pi}_{ij}$ in eqn 4.83 form a POM.

(4.31) Three measurements are performed in succession on a system prepared in a state with density operator $\hat{\rho}$. The results for the first, second, and third measurements are labelled i, j, and k, respectively, and the associated effect operators are $\hat{A}_i$, $\hat{B}_j$, and $\hat{C}_k$.

   (a) Write down expressions for the following probabilities:
      (i) $P(i, j, k)$;
      (ii) $P(j)$;
      (iii) $P(k|i, j)$;
      (iv) $P(j|k)$.

   (b) Write down the form of $\hat{\rho}$ following the third measurement given that the following results are known:
      (i) i, j, and k;
      (ii) none of the results;
      (iii) i and j;
      (iv) i and k.

(4.32) A qubit has one stable state $|0\rangle$ and one unstable state $|1\rangle$. Its density operator matrix elements satisfy the equations

$$\frac{d}{dt}\langle 1|\hat{\rho}(t)|1\rangle = -2\Gamma\langle 1|\hat{\rho}(t)|1\rangle = -\frac{d}{dt}\langle 0|\hat{\rho}(t)|0\rangle,$$

$$\frac{d}{dt}\langle 1|\hat{\rho}(t)|0\rangle = -\gamma\langle 1|\hat{\rho}(t)|0\rangle,$$

$$\frac{d}{dt}\langle 0|\hat{\rho}(t)|1\rangle = -\gamma\langle 0|\hat{\rho}(t)|1\rangle.$$

Show that the positivity of the evolved density operator requires that $\gamma \geq \Gamma$.

(4.33) Construct the analogue of eqn 4.91 if the spin flips occur independently; that is, flips about none, one, two, or three of the axes can occur in sequence. Let the probabilities that a spin flip occurs about the x, y, and z axis be q_x, q_y, and q_z respectively.

(4.34) Show that the transformation in eqn 4.91 either reduces $\text{Tr}(\hat{\rho}^2)$ or leaves it unchanged. Further, show that the transformed density operator is $\hat{I}/2$ for all $\hat{\rho}$ if and only if $p_x = p_y = p_z = 1/4$.

(4.35) Show that it is always possible to have an operation that *increases* the overlap between a pair of quantum states, that is, a process that changes $|\psi_{1,2}\rangle$ into $|\psi'_{1,2}\rangle$, where

$$|\langle\psi'_1|\psi'_2\rangle| > |\langle\psi_1|\psi_2\rangle|.$$

(4.36) Show that for optimal state separation of a pair of quantum states $|\psi_1\rangle$ and $|\psi_2\rangle$, *failure* to separate the states necessarily causes the states to be transformed into a single common state.

(4.37) Use the bound on the probability for successful state separation in eqn 4.96 to limit the probability for making N copies of a system known to have been prepared in either the state $|\psi_1\rangle$ or the state $|\psi_2\rangle$. How might you interpret the $N \to \infty$ limit of this probability?

(4.38) Consider the single-qubit transformation

$$\rho \to \rho' = \frac{1}{2}\hat{I} + \kappa\begin{pmatrix} 0 & \rho_{01} \\ \rho_{10} & 0 \end{pmatrix},$$

where κ is a real constant.

   (a) Show that the resulting density operator is positive if $|\kappa| \leq 1$.
   (b) Find the values of κ for which the transformation is completely positive and therefore physically implementable.
   [*Hint:* you might proceed by calculating the matrix form of the associated transformation, as in Appendix J.]

(4.39) Find, for the transformation in the previous question, a suitable set of effect operators $\{\hat{A}_i\}$. Is this set unique?

Entanglement

We have seen, in Section 2.5, how the superposition principle leads to the existence of entangled states of two or more quantum systems. Such states are characterized by the existence of correlations between the systems, the form of which cannot be satisfactorily accounted for by any classical theory. These have played a central role in the development of quantum theory since early in its development, starting with the famous paradox or dilemma of Einstein, Podolsky, and Rosen (EPR). No less disturbing than the EPR dilemma is the problem of Schrödinger's cat, an example of the apparent absurdity of following entanglement into the macroscopic world. It was Schrödinger who gave us the name *entanglement*; he emphasized its fundamental significance when he wrote, 'I would call this not *one* but *the* characteristic trait of quantum mechanics, the one that enforces the entire departure from classical thought'.

The EPR dilemma represents a profound challenge to classical reasoning in that it seems to present a conflict between the ideas of the reality of physical properties and the locality imposed by the finite velocity of light. This challenge and the developments that followed have served to refine the concept of entanglement and will be described in the first section of this chapter.

In the discipline of quantum information, entanglement is viewed as a resource to be exploited. We shall find, both here and in the subsequent chapters, that our subject owes much of its distinctive flavour to the utilization of entanglement.

5.1 Non-locality

We start by recalling that a state of two quantum systems is entangled if its density operator cannot be written as a product of density operators for the two systems, or as a probability-weighted sum of such products. For pure states, the condition for entanglement can be stated more simply: a pure state of two quantum systems is not entangled only if the state vector can be written as a product of state vectors for the two systems. Consider the two-qubit state

$$|\psi\rangle = \cos\theta|0\rangle \otimes |0\rangle + \sin\theta|1\rangle \otimes |1\rangle. \tag{5.1}$$

This will be the unentangled or product state $|0\rangle \otimes |0\rangle$ if $\theta = 0, \pi$ (or $|1\rangle \otimes |1\rangle$ if $\theta = \pi/2, 3\pi/2$), but for other values of θ the state is entangled. The state is most strongly entangled, that is, furthest from a product state, if $\cos\theta$ and $\sin\theta$ are equal in magnitude ($\theta = \pi/4, 3\pi/4$). We shall return to the question of quantifying entanglement in Section 8.4, but we note here that a maximally entangled state of two quantum systems is a pure state for which the reduced density operator of one of the two systems is proportional to the identity operator. For the pure state in eqn 5.1, the reduced density operator for the first qubit is

$$\hat{\rho} = \begin{pmatrix} \cos^2\theta & \\ 0 & \sin^2\theta \end{pmatrix}, \tag{5.2}$$

so the two-qubit state is maximally entangled for $\cos^2\theta = \frac{1}{2}$.

If two distant parties, whom we shall call Alice and Bob, each have one of a pair of entangled quantum systems, then the actions of Alice on her system can have a remarkable effect on the state of Bob's system. Any measurement that Alice might perform will reveal information about the state of her qubit, and in doing so will change the state of Bob's. This is true irrespective of the distance between Alice and Bob. If, for example, Alice and Bob's qubits were prepared in the state of eqn 5.1 and Alice measures the observable corresponding to the operator $\hat{\sigma}_z$, then on her finding the result $+1$ (or -1), corresponding to the state $|0\rangle$ (or $|1\rangle$), the state of the *two* qubits is instantaneously changed into $|0\rangle \otimes |0\rangle$ (or $|1\rangle \otimes |1\rangle$). The surprising feature of this, embodied in the EPR dilemma, is that the state of Bob's particle changes immediately after Alice's measurement and that this change occurs instantaneously, irrespective of the distance between Alice and Bob.

There is no dilemma, of course, in the existence of pre-established values for the measured observables, as would be the case if the state in eqn 5.1 represented a pair of qubits which were prepared in the state $|0\rangle \otimes |0\rangle$ with probability $\cos^2\theta$ or the state $|1\rangle \otimes |1\rangle$ with probability $\sin^2\theta$. This does not describe fully the correlations associated with the entangled state, however, as Alice and Bob can measure observables other than $\hat{\sigma}_z$. It is in the possibility of measuring incompatible observables on each of the entangled systems that the EPR paradox arises. This paradox has played an important role in the development of quantum theory, and it is interesting and instructive to follow the EPR argument in detail and, in particular, its challenge to the completeness of quantum theory. Such an analysis lies beyond the scope of this book, but may be found in some of the titles suggested for further reading.

Bohm presented the EPR dilemma in terms of the spins of a pair of spin-1/2 particles prepared in a state of zero total angular momentum. When expressed in the language of quantum information, this corresponds to a pair of qubits prepared in the Bell state

$$|\Psi^-\rangle_{AB} = \frac{1}{\sqrt{2}} \left(|0\rangle_A |1\rangle_B - |1\rangle_A |0\rangle_B \right). \tag{5.3}$$

If Alice measures $\hat{\sigma}_z$ on her qubit then she immediately establishes that the state of Bob's qubit is $|0\rangle_B$ or $|1\rangle_B$, corresponding, respectively, to her measurement results -1 and $+1$. This could be demonstrated were Bob to perform a measurement of $\hat{\sigma}_z$ on his qubit. If Alice chooses to measure $\hat{\sigma}_x$, however, then she immediately establishes that the state of

Bob's qubit is $|0'\rangle_B = 2^{-1/2}(|0\rangle_B + |1\rangle_B)$ or $|1'\rangle_B = 2^{-1/2}(|0\rangle_B - |1\rangle_B)$ corresponding, respectively, to her measurement results -1 and $+1$. Again, this could be demonstrated were Bob to measure $\hat{\sigma}_x$. By her choice of observable, $\hat{\sigma}_z$ or $\hat{\sigma}_x$, Alice can establish either of two *incompatible* properties of Bob's qubit. It is clear that there is no quantum state having well-defined values for both $\hat{\sigma}_z$ and $\hat{\sigma}_x$, as these operators have no common eigenstates. This means that the values of $\hat{\sigma}_z$ and $\hat{\sigma}_x$ for Bob's qubit could not *both* have been established at the source of the entangled qubits. It seems, therefore, that the effect of Alice's measurement must have changed, instantaneously, Bob's qubit. This conflicts with ideas from special relativity, however, which require that no signal or other physical influence can propagate faster than the speed of light. This combination of realism (that the properties of Bob's qubit exist whether or not they are measured) and of locality (that physical influences cannot propagate from Alice to Bob at a speed greater than that of light) is called local realism. The remarkable conflict between local realism and the properties of entangled quantum states has been given the name non-locality.

The EPR paradox is resolved, at least in part, by the no-signalling theorem of Ghirardi, Rimini, and Weber. This proves that Alice's choice of measurement has no observable consequences for Bob. The probabilities for the possible outcomes of Bob's measurements are not affected by Alice's choice of measurement or, indeed, whether or not Alice makes a measurement. Consider, for example, the situation that arises in Bohm's version of the EPR paradox. If Alice and Bob both choose to measure $\hat{\sigma}_z$ on their qubits then the probabilities that their respective measurements give the results $+1$ and -1 are

$$P_{AB}^{zz}(+1,+1) = 0 = P_{AB}^{zz}(-1,-1),$$
$$P_{AB}^{zz}(+1,-1) = \frac{1}{2} = P_{AB}^{zz}(-1,+1). \tag{5.4}$$

The result of Alice's measurement is not known to Bob, and the probabilities for Bob's two measurement results are

$$P_B^z(+1) = P_{AB}^{zz}(+1,+1) + P_{AB}^{zz}(-1,+1) = \frac{1}{2},$$
$$P_B^z(-1) = P_{AB}^{zz}(+1,-1) + P_{AB}^{zz}(-1,-1) = \frac{1}{2}. \tag{5.5}$$

If Alice measures $\hat{\sigma}_x$ and Bob measures $\hat{\sigma}_z$, then the joint-measurement probabilities are

$$P_{AB}^{xz}(+1,+1) = \frac{1}{4} = P_{AB}^{xz}(+1,-1),$$
$$P_{AB}^{xz}(-1,+1) = \frac{1}{4} = P_{AB}^{xz}(-1,-1). \tag{5.6}$$

Bob's measurement probabilities in this situation are precisely the same as in the case in which Alice measured $\hat{\sigma}_z$:

$$P_B^z(+1) = P_{AB}^{xz}(+1, +1) + P_{AB}^{xz}(-1, +1) = \frac{1}{2},$$

$$P_B^z(-1) = P_{AB}^{xz}(+1, -1) + P_{AB}^{xz}(-1, -1) = \frac{1}{2}. \qquad (5.7)$$

Alice's choice of observation has no effect on the outcome of Bob's measurement, and so Alice cannot send a signal to Bob by simply choosing to measure $\hat{\sigma}_z$ or $\hat{\sigma}_x$.

The no-signalling theorem is more general than this and can be proven for any measurements carried on any state of two systems shared by Alice and Bob. Consider such a state represented by the density operator $\hat{\rho}_{AB}$. The most general measurement Bob can perform is described by a POM, and we consider such a measurement with the associated probability operators $\{\hat{\pi}_i^B\}$. Alice chooses between two possible measurements with the probability operators $\{\hat{\pi}_j^{A1}\}$ and $\{\hat{\pi}_k^{A2}\}$, respectively. The joint probabilities for the outcomes of Alice's and Bob's measurements are

$$P_{AB}^{A1}(j, i) = \text{Tr}\left(\hat{\rho}_{AB}\hat{\pi}_j^{A1} \otimes \hat{\pi}_i^B\right) \qquad (5.8)$$

for Alice's $A1$ measurement and

$$P_{AB}^{A2}(k, i) = \text{Tr}\left(\hat{\rho}_{AB}\hat{\pi}_k^{A2} \otimes \hat{\pi}_i^B\right) \qquad (5.9)$$

for Alice's $A2$ measurement. The probabilities for Bob's measurement outcomes are independent of Alice's choice of measurement because the sum of the elements of a POM is the identity operator:

$$\sum_j P_{AB}^{A1}(j, i) = \sum_j \text{Tr}\left(\hat{\rho}_{AB}\hat{\pi}_j^{A1} \otimes \hat{\pi}_i^B\right) = \text{Tr}\left(\hat{\rho}_{AB}\hat{\pi}_i^B\right),$$

$$\sum_k P_{AB}^{A2}(k, i) = \sum_k \text{Tr}\left(\hat{\rho}_{AB}\hat{\pi}_k^{A2} \otimes \hat{\pi}_i^B\right) = \text{Tr}\left(\hat{\rho}_{AB}\hat{\pi}_i^B\right). \qquad (5.10)$$

The properties of Alice's system alone can be described purely in terms of the reduced density operator $\hat{\rho}_A = \text{Tr}_B\left(\hat{\rho}_{AB}\right)$, and those of Bob's system by the reduced density operator $\hat{\rho}_B = \text{Tr}_A\left(\hat{\rho}_{AB}\right)$. Another way to understand the no-signalling theorem is to realize that no action carried out by Alice on her system can possibly change the reduced density matrix for Bob's qubit. Alice's actions cannot have any observable effect on Bob's measurement results.

It is certainly possible to view the no-signalling theorem as a resolution of the EPR paradox. To do so, however, would be to overlook a more subtle problem revealed by Bell and expressed in terms of his famous inequality. Bell's inequality is an experimentally testable consequence of the combination of locality and realism and it follows that violation of this inequality constitutes an explicit demonstration of non-locality.

We derive Bell's inequality here in the form in which it is usually expressed, which was first given by Clauser, Horne, Shimony, and Holt. In the experimental situation, there is a source of pairs of spin-half particles, one particle of which is sent to Alice and one to Bob. They each make a measurement of a component of spin, and we let the unit

No signalling at any speed The EPR paradox is at its most worrying when Alice and Bob are space-like separated so that, according to the requirements of special relativity, no signal can possibly pass between them. The speed of light does not enter into the no-signalling theorem nor, indeed, does it appear in non-relativistic quantum theory. It follows that the no-signalling theorem prohibits signalling at any speed unless, of course, it is accompanied by further communications between Alice and Bob.

vectors $\vec{a}$ and $\vec{b}$, respectively, denote the directions of Alice's and Bob's components. In quantum theory, the measurements are described by the operators $\vec{a} \cdot \vec{\sigma}$ and $\vec{b} \cdot \vec{\sigma}$ and the result of each of the measurements is $+1$ or -1. The question is whether or not these values *exist*, even if they are not measured. If these values do exist then we also need to ask whether they were determined by the source, or perhaps depend on the choice of measurement made by the distant observer. The idea that properties exist even if they are not measured embodies realism, and the idea that they are independent of the measurement choices of the distant observer is a consequence of locality. We can build a local realistic theory of this arrangement by introducing values for each of the possible observables that Alice and Bob might measure. To this end, we denote by A the result of a measurement Alice might perform; this will depend on her choice of measurement direction, $\vec{a}$, and also on the statistical and unknown properties of the source. We describe these properties of the source by a set of hidden variables λ and an associated probability density $\rho(\lambda)$. Locality is imposed by not allowing A to depend on Bob's choice of measurement direction, $\vec{b}$. Similarly, we denote the results of Bob's possible measurements by B and allow these to depend on $\vec{b}$ and λ but not on $\vec{a}$. Naturally, A and B can take the values $+1$ or -1, but without performing the measurement or accessing the values λ we cannot determine which.

Any correlations between the measurements carried out by Alice and Bob will be revealed in the joint probabilities for their measurement outcomes. It suffices, for our purposes, to consider the probability that their measurement results are the same (both $+1$ or both -1) minus the probability that they are different. If Alice and Bob measure their spins along the $\vec{a}$ and $\vec{b}$ directions, respectively then, in a local realistic theory, this quantity can be written in the form

$$E(\vec{a}, \vec{b}) = \int d\lambda \, \rho(\lambda) A(\vec{a}, \lambda) B(\vec{b}, \lambda). \qquad (5.11)$$

We let $\vec{a}'$ and $\vec{b}'$ be two other directions along which Alice and Bob can make measurements. It then follows from eqn 5.11 that

$$E(\vec{a}, \vec{b}) - E(\vec{a}, \vec{b}') = \int d\lambda \, \rho(\lambda) \left[A(\vec{a}, \lambda) B(\vec{b}, \lambda) - A(\vec{a}, \lambda) B(\vec{b}', \lambda) \right]$$
$$= \int d\lambda \, \rho(\lambda) A(\vec{a}, \lambda) B(\vec{b}, \lambda) \left[1 \pm A(\vec{a}', \lambda) B(\vec{b}', \lambda) \right]$$
$$- \int d\lambda \, \rho(\lambda) A(\vec{a}, \lambda) B(\vec{b}', \lambda) \left[1 \pm A(\vec{a}', \lambda) B(\vec{b}, \lambda) \right]. \qquad (5.12)$$

Here we have made explicit use of the idea of realism by including in the same product the values $A(\vec{a}, \lambda)$ and $A(\vec{a}', \lambda)$ (and indeed $B(\vec{b}, \lambda)$ and $B(\vec{b}'\lambda)$), even though the corresponding observables in quantum theory are incompatible. The fact that $|A| = 1$ and $|B| = 1$ for both pairs of

Hidden variables   The supposed statistical and unknown properties of the source are referred to as hidden variables. These were introduced in the hope that the probabilities and indeterminacy of quantum theory might hide a more fundamental theory, in much the same way that statistical mechanics is less fundamental than the underlying mechanics. Testing theories based on hidden variables against quantum theory tells us much about the subtlety of the quantum world.

measurement directions and for all values of λ leads to the inequality

$$\left| E(\vec{a}, \vec{b}) - E(\vec{a}, \vec{b}') \right| \leq \int d\lambda\, \rho(\lambda) \left[1 \pm A(\vec{a}', \lambda)B(\vec{b}', \lambda) \right]$$
$$+ \int d\lambda\, \rho(\lambda) \left[1 \pm A(\vec{a}', \lambda)B(\vec{b}, \lambda) \right]$$
$$= 2 \pm \left[E(\vec{a}', \vec{b}') + E(\vec{a}', \vec{b}) \right]. \tag{5.13}$$

This is Bell's inequality, which is usually written in the more symmetrical form

$$S = \left| E(\vec{a}, \vec{b}) - E(\vec{a}, \vec{b}') \right| + \left| E(\vec{a}', \vec{b}') + E(\vec{a}', \vec{b}) \right| \leq 2. \tag{5.14}$$

It is remarkable that the correlations associated with a pair of spins (or qubits) prepared in the Bell state $|\Psi^-\rangle_{AB}$ can violate Bell's inequality. To see this we note that, for this state, quantum mechanics predicts that

$$E(\vec{a}, \vec{b}) = {}_{AB}\langle \Psi^- | \vec{a} \cdot \hat{\vec{\sigma}} \otimes \vec{b} \cdot \hat{\vec{\sigma}} | \Psi^- \rangle_{AB} = -\vec{a} \cdot \vec{b}. \tag{5.15}$$

If we put these into Bell's inequality then we find a maximum value for the left-hand side when the vectors $\vec{b}$ and $\vec{b}'$ are mutually perpendicular and $\vec{a}$ and $\vec{a}'$ are parallel to $\vec{b} - \vec{b}'$ and $\vec{b} + \vec{b}'$, respectively. This arrangement is depicted in Fig. 5.1. For these observables we find $S = 2\sqrt{2}$, in clear violation of Bell's inequality. Experiments clearly suggest that Bell's inequality is indeed violated for entangled states and hence that non-local phenomena are part of the physical world. In quantum information, we view entanglement as a resource and we can use the violation of Bell's inequality as evidence for the existence of entanglement for a given physical system.

It is possible to demonstrate a contradiction between the predictions of local realism and of quantum theory without resorting to an inequality. There exist a number of such demonstrations, but the simplest is one for three qubits prepared in the Greenberger–Horne–Zeilinger state,

$$|\text{GHZ}\rangle = \frac{1}{\sqrt{2}} \left(|000\rangle + |111\rangle \right). \tag{5.16}$$

We suppose that three parties, Alice, Bob, and Claire, each have one of the qubits and that they each measure one of the pair of observables corresponding to the operators $\hat{\sigma}_x$ and $\hat{\sigma}_y$. Naturally, the results of their measurements, which we denote m_x or m_y, will be $+1$ or -1. The state in eqn 5.16 is an eigenstate of the three operators $\hat{\sigma}_x \otimes \hat{\sigma}_y \otimes \hat{\sigma}_y$, $\hat{\sigma}_y \otimes \hat{\sigma}_x \otimes \hat{\sigma}_y$, and $\hat{\sigma}_y \otimes \hat{\sigma}_y \otimes \hat{\sigma}_x$, with the eigenvalue in each case being -1. This means that if two of Alice, Bob, and Claire measure $\hat{\sigma}_y$ and the other measures $\hat{\sigma}_x$ then the product of their measurement results will certainly be -1. From the local realistic viewpoint, the values of m_x and m_y for each system exist whether or not they are measured, and each of these is independent of the observations carried out on the other systems. The eigenvalue property described above then requires

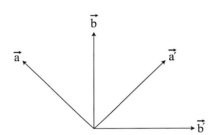

Fig. 5.1 Relative orientations of the spin-measurement directions for maximum violation of Bell's inequality.

Loopholes   It is still just about possible, but for most physicists unreasonable, to hold to a local realistic view in the face of existing experimental evidence. The combination of high-efficiency detection, space-like separation of the observers' qubits, and random and independent choices of observables, required to close all possible loopholes, has yet to be realized in a *single* experiment.

that these values satisfy the equations

$$m_x^A m_y^B m_y^C = -1,$$
$$m_y^A m_x^B m_y^C = -1,$$
$$m_y^A m_y^B m_x^C = -1. \qquad (5.17)$$

Each of the quantities m_x and m_y has the value $+1$ or -1, and if we multiply together these three products then we find

$$m_x^A m_y^B m_y^C \times m_y^A m_x^B m_y^C \times m_y^A m_y^B m_x^C = m_x^A m_x^B m_x^C = -1. \qquad (5.18)$$

The local realistic description leads us to conclude that if Alice, Bob, and Claire all measure $\hat{\sigma}_x$ then the product of their results will be -1. The state $|\text{GHZ}\rangle$ is indeed an eigenstate of $\hat{\sigma}_x \otimes \hat{\sigma}_x \otimes \hat{\sigma}_x$, but with the eigenvalue $+1$, a direct *contradiction* of the value required by a local realistic theory. The origin of this contradiction is in the anticommutation property of the Pauli operators in eqn 2.88:

$$(\hat{\sigma}_x \otimes \hat{\sigma}_y \otimes \hat{\sigma}_y)(\hat{\sigma}_y \otimes \hat{\sigma}_x \otimes \hat{\sigma}_y)(\hat{\sigma}_y \otimes \hat{\sigma}_y \otimes \hat{\sigma}_x) = \hat{\sigma}_x \otimes (\hat{\sigma}_y \hat{\sigma}_x \hat{\sigma}_y) \otimes \hat{\sigma}_x$$
$$= -\hat{\sigma}_x \otimes \hat{\sigma}_x \otimes \hat{\sigma}_x. \qquad (5.19)$$

It follows that $|\text{GHZ}\rangle$ is an eigenstate of $-\hat{\sigma}_x \otimes \hat{\sigma}_x \otimes \hat{\sigma}_x$ with eigenvalue -1. A further example of a violation of local realism without an inequality is Hardy's theorem, which is described in Appendix K.

5.2   Indirect measurements

The EPR paradox encapsulates the idea that a measurement performed on one of a pair of entangled systems provides information about its partner. As such, it can be thought of as an indirect measurement of the otherwise unobserved system. For the entangled state $|\Psi^-\rangle_{AB}$, for example, Alice can measure any spin component and thereby simultaneously determine the state of Bob's qubit; if she measures $\hat{\sigma}_x$ and finds the result 0 then she also determines that Bob's qubit is in the state $|1'\rangle$.

If we can tailor an interaction between two quantum systems then we can exploit this idea to perform indirect measurements. Suppose, for example, that we wish to perform an indirect measurement of $\hat{\sigma}_z$ by observing an ancillary qubit. One way to achieve this is to prepare the ancilla in the state $2^{-1/2}(|0\rangle + i|1\rangle)$ and to change this state in a way that depends on the state of the qubit to be observed indirectly. A suitable Hamiltonian for this purpose is

$$\hat{H} = g\hat{\sigma}_z \otimes \hat{\sigma}_x, \qquad (5.20)$$

where the first operator acts on the state of the qubit of interest and the second acts on the ancilla. The Bloch vector for our ancilla will rotate under the action of this Hamiltonian from the along the y-axis towards the z-axis, with the choice of poles depending on whether our initial qubit was prepared in the state $|0\rangle$ or $|1\rangle$. If we choose the interaction time T such that $gT/\hbar = \pi/4$ then the unitary evolution produces the states

$$\exp\left(-i\frac{\hat{H}T}{\hbar}\right)|0\rangle \otimes \frac{1}{\sqrt{2}}\left(|0\rangle + i|1\rangle\right) = |0\rangle \otimes |0\rangle,$$

$$\exp\left(-i\frac{\hat{H}T}{\hbar}\right)|1\rangle \otimes \frac{1}{\sqrt{2}}\left(|0\rangle + i|1\rangle\right) = i|1\rangle \otimes |1\rangle. \tag{5.21}$$

Clearly, the eigenvalue of $\hat{\sigma}_z$ for our qubit has been copied onto the state of the ancilla, and a measurement of $\hat{\sigma}_z$ carried out on the ancilla will reveal the same result as would have been found by means of a direct measurement. For a more general initial state, the interaction will produce an entangled state of the two qubits:

$$\exp\left(-i\frac{\hat{H}T}{\hbar}\right)(c_0|0\rangle + c_1|1\rangle) \otimes \frac{1}{\sqrt{2}}\left(|0\rangle + i|1\rangle\right) = c_0|0\rangle \otimes |0\rangle + ic_1|1\rangle \otimes |1\rangle. \tag{5.22}$$

The interaction clearly modifies a general state of the qubit but, by design, leaves unchanged the probabilities for the observable of interest ($|c_0|^2$ and $|c_1|^2$).

If the Hamiltonian in eqn 5.20 is replaced by

$$\hat{H} = -g\hat{\sigma}_x \otimes \hat{\sigma}_z, \tag{5.23}$$

with the second qubit again prepared in the state $2^{-1/2}\left(|0\rangle + i|1\rangle\right)$, then choosing the interaction time such that $gT/\hbar = \pi/4$ leads to an entangled state in which both qubits are in the same eigenstate of $\hat{\sigma}_x$. A measurement of $\hat{\sigma}_x$ for the second (ancillary) qubit would then constitute an indirect measurement of the first qubit. We can combine the two interactions of eqns 5.20 and 5.23 into a single Hamiltonian of the form

$$\hat{H} = g\left(\hat{\sigma}_z \otimes \hat{\sigma}_x - \hat{\sigma}_x \otimes \hat{\sigma}_z\right). \tag{5.24}$$

The interaction between the two qubits then imprints information about both $\hat{\sigma}_z$ and $\hat{\sigma}_x$ for the first particle onto the state of the ancilla. Naturally, the incompatibility of these observables ensures that the probabilities for the eigenvalues of both $\hat{\sigma}_z$ and $\hat{\sigma}_x$ will, in general, be changed by such an interaction.

It is by no means necessary to employ a qubit as our ancilla. We might, for example, couple our qubit to the position of a particle through the Hamiltonian

$$\hat{H} = g\hat{\sigma}_z \otimes \hat{x}. \tag{5.25}$$

The resulting dynamics will result in a shift of the momentum of our second particle, with the sign of the shift determined by whether our qubit was prepared in the state $|0\rangle$ or $|1\rangle$. If our second particle is prepared in the motional state $|\phi\rangle$ and if the interaction time T is sufficiently short

Quantum non-demolition measurements   It is often useful to design measurements which leave a desired property unchanged, as the observable corresponding to $\hat{\sigma}_z$ is here. In this way, for example, a quantum property can be monitored repeatedly in order to detect a change due to some external influence. Measurements of this kind are commonly referred to as quantum non-demolition, or QND, measurements.

for us to ignore its free motion, then the unitary evolution produces the
state

$$\exp\left(-i\frac{\hat{H}T}{\hbar}\right)(c_0|0\rangle + c_1|1\rangle) \otimes |\phi\rangle = c_0|0\rangle \otimes \exp\left(-i\frac{gT\hat{x}}{\hbar}\right)|\phi\rangle$$
$$+c_1|1\rangle \otimes \exp\left(+i\frac{gT\hat{x}}{\hbar}\right)|\phi\rangle. \tag{5.26}$$

If the momentum wavefunction for the motional state is $\phi(p) = \langle p|\phi\rangle$
then the interaction induces a shift in the momentum of the form

$$\langle p|\exp\left(\mp i\frac{gT\hat{x}}{\hbar}\right)|\phi\rangle = \exp\left(\pm gT\frac{\partial}{\partial p}\right)\phi(p)$$
$$= \phi(p \pm gT). \tag{5.27}$$

If the interaction is sufficiently strong for gT to greatly exceed the mo-
mentum uncertainty for the state $|\phi\rangle$ then a measurement of the mo-
mentum of the ancilla particle after the interaction will constitute an
indirect measurement of $\hat{\sigma}_z$ for our qubit. It is entirely possible to realize a measurement of this type using an atomic nuclear spin as our
qubit and coupling this to the atomic position. This is, of course, how a
Stern–Gerlach measurement of spin is performed. A closely related example from optics is the polarizing beam splitter, which acts to transmit
horizontally polarized light and reflect vertically polarized light. For a
single photon in a general polarization state, this results in an entanglement between polarization and direction of propagation. Detecting
the photon in the transmitted or reflected mode constitutes an indirect
determination of the linear polarization as horizontal or vertical.

It should be emphasized that the process of entangling our qubit with
the ancilla does not in itself constitute an indirect measurement. The
transformation creating the entangled state is unitary and can be reversed, at least in principle; the indirect measurement occurs only on
observing the ancilla. In indirect measurements, therefore, it is possible
to separate in time the measurement from the interaction with the ancilla. We can induce an interaction between our quantum system and an
ancilla and then, at a later time, decide which measurement to perform.
As a simple example, consider the optical interferometer depicted in Fig.
5.2. At the input is a polarizing beam splitter, and this means, for example, that an input photon with left circular polarization is transformed
into the entangled state

$$\frac{1}{\sqrt{2}}(|H\rangle + i|V\rangle) \rightarrow \frac{1}{\sqrt{2}}(|H\rangle \otimes |L\rangle + i|V\rangle \otimes |U\rangle), \tag{5.28}$$

where the states $|L\rangle$ and $|U\rangle$ correspond, respectively, to the upper and
lower paths through the interferometer. These paths play the role, in
this example, of the ancilla. Included within the interferometer is a
relative phase shift, the action of which transforms our entangled state

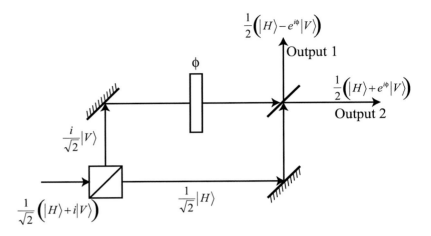

Fig. 5.2 In this interferometer, the input polarizing beam splitter entangles the polarization with the optical path. A measurement of the path at this point would be an indirect measurement of polarization. Here, however, we demonstrate the coherence of the superposition state by recombining the beams at the output beam splitter.

into

$$\frac{1}{\sqrt{2}} \left(|H\rangle \otimes |L\rangle + i|V\rangle \otimes |U\rangle \right) \rightarrow \frac{1}{\sqrt{2}} \left(|H\rangle \otimes |L\rangle + ie^{i\phi}|V\rangle \otimes |U\rangle \right).$$

(5.29)

Finally, recombining the fields at a symmetric output beam splitter (with $t = 1/\sqrt{2}$ and $r = i/\sqrt{2}$) produces the entangled state

$$|\psi\rangle = \frac{1}{2} \left[\left(|H\rangle - e^{i\phi}|V\rangle \right) \otimes |\text{Output1}\rangle + i \left(|H\rangle + e^{i\phi}|V\rangle \right) \otimes |\text{Output2}\rangle \right].$$

(5.30)

It is clear that our photon is equally likely to be found in either of the two output ports. This is true also if we perform a polarization-sensitive measurement, determining both the output port and whether the photon is horizontally or vertically polarized. The relative phase ϕ was imposed after the linear polarization was correlated with the optical path but it still appears in the state in eqn 5.30. It can be recovered only if we are prepared to give up the information about the linear polarization. One way to do this is to measure both the output port and the circular polarization. The probabilities for each of the four possible outcomes are

$$P\left(\text{Output1}, L\right) = \frac{1}{4}(1 - \sin\phi) = P\left(\text{Output2}, R\right)$$

$$P\left(\text{Output1}, R\right) = \frac{1}{4}(1 + \sin\phi) = P\left(\text{Output2}, L\right).$$

(5.31)

By varying the phase shift it is possible to correlate, perfectly, the output port with the degree of circular polarization. This is possible because it is the linear polarization that is correlated with the path through the interferometer; measuring the circular polarization destroys any 'which-way' information about the path the photon took through the interferometer. Had the interaction at the input beam splitter constituted a measurement of the horizontal or vertical polarization, then no such interference would be observed.

5.3   Ebits and shared entanglement

If two parties, Alice and Bob, have in their possession quantum systems in a common entangled state then we say that they share some entanglement. The amount of this shared entanglement will depend on the form of the entangled state but not on the nature of the component quantum systems. The entanglement allows Alice and Bob to perform a number of tasks which, without entanglement, would be either impossible or less efficient. For this reason it is useful to think of shared entanglement as a resource, both for quantum communication and for quantum information processing. We shall discuss how to quantify this resource in Chapter 8, but for the present it is sufficient to introduce the ebit as a unit of shared entanglement. Alice and Bob share one ebit if they each have a qubit and these two qubits have been prepared in a maximally entangled state. If Alice's and Bob's qubits have been prepared in the pure state $|\psi\rangle_{AB}$ and

$$\hat{\rho}_A = \text{Tr}_B \left(|\psi\rangle_{AB\,AB}\langle\psi| \right) = \frac{1}{2}\hat{I}_A$$
$$\hat{\rho}_B = \text{Tr}_A \left(|\psi\rangle_{AB\,AB}\langle\psi| \right) = \frac{1}{2}\hat{I}_B, \tag{5.32}$$

then the two qubits constitute one ebit. Simple examples of such maximally entangled states are the Bell states (eqns 2.108).

The discussion of non-locality in the first section of this chapter is useful in establishing some important properties of ebits. In particular, ebits can only be produced by use of a quantum communication channel. The most straightforward way to do this is for Alice to prepare two qubits in the maximally entangled state $|\psi\rangle_{AB}$ and then to send qubit B to Bob. Let us suppose, for the moment, that an ebit can be produced using classical communications. A classical channel is described entirely in terms of probabilities and, in particular, joint probabilities exist for every possible signal preparation event associated with Alice, and for all possible detection events observed by Bob. It necessarily follows that the correlations produced by such classical means must satisfy Bell's inequality. As ebits can produce an observable violation of this, they cannot be produced using only classical communications. This simple conclusion has profound consequences for our subject; if ebits cannot be generated using classical channels but this task can be performed using a quantum channel, then quantum channels must be fundamentally different from classical channels. It then follows that quantum communication is distinct from its classical counterpart and that quantum information is very different from classical information.

The requirement for quantum communications if one wishes to produce ebits is an example of an important principle in quantum information. If a desired task can be completed using only classical communications together with local transformations or other operations on local quantum systems, then this is likely to be far simpler and more reliable to perform than a task that requires a quantum channel. Tasks which can be performed without employing quantum communications are commonly referred to as LOCC tasks (local operations and classical communications). Clearly, generating ebits is not an LOCC task.

One of the remarkable properties of ebits is that the associated correlations cannot reliably be mimicked by any other quantum system. Let us suppose, by way of illustration, that Alice and Bob each have a qubit and that these have been prepared in the Bell state $|\Psi^-\rangle_{AB}$. Bob can subsequently identify himself to Alice by giving her his qubit so that she can perform a Bell measurement on the two qubits. If she finds a measurement result corresponding to the two-qubit state $|\Psi^-\rangle_{AB}$ then Bob may indeed be who he claims to be. A result corresponding to one of the other three Bell states will reveal an impostor. Suppose that Claire tries to impersonate Bob by giving Alice a qubit prepared in the state $|0\rangle_C$. The resulting density for the two qubits is then

$$\hat{\rho}_{AC} = \text{Tr}_B\left(|\Psi^-\rangle_{AB\,AB}\langle\Psi^-|\right) \otimes |0\rangle_{CC}\langle 0| = \frac{1}{2}\hat{I}_A \otimes |0\rangle_{CC}\langle 0|. \quad (5.33)$$

The probability that the qubits will pass Alice's test is

$$P(\text{pass}) = {}_{AC}\langle\Psi^-|\hat{\rho}_{AC}|\Psi^-\rangle_{AC} = \frac{1}{4}. \quad (5.34)$$

If Alice and Bob share n ebits, with each pair prepared in the state $|\Psi^-\rangle$, then the probability that Claire will be able to pass as Bob using the above strategy is 4^{-n}, which can be made arbitrarily small for sufficiently large n. We can think of this use of ebits as a lock with only one key; Alice's n qubits are the lock and Bob's the key. This idea is reminiscent, of course, of Wiesner's quantum money, as described in Section 3.4.

The ebit is a very useful idea in analysing the entanglement between two parties, but the quantitative description of more complicated states can be problematic. Consider, for example, the GHZ state shared between three parties, Alice, Bob, and Claire,

$$|\text{GHZ}\rangle = \frac{1}{\sqrt{2}}\left(|0\rangle_A|0\rangle_B|0\rangle_C + |1\rangle_A|1\rangle_B|1\rangle_C\right). \quad (5.35)$$

Do Alice and Bob share an ebit in this case? As it stands, the answer would appear to be no as, without the intervention of Claire, their best description of the state of their qubits is the unentangled mixed state

$$\hat{\rho}_{AB} = \frac{1}{2}\left(|0\rangle_{AA}\langle 0| \otimes |0\rangle_{BB}\langle 0| + |1\rangle_{AA}\langle 1| \otimes |1\rangle_{BB}\langle 1|\right). \quad (5.36)$$

The only satisfactory answer, however, is that it depends on what Claire does. If Claire makes a measurement on her qubit of the observable corresponding to $\hat{\sigma}_z$ then the resulting state cannot be entangled and so Alice and Bob will not share an ebit. This suggests that we should not associate an ebit with the state. If, however, Claire measures $\hat{\sigma}_x$ then the resulting state of Alice and Bob's qubit will be

$$|\psi\rangle_{AB} = \frac{1}{\sqrt{2}}\left(|0\rangle_A|0\rangle_B \pm |1\rangle_A|1\rangle_B\right), \quad (5.37)$$

with the + and − signs corresponding, respectively, to Claire's measurement results +1 and −1. If Claire sends, through a classical channel, her measurement result to Alice and Bob then they know that the state of their qubits is one of the Bell states ($|\Phi^+\rangle_{AB}$ or $|\Phi^-\rangle_{AB}$) and they then share an ebit. Claire can generate an ebit for Alice and Bob in this way by performing an LOCC task and this suggests that there is, in fact, an ebit for Alice and Bob embedded in the GHZ state. This situation is somewhat confusing, and uncovering a satisfactory method for quantifying and characterizing entanglement of multiple quantum systems remains an active area of research.

5.4   Quantum dense coding

One rather obvious and fundamental question which we have not yet addressed is the amount of information, or number of bits, that we can encode on a single qubit. The natural answer is one bit, and this can be achieved if Alice selects two orthogonal states, for example $|0\rangle$ and $|1\rangle$, and choosing either of these with equal probability. A measurement of $\hat{\sigma}_z$ will then reveal the bit value. In this scheme, the qubit is simply acting as a physical implementation of a classical bit. One bit does indeed turn out to be the maximum for a single qubit but to prove it is not straightforward, given the wide variety of possible states which can be prepared and measurements that can be performed. A full proof of this will have to wait until the treatment of quantum information theory in Chapter 8.

If we have at our disposal two qubits then we can encode a maximum of two bits of information on these. The natural way to do this is to use the four states $|0\rangle|0\rangle$, $|0\rangle|1\rangle$, $|1\rangle|0\rangle$, and $|1\rangle|1\rangle$ to represent the binary values 00, 01, 10, and 11. We do not have to use these states, however, and any four orthogonal states will be equally appropriate. In particular, we could employ the four Bell states

$$|\Psi^-\rangle = \frac{1}{\sqrt{2}}\left(|0\rangle|1\rangle - |1\rangle|0\rangle\right),$$

$$|\Psi^+\rangle = \frac{1}{\sqrt{2}}\left(|0\rangle|1\rangle + |1\rangle|0\rangle\right),$$

$$|\Phi^-\rangle = \frac{1}{\sqrt{2}}\left(|0\rangle|0\rangle - |1\rangle|1\rangle\right),$$

$$|\Phi^+\rangle = \frac{1}{\sqrt{2}}\left(|0\rangle|0\rangle + |1\rangle|1\rangle\right). \tag{5.38}$$

If Alice wishes to use these states to send two classical bits to Bob then she must first develop the means to prepare a pair of qubits in each of these possible states. One interesting way to do this is to prepare the qubits in the state $|\Psi^-\rangle$ and then to apply one of the four Pauli operators $\hat{I}$, $\hat{\sigma}_x$, $\hat{\sigma}_y$, and $\hat{\sigma}_z$ to the first qubit:

$$\hat{I} \otimes \hat{I} |\Psi^-\rangle = |\Psi^-\rangle,$$
$$\hat{\sigma}_x \otimes \hat{I} |\Psi^-\rangle = -|\Phi^-\rangle,$$
$$\hat{\sigma}_y \otimes \hat{I} |\Psi^-\rangle = i|\Phi^+\rangle,$$
$$\hat{\sigma}_z \otimes \hat{I} |\Psi^-\rangle = |\Psi^+\rangle. \tag{5.39}$$

The global phases of these states are, of course, unobservable so it is clear that the Bell state $|\Psi^-\rangle$ can be transformed into any of the other three by means of a unitary transformation acting on the first qubit alone.

The fact that any of the four orthogonal Bell states can be selected by action on just one of the qubits means that it is not necessary for Alice to prepare the two qubits for transmission to Bob if they already share an ebit. Suppose, for example, that Alice and Bob share an ebit in the form of a pair of qubits prepared in the Bell state $|\Psi^-\rangle_{AB}$. Alice can encode two bits of information onto this state by acting on her qubit with one of the four Pauli operators. If she then sends her single qubit to Bob then, by performing a measurement in the Bell-state basis on the two qubits, he will be able to recover the two bits of information. This phenomenon, whereby we encode four bits of information on a single qubit, was discovered by Bennett and Wiesner and is known as quantum dense coding (or sometimes superdense coding). We should emphasize that we really need two qubits for quantum dense coding, but that these comprise a pre-established ebit shared between Alice and Bob. The remarkable property is that after the message has been selected, Alice needs to send only one qubit to Bob.

It is interesting to ask where the two bits of information reside in quantum dense coding. With classical bits, of course, each one of the two bits of information has to reside on one of the two systems. Dense coding, exploiting pre-established correlations between such bits, would then not be possible as Alice cannot change Bob's bit value. We can use correlations between a pair of classical bits to send information, but then the maximum amount of information conveyed would be just one bit. To see this we note that the two possible (perfect) correlations between a pair of bits are that the bit values are the same (00 or 11) or that they are different (01 or 10). If Alice and Bob share a pair of bits correlated in this way then Alice can convey *just one bit* of information to Bob, by either flipping her bit value or leaving it unchanged and then transmitting it.

In quantum dense coding, a measurement by Bob on his single qubit will give a completely random result, entirely consistent with its reduced density operator

$$\hat{\rho}_B = \frac{1}{2}\hat{I}. \tag{5.40}$$

This density operator is unchanged by Alice's performance of a unitary transformation. Were this not the case, of course, then we would violate the no-signalling theorem proven in Section 5.1. Similarly, the bits cannot be recovered or checked by Alice, even though she chose the transformation. This is because the reduced density operator for her qubit is

$$\hat{\rho}_A = \frac{1}{2}\hat{I} \tag{5.41}$$

and this is unchanged by unitary transformations, such as those associated with the Pauli operators. Neither Alice nor Bob can recover the encoded bits by acting alone. It is only by combining the qubits, after Alice has transmitted her qubit, that Bob can recover the two bits of information. The four bits clearly reside in the entangled state of the two qubits but only non-locally; all of the information resides in the quantum correlations between the qubits.

It is reasonable to ask whether it might be possible to encode more than two bits of information on a single qubit by preparing it in a more complicated entangled state of more than two qubits. In order to answer this question, let us suppose that Alice has a single qubit while Bob has multiple qubits and that these have been prepared in the pure state

$$|\psi\rangle_{AB} = \frac{1}{\sqrt{2}}\left(|0\rangle_A|\phi_0\rangle_B - |1\rangle_A|\phi_1\rangle_B\right). \tag{5.42}$$

There will exist, of course, a Schmidt decomposition for this state between Alice's qubit and Bob's system and we shall assume that this state has been written in the Schmidt form so that $\langle\phi_0|\phi_1\rangle = 0$. If Alice applies one of the four Pauli operators to her qubit then the result is one of four orthogonal states. It is clear, moreover, that these four states exhaust the state space spanned by the four states $|0\rangle_A|\phi_0\rangle$, $|0\rangle|\phi_1\rangle$, $|1\rangle|\phi_0\rangle$, and $|1\rangle|\phi_1\rangle$ and so represent the maximum amount of information that Alice can send to Bob by transmitting her qubit. Bob's system lives in a state space which is larger than the two dimensions spanned by $|\phi_0\rangle$ and $|\phi_1\rangle$, but Alice is unable to utilize this as the no-signalling theorem requires that Bob's system remains a mixture of just the two states $|\phi_0\rangle$ and $|\phi_1\rangle$. The state in eqn 5.42, in fact, constitutes just one ebit as it is a maximally entangled state between Alice's qubit and an effective two-state system in Bob's domain spanned by the two orthogonal states $|\phi_0\rangle$ and $|\phi_1\rangle$. Sending two bits per qubit is the maximum that can be achieved, and this requires the qubit to be part of a pre-established ebit shared between Alice and Bob.

5.5   Teleportation

In quantum dense coding, we use two qubits to send two bits of classical information, but one of these qubits has been sent to Bob before Alice has selected the values of the two bits for transmission. It is almost as if one of the bits has been sent 'backwards in time' from Alice to Bob via the source of the entangled qubits. Is it also possible to use this strange method of communication as a quantum channel to send quantum information from Alice to Bob? Remarkably, the answer is yes; Alice can send a qubit to Bob using just one ebit and two bits of classical information. The method, discovered by Bennett, Brassard, Crépeau, Jozsa, Peres, and Wootters, is quantum teleportation. The term 'teleportation' conjures up images from science fiction of objects dematerializing at one location before reappearing at another. This idea is misleading, however, as the physical object does not move but rather it is the quantum information, encoded in its state vector, that is transferred from a local quantum system to a distant one. It might be better to think of a teleportation device as a fax machine for quantum information; a fax machine link reads the information from one piece of paper and prints it on another at a distant location.

Suppose that Alice needs to send to Bob a qubit prepared in the pure state

$$|\psi\rangle = \alpha|0\rangle + \beta|1\rangle, \qquad (5.43)$$

where α and β are a pair of complex amplitudes. If Alice knows the state then she could send to Bob, via a classical communication channel, instructions for creating a copy of the state. If the state is needed very precisely then this will require a large number of bits. A more serious problem is that Alice might not know the state of the qubit that she needs to send. One way to overcome both problems is to send the qubit to Bob using a quantum channel. For example, she could encode the qubit onto the polarization state of a single photon and transmit this through free space to Bob. If no reliable quantum channel exists, however, then Alice can still teleport the qubit to Bob if they share an ebit. To see how this works, consider the three-qubit state

$$|\psi\rangle_a \otimes |\Psi^-\rangle_{AB} = (\alpha|0\rangle_a + \beta|1\rangle_a)\frac{1}{\sqrt{2}}\left(|0\rangle_A|1\rangle_B - |1\rangle_A|0\rangle_B\right), \qquad (5.44)$$

in which Alice has the qubit a to be transmitted, while Alice and Bob have qubits A and B in the form of an ebit prepared in the state $|\Psi^-\rangle_{AB}$. We can rewrite this state in terms of the Bell-state basis for Alice's qubits in the form

$$\begin{aligned}
|\psi\rangle_a \otimes |\Psi^-\rangle_{AB} = \frac{1}{2}\big[&|\Psi^-\rangle_{aA}\left(-\alpha|0\rangle_B - \beta|1\rangle_B\right) \\
+&|\Psi^+\rangle_{aA}\left(-\alpha|0\rangle_B + \beta|1\rangle_B\right) \\
+&|\Phi^-\rangle_{aA}\left(\alpha|1\rangle_B + \beta|0\rangle_B\right) \\
+&|\Phi^+\rangle_{aA}\left(\alpha|1\rangle_B - \beta|0\rangle_B\right)\big].
\end{aligned} \qquad (5.45)$$

Written in this form, it appears as though Bob already has the qubit, in that the state of his qubit appears to depend on the parameters α and β. This is not correct, of course, as the reduced density operator for Bob's qubit is the mixed state with density operator $\frac{1}{2}\hat{I}$. Alice can prepare Bob's qubit in a state depending on α and β, however, by measuring her two qubits in the Bell-state basis. The four equally probable outcomes, associated with the states $|\Psi^-\rangle_{aA}$, $|\Psi^+\rangle_{aA}$, $|\Phi^-\rangle_{aA}$, and $|\Phi^+\rangle_{aA}$, leave Bob's qubit in the four corresponding states $-\alpha|0\rangle_B - \beta|1\rangle_B$, $-\alpha|0\rangle_B + \beta|1\rangle_B$, $\alpha|1\rangle_B + \beta|0\rangle_B$ and $\alpha|1\rangle_B - \beta|0\rangle_B$. These four states are related to the original qubit state in eqn 5.43 by the action of one of the four Pauli operators:

$$\hat{I}\left(-\alpha|0\rangle_B - \beta|1\rangle_B\right) = -\left(\alpha|0\rangle_B + \beta|1\rangle_B\right),$$
$$\hat{\sigma}_z\left(-\alpha|0\rangle_B + \beta|1\rangle_B\right) = -\left(\alpha|0\rangle_B + \beta|1\rangle_B\right),$$
$$\hat{\sigma}_x\left(\alpha|1\rangle_B + \beta|0\rangle_B\right) = \left(\alpha|0\rangle_B + \beta|1\rangle_B\right),$$
$$\hat{\sigma}_y\left(\alpha|1\rangle_B - \beta|0\rangle_B\right) = -i\left(\alpha|0\rangle_B + \beta|1\rangle_B\right). \qquad (5.46)$$

All that Alice needs to do is to tell Bob which of these four equiprobable operators to apply to his qubit and the result will be a copy, at Bob's location, of Alice's original qubit. With four a priori equally likely outcomes of Alice's measurement, she needs to send to Bob only two (classical) bits of information so that he knows which Pauli operator to apply to his qubit. A schematic representation of this teleportation procedure is given in Fig. 5.3.

At the heart of the teleportation protocol described above is the Bell state $|\Psi^-\rangle_{AB}$ shared by Alice and Bob, and its defining property that each spin component for the two qubits is anticorrelated. This is most simply expressed in the fact that the state $|\Psi^-\rangle$ is a simultaneous eigenstate of the three operators $\hat{\sigma}_x \otimes \hat{\sigma}_x$, $\hat{\sigma}_y \otimes \hat{\sigma}_y$, and $\hat{\sigma}_z \otimes \hat{\sigma}_z$, with the eigenvalue -1 in each case. This means that if we can establish the value of a spin component of the qubit A in the state in eqn 5.44 then it necessarily follows that we have also established that the qubit B has the opposite spin. This is, of course, the starting point for Bohm's presentation of the EPR dilemma. We can understand the Bell measurement in teleportation as a comparison of the spins of Alice's two qubits. If she finds a result corresponding to the state $|\Psi^-\rangle_{aA}$ then she knows that the state of qubit A is orthogonal to that of the qubit to be teleported. It immediately follows, from the spin anticorrelation properties of the ebit state $|\Psi^-\rangle_{AB}$, that Alice's measurement leaves Bob's qubit in the desired qubit state $|\psi\rangle$. If the result of Alice's Bell measurement is the state $|\Psi^+\rangle_{aA}$, then this corresponds to a simultaneous eigenstate of $\hat{\sigma}_x \otimes \hat{\sigma}_x$, $\hat{\sigma}_y \otimes \hat{\sigma}_y$, and $\hat{\sigma}_z \otimes \hat{\sigma}_z$ with eigenvalues $+1$, $+1$, and -1, respectively. It then follows that the state of the qubit A differs from that which is orthogonal to $|\psi\rangle$ by a rotation through π about the z-axis on the Bloch sphere. The same rotation, enacted by the action of $\hat{\sigma}_z$, on Bob's qubit will leave it in the state $|\psi\rangle$. Similarly, if Alice's Bell measurement gives results corresponding to the state $|\Phi^-\rangle_{aA}$ or $|\Phi^+\rangle_{aA}$, then application of the operator $\hat{\sigma}_x$ or $\hat{\sigma}_y$, respectively, to Bob's qubit will leave it in the state $|\psi\rangle$. The possible operations involved in teleportation and these chains of inference are summarized in Table 5.1.

The no-signalling theorem requires that Bob knows nothing about Alice's qubit before receiving the two bits of information telling him which of the four Pauli operators he has to apply in order to recover the state $|\psi\rangle$. In the absence of this information, Bob's best description of the state of his qubit is as an equally weighted mixture of states generated by the action of the four Pauli operators on $|\psi\rangle$:

$$\hat{\rho}_B = \frac{1}{4}\left(|\psi\rangle\langle\psi| + \hat{\sigma}_z|\psi\rangle\langle\psi|\hat{\sigma}_z + \hat{\sigma}_x|\psi\rangle\langle\psi|\hat{\sigma}_x + \hat{\sigma}_y|\psi\rangle\langle\psi|\hat{\sigma}_y\right). \qquad (5.47)$$

It is straightforward to show (see Exercise (4.33)) that this density oper-

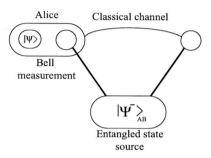

Fig. 5.3 A representation of the ebit-based teleportation scheme.

ator is $\frac{1}{2}\hat{\mathrm{I}}$ irrespective of the form of $|\psi\rangle$, as required by the no-signalling theorem.

Table 5.1 The possible events in a teleportation protocol.

Alice's Bell meas.	Inferred spin properties			State of Bob's qubit	Required unitary transf.
	$\hat{\sigma}_x \otimes \hat{\sigma}_x$	$\hat{\sigma}_y \otimes \hat{\sigma}_y$	$\hat{\sigma}_z \otimes \hat{\sigma}_z$		
$|\Psi^-\rangle_{aA}$	-1	-1	-1	$|\psi\rangle$	$\hat{\mathrm{I}}$
$|\Psi^+\rangle_{aA}$	$+1$	$+1$	-1	$\hat{\sigma}_z|\psi\rangle$	$\hat{\sigma}_z$
$|\Phi^-\rangle_{aA}$	-1	$+1$	$+1$	$\hat{\sigma}_x|\psi\rangle$	$\hat{\sigma}_x$
$|\Phi^+\rangle_{aA}$	$+1$	-1	$+1$	$\hat{\sigma}_y|\psi\rangle$	$\hat{\sigma}_y$

At the end of the teleportation protocol Bob's qubit is left in the state $|\psi\rangle_B$, but what has happened to Alice's original qubit prepared in the state $|\psi\rangle_a$? If Alice performed her Bell-state measurement as an ideal von Neumann measurement then the two qubits a and A are left in one of the four Bell states, and all dependence on the initial state $|\psi\rangle_a$ is lost. Were this not the case, of course, then we would run into a conflict with the no-cloning theorem, which tells us that it is impossible to make more than one copy of an unknown qubit. A teleportation device can transmit the state of any qubit from Alice to Bob but only at the expense of erasing the quantum information in the original state of Alice's qubit.

It is interesting to ask how well we can approximate teleportation using a classical communication channel, that is, to perform teleportation using only LOCC operations. In order to address this question, we need to introduce a figure of merit for the process, and a convenient quantity is the fidelity. Consider an operation, such as teleportation, which has been designed to generate a state $|\psi\rangle$ but because of some imperfection, whether practical or fundamental, produces a mixed state with density operator $\hat{\rho}$. The fidelity F is then defined to be

$$F = \langle\psi|\hat{\rho}|\psi\rangle. \tag{5.48}$$

This quantity has a simple meaning; it is the probability that the state produced will pass a test to identify it as the desired state $|\psi\rangle$. The fidelity clearly takes the maximum value of unity only if the state produced is $\hat{\rho} = |\psi\rangle\langle\psi|$. One way in which we might attempt to perform teleportation would be for Alice to measure one spin component of the qubit to be 'teleported' and to send the bit value corresponding to her measurement result, to Bob. Bob would then prepare a qubit in the eigenstate corresponding to Alice's measurement result. Suppose, for example, that Alice and Bob agree to use a measurement of $\hat{\sigma}_z$. For the general state in eqn 5.43, this will lead Alice to identify the state $|0\rangle$ with probability $|\alpha|^2$ and the state $|1\rangle$ with probability $|\beta|^2$. It follows

Definition of the fidelity The square root of the quantity in eqn 5.48 is also, confusingly, referred to as the fidelity. It is also useful to be able to define a fidelity for a process designed to produce a mixed state. We shall discuss the fidelity for such mixed states, and at greater length, in Chapter 8.

that Bob will prepare the state $|0\rangle$ with probability $|\alpha|^2$ or the state $|1\rangle$ with probability $|\beta|^2$, so that, on average, the state of Bob's qubit will be $\hat{\rho} = |\alpha|^2|0\rangle\langle 0| + |\beta|^2|1\rangle\langle 1|$. The resulting fidelity for Bob's qubit is

$$F = (\alpha^*\langle 0| + \beta^*\langle 0|)\,\hat{\rho}\,(\alpha|0\rangle + \beta|1\rangle))$$
$$= |\alpha|^4 + |\beta|^4. \tag{5.49}$$

This LOCC scheme works well if either $|\alpha|$ or $|\beta|$ is very small so that the state to be transmitted has a large overlap with $|1\rangle$ or $|0\rangle$. For states with $|\alpha| = |\beta|$, however, the fidelity is only $\frac{1}{2}$ and Bob's fidelity is no better than he could have achieved by guessing. High-fidelity communication of a general qubit requires the use of a quantum communication channel, either to send or to teleport the qubit.

The quantum nature of teleportation is clearly illustrated if the qubit to be teleported is itself part of an entangled state of two qubits, for example the state

$$|\Psi^-\rangle_{Ca}|\Psi^-\rangle_{AB} = \frac{1}{2}\left(|0\rangle_C|1\rangle_a - |1\rangle_C|0\rangle_a\right)\left(|0\rangle_A|1\rangle_B - |1\rangle_A|0\rangle_B\right), \tag{5.50}$$

which corresponds to an ebit shared by Claire and Alice and a second ebit shared by Alice and Bob. Alice can perform a Bell measurement on her two qubits, a and A, and, depending on the outcome, tell Bob to apply to his qubit one of the four Pauli operators. Suppose, for example, that Alice performs her Bell measurement and finds the result corresponding to the state $|\Phi^+\rangle_{aA}$. The resulting state of Alice's and Bob's qubits is then

$$_{aA}\langle\Phi^+||\Psi^-\rangle_{Ca}|\Psi^-\rangle_{AB} \propto |\Phi^+\rangle_{BC}, \tag{5.51}$$

which, apart from an unimportant prefactor, is one of the Bell states in eqn 5.38. The teleportation protocol requires Alice, having measured the state $|\Phi^+\rangle_{aA}$, to instruct Bob to apply to his qubit the operator $\hat{\sigma}_y$, which produces the required entangled state for Bob and Claire:

$$\hat{\sigma}_y^B|\Psi^+\rangle_{BC} = \frac{1}{\sqrt{2}}\left(i|1\rangle_B|0\rangle_C - i|0\rangle_B|1\rangle_C\right) = i|\Psi^-\rangle_{BC}. \tag{5.52}$$

This process, whereby teleportation is employed to establish an ebit between two parties, Bob and Claire, who share neither a quantum channel nor an ebit, is called entanglement swapping. The pre-existing entanglement between Claire and Alice is swapped for entanglement between Bob and Claire at the cost of one ebit shared between Alice and Bob. The fact that a non-local correlation can be established between Bob and Claire, by means of teleportation from Alice to Bob, is a further demonstration of the fact that teleportation realizes a quantum channel and so is not an LOCC task.

Teleportation is not limited to qubits, and can be performed for any quantum system if a maximally entangled state in a sufficiently large state space is shared by Alice and Bob. Suppose, for example, that they share an entangled state of two systems, each described by a state-space of dimension d, of the form

$$|\Psi^{00}\rangle_{AB} = \frac{1}{\sqrt{d}} \sum_{j=0}^{d-1} |j\rangle_A |j\rangle_B,$$ (5.53)

where the states $\{|j\rangle_{A,B}\}$ form a complete orthogonal set for Alice and for Bob. Alice can use this entangled state to send to Bob the state of a d-dimensional quantum system,

$$|\psi\rangle_a = \sum_{i=0}^{d-1} \alpha_i |i\rangle_a.$$ (5.54)

The teleportation protocol proceeds in very much the same way as for qubits. Alice first performs a measurement in the basis of states constructed by extending the Bell states to the d^2-dimensional product space,

$$|\Psi^{nm}\rangle = \frac{1}{\sqrt{d}} \sum_{j=0}^{d-1} e^{i2\pi jn/d} |j\rangle \otimes |(j+m)\bmod d\rangle, \quad n, m = 0, \cdots, d-1.$$ (5.55)

Alice's measurement result is equally likely to identify any one of these d^2 Bell states. If the result corresponds to the state $|\Psi^{nm}\rangle$ then Bob's system will be projected into the state

$$_{aA}\langle \Psi^{nm}||\psi\rangle_a |\Psi^{00}\rangle_{AB} \propto \sum_{j=0}^{d-1} \alpha_j e^{-i2\pi jn/d} |(j+m)\bmod d\rangle_B.$$ (5.56)

If Alice tells Bob her measurement result, which is equivalent to sending him the two numbers n and m, then he can apply to his system the unitary transformation

$$\hat{U}_{nm} = \sum_{k=0}^{d-1} e^{i2\pi kn/d} |k\rangle \langle (k+m)\bmod d|,$$ (5.57)

which transforms the state of his qubit into the original state of Alice's system given in eqn 5.54, as required.

Teleportation can even be performed, at least in principle, for quantum systems in an infinite-dimensional state space. A simple demonstration of this was given by Vaidman, who showed how the motional state of a particle could be transferred from Alice to Bob using *continuous-variable teleportation*. We suppose that Alice and Bob each have a particle, the paricles being labelled A and B, and that these have been prepared in a perfectly correlated state with position and momentum representations

$$\psi_{AB}(x_A, x_B) = \delta(x_A - x_B + L),$$
$$\phi_{AB}(p_A, p_B) = e^{ip_B L/\hbar} \delta(p_A + p_B),$$ (5.58)

where L is the distance between Alice and Bob. Alice also has a second particle, labelled a, prepared in the state $\psi_a(x_a)$, and wishes to transfer the state of this particle onto Bob's particle. Instead of a Bell measurement, Alice can measure the relative position and total momentum of the particles a and A. These are compatible observables, as the operators $\hat{x}_a - \hat{x}_A$ and $\hat{p}_a + \hat{p}_A$ commute. Let us suppose that these measurements have been performed and that Alice finds the results

$$x_a - x_A = X,$$
$$p_a + p_A = P. \tag{5.59}$$

Following this measurement, the state of Bob's particle will be

$$\psi_B(x_B) = \psi_a(x_B - L + X)\, e^{-i(x_B - L + X)P/\hbar}. \tag{5.60}$$

This has the same form as the original state, but with the position shifted by $L - X$ and the momentum by $-P$. It only remains for Alice to send to Bob his measurement results, X and P, so that Bob can shift the position and momentum of his particle. This shift corresponds to acting on his state with the unitary displacement operator (discussed in Appendix H)

$$\hat{D}(-X, -P) = \exp\left[i\left(-X\hat{p}_B + P\hat{x}_B\right)/\hbar\right]$$
$$= \exp\left(iXP/\hbar\right)\exp\left(-iX\hat{p}_B/\hbar\right)\exp\left(iP\hat{x}_B/\hbar\right). \tag{5.61}$$

By performing this unitary transformation on the state of his particle, Bob produces the state

$$\hat{D}(-X, -P)\psi_B(x_B) = \psi_a(x_B - L)\, e^{iLP/\hbar}, \tag{5.62}$$

which, apart from an unimportant phase factor, is the state of Alice's original particle shifted by the distance L separating Alice and Bob.

One inevitable problem with continuous-variable teleportation is that the maximally entangled state in eqn 5.58 is unphysical and cannot be prepared. The best we can do is to prepare a state with well-localized but not precisely defined values of $x_A - x_B$ and $p_A + p_B$. One such state is the Gaussian wavefunction

$$\psi_{AB}(x_A, x_B) = (\pi\sigma)^{-1/2}\exp\left[-\frac{k(x_A - x_B + L)^2}{4\sigma^2} - \frac{(x_A + x_B)^2}{4k\sigma^2}\right], \tag{5.63}$$

where k is a large positive constant. Teleportation with this state is still possible but the imperfect nature of the position and momentum correlations leads to a loss of fidelity in the teleportation.

Suggestions for further reading

Bell, J. S. (1987). *Speakable and unspeakable in quantum mechanics.* Cambridge University Press, Cambridge.

Bohm, D. (1951). *Quantum theory.* Prentice Hall, Upper Saddle River, NJ. (Reprinted by Dover, New York, 1989.)

Bouwmeester, D., Ekert, A., and Zeilinger, A. (eds) (2000). *The physics of quantum information.* Springer-Verlag, Berlin.

Bruß, D. and Leuchs, G. (eds) (2007). *Lectures on quantum information theory.* Wiley-VCH, Weinheim.

Jaeger, G. (2007). *Quantum information: an overview.* Springer, New York.

Macchiavello, C., Palma, G. M., and Zeilinger, A. (eds) (2000). *Quantum computation and quantum information theory.* World Scientific, Singapore.

Loepp, S. and Wootters, W. K. (2006). *Protecting information: from classical error correction to quantum cryptography.* Cambridge University Press, Cambridge.

Peres, A. (1995). *Quantum theory: concepts and methods.* Kluwer Academic, Dordrecht.

Redhead, M. (1987). *Incompleteness, nonlocality and realism.* Oxford University Press, Oxford.

Vedral, V. (2006). *Introduction to quantum information science.* Oxford University Press, Oxford.

Wheeler, J. A. and Zurek, W. H. (eds) (1983). *Quantum theory and measurement.* Princeton University Press, Princeton.

Whitaker, A. (2006). *Einstein, Bohr and the quantum dilemma* (2nd edn). Cambridge University Press, Cambridge.

Exercises

(5.1) Show that each of the four Bell states in eqns 2.108 is maximally entangled.

(5.2) Show that the Bell state $|\Psi^-\rangle$, considered as a state of two spin-1/2 particles, is the eigenstate with zero total angular momentum.

(5.3) Two spin-1 particles have orthogonal basis states $|1\rangle, |0\rangle$, and $|-1\rangle$. In this basis, the Cartesian components of the angular momentum correspond to the operators

$$\hat{J}_z = \hbar \begin{pmatrix} 1 & 0 & 0 \\ 0 & 0 & 0 \\ 0 & 0 & -1 \end{pmatrix},$$

$$\hat{J}_x = \frac{\hbar}{\sqrt{2}} \begin{pmatrix} 0 & 1 & 0 \\ 1 & 0 & 1 \\ 0 & 1 & 0 \end{pmatrix},$$

$$\hat{J}_y = \frac{\hbar}{\sqrt{2}} \begin{pmatrix} 0 & -i & 0 \\ i & 0 & -i \\ 0 & i & 0 \end{pmatrix}.$$

Find the eigenstate of zero total angular momentum of the two particles and show that this is a maximally entangled state.

(5.4) Show that, for the Bell state $|\Psi^-\rangle$, a measurement of $\vec{a} \cdot \hat{\vec{\sigma}}$ for any unit vector $\vec{a}$ on both particles will always lead to anticorrelated results.

(5.5) Show that the no-signalling theorem applies for any *operations* that Alice might choose to perform. You can consider a pair of operations associated with the effect operators $\{\hat{A}_j^{A1}\}$ and $\{\hat{A}_k^{A2}\}$.

(5.6) Consider the entangled two-qubit state

$$|\Psi\rangle = \frac{1}{\sqrt{2}} \left(|\psi_1\rangle_A |0\rangle_B + |\psi_2\rangle_A |1\rangle_B \right),$$

where

$$|\psi_1\rangle = \cos\theta |0\rangle + \sin\theta |1\rangle,$$
$$|\psi_2\rangle = \cos\theta |0\rangle - \sin\theta |1\rangle.$$

(a) Obtain the Schmidt decomposition for the state $|\Psi\rangle$ and hence calculate the reduced density operator $\hat{\rho}_B$.

(b) If Alice can determine that her qubit is in the state $|\psi_1\rangle$ and *not* in the state $|\psi_2\rangle$ then she also determines that Bob's qubit is in the state $|0\rangle$. Similarly, if she can determine that her qubit is in the state $|\psi_2\rangle$ and *not* in the state $|\psi_1\rangle$ then she also determines that Bob's qubit is in the state $|1\rangle$. Failure to achieve this unambiguous state discrimination will leave Bob's qubit in some other state. Use this idea, together with the no-signalling theorem, to obtain the upper bound on the probability for unambiguous state discrimination.

(5.7) Confirm that, in a local realistic theory, the probability that Alice's and Bob's measurements are both +1 or both −1 minus the probability that they are different can be written in the form of eqn 5.11.

(5.8) Alice and Bob each measure $\hat{\sigma}_x$ or $\hat{\sigma}_z$ on each member of an ensemble of pairs of qubits, each of which is prepared in the state $|\Psi^-\rangle_{AB}$. Will Bell's inequality (eqn 5.14) be violated for these results or not?

(5.9) Confirm that for the singlet state, $E(\vec{a}, \vec{b}) = -\vec{a} \cdot \vec{b}$. Calculate $E(\vec{a}, \vec{b})$ for each of the other three Bell states.

(5.10) Show that the arrangement of measurement directions depicted in Fig. 5.1 is the only one for which the singlet state leads to a maximum violation of Bell's inequality.

(5.11) Show that Bell's inequality is violated, for appropriate choices of observables, for all pure entangled states of two qubits.

[*Hint:* you might start by defining the states $|0\rangle$ and $|1\rangle$ for each qubit so that the Schmidt decomposition of the general entangled state is $c_1|0\rangle \otimes |1\rangle - c_2|1\rangle \otimes |0\rangle$.]

(5.12) Find all of the simultaneous eigenstates of $\hat{\sigma}_x \otimes \hat{\sigma}_y \otimes \hat{\sigma}_y$, $\hat{\sigma}_y \otimes \hat{\sigma}_x \otimes \hat{\sigma}_y$, and $\hat{\sigma}_y \otimes \hat{\sigma}_y \otimes \hat{\sigma}_x$. Show that preparing any of these will allow a demonstration of a conflict with local realism.

(5.13) (a) Show, by a suitable labelling of the basis states, that any non-maximally entangled pure state of two qubits can be written in the form

$$|\psi\rangle = a|0\rangle_A|0\rangle_B + b|1\rangle_A|0\rangle + c|0\rangle_A|1\rangle_B,$$

where none of the coefficients a, b, and c are zero.

(b) Demonstrate a contradiction between local realism and quantum theory for this state in the spirit of Appendix K.

(c) Why does the contradiction fail for maximally entangled states?

(5.14) Confirm that the Hamiltonian in eqn 5.23 does indeed induce dynamics which can be used to realize an indirect measurement of $\hat{\sigma}_x$.

(5.15) For the Hamiltonian (5.24) show that the corresponding unitary time evolution operator is

$$\hat{U} = \cos^2\theta\, \hat{I} \otimes \hat{I} - i \sin\theta \cos\theta \left(\hat{\sigma}_z \otimes \hat{\sigma}_x - \hat{\sigma}_x \otimes \hat{\sigma}_z \right)$$
$$+ \sin^2\theta\, \hat{\sigma}_y \otimes \hat{\sigma}_y,$$

where $\theta = gT/\hbar$.

(a) Is it possible to choose θ so that we can determine the value of either $\hat{\sigma}_z$ or $\hat{\sigma}_x$ with certainty?

(b) An interaction of this kind has been analysed by Fuchs *et al.* in the context of quantum key distribution. The idea is that Eve might use an interaction of this kind and need only measure her ancilla after the public discussion between Alice and Bob. How does such a strategy compare with intercepting the qubits, measuring them, and then sending a freshly prepared qubit to Bob?

(5.16) An indirect measurement of $\hat{\sigma}_z$ is performed so as to produce the state in eqn 5.26. The momentum wavefunction for the initial motional state is

$$\phi(p) = \left(\frac{a}{\pi}\right)^{1/4} \exp\left(-\frac{ap^2}{2}\right),$$

where a is a positive constant.

   (a) A subsequent measurement of momentum associates positive results with the spin value 0 and negative results with the value 1. What is the probability of error in performing such a measurement?

   (b) How does the value obtained in part (a) compare with the minimum-error strategy for discriminating between the two states in eqn 5.27?

As an additional challenge, consider the effect of including the particle kinetic energy in the Hamiltonian so that

$$\hat{H} = \hat{I} \otimes \frac{1}{2m}\hat{p}^2 + g\hat{\sigma}_z \otimes \hat{x}.$$

(5.17) Is quantum key distribution an LOCC task?

(5.18) Consider the 'lock with only one key' problem described in Section 5.3. Which state gives Claire the best chance of deceiving Alice so as to pass as Bob? What is the corresponding probability that Claire's key will open the lock if it is based on n ebits?

(5.19) Show that by a suitable set of LOCC operations (including instructions sent to Alice and Bob to perform local operations on their qubits), Claire can transform the GHZ state in eqn 5.35 into any desired pure state for Alice and Bob.

(5.20)   (a) Show that each of the four Bell states can be transformed into any of the others by the action of a Pauli operator on one of the component qubits.

   (b) Confirm the more general result that any maximally entangled state of two qubits is related to any other by the action of a unitary transformation on just one of the qubits.

(5.21) Alice wishes to employ dense coding to send classical information to Bob, but they have only a pair of qubits prepared in the mixed Werner state given in eqn 2.120.

   (a) Calculate the mutual information between Alice and Bob if Alice selects each of the four possible transformations with equal probability.

   (b) How much information can be transmitted for a non-entangled state? Why is this value less than the one-bit maximum value using classical correlations?

(5.22) Alice wishes to send to Bob instructions, via a classical channel, for preparing the state in eqn 5.43 which is known to her. Calculate the number of bits required if Bob is to be able to reconstruct the state with a fidelity of at least $1 - 10^{-6}$.

(5.23) Confirm that the states in eqns 5.44 and 5.45 are indeed equivalent.

(5.24) Show that a qubit can be teleported using any of the three Bell states $|\Psi^+\rangle_{AB}$, $|\Phi^-\rangle_{AB}$ and $|\Phi^+\rangle_{AB}$ in place of an ebit prepared in the state $|\Psi^-\rangle_{AB}$.

(5.25) The teleportation process transfers the state of Alice's qubit to Bob and, in doing so, erases any memory of the state at Alice's location. Consider a process which enacts the transformation

$$(\alpha|0\rangle + \beta|1\rangle) \otimes |B\rangle \rightarrow |B'\rangle \otimes (\alpha|0\rangle + \beta|1\rangle),$$

where the state $|B\rangle$ is a 'blank' state onto which the state of the left qubit is to be teleported. Show that for this process to work for any choice of α and β, we require the state $|B'\rangle$ to be independent of α and β.

(5.26) Find the average of the fidelity for the LOCC version of teleportation by averaging the fidelity in eqn 5.49 for all possible pure states.

[*Hint:* you could do this by writing $\alpha = \cos(\theta/2)$, $\beta = \sin(\theta/2)e^{i\phi}$ and then integrating the fidelity over the surface of the Bloch sphere.]

(5.27) Alice wishes to teleport the pure state of a qubit to Bob but they share only an imperfect ebit in the form of the mixed Werner state of eqn 2.120. Calculate the fidelity for the qubit Bob receives.

(5.28) Calculate the average fidelity for teleportation of an unknown state using the non-maximally entangled state $\cos\varphi|0\rangle_A|1\rangle_B - \sin\varphi|1\rangle_A|0\rangle_B$.

(5.29) Demonstrate entanglement swapping explicitly by rewriting eqn 5.50 in terms of Bell states for the qubits a and A and for the qubits B and C and then applying the requisite Pauli operator to Bob's qubit.

(5.30) Alice and Bob each have a qubit in an unknown state and wish to swap these qubits. They have at their disposal a number of ebits and the ability to perform LOCC operations. One way to achieve the desired swap would be for Alice to teleport her qubit to Bob and for Bob to teleport his qubit to Alice. This requires the use of two ebits. Is it possible to achieve the desired swap using just *one* ebit?

[*Hint:* suppose that each of the qubits to be swapped is itself a part of a locally prepared maximally entangled state and consider how many ebits Alice and Bob would then share after the swap.]

(5.31) Prove that the state in eqn 5.53 is maximally entangled.

(5.32) Show that for $d = 2$ the states in eqn 5.55 reduce to the Bell states.

(5.33) Show that the d^2 states in eqn 5.55 form a complete orthonormal set for the d^2-dimensional product space spanned by the states $|i\rangle \otimes |j\rangle$.

(5.34) Teleporting the state of a d-dimensional quantum system requires Alice to send to Bob one of d^2 equally likely messages. How many bits are required to do this?

(5.35)   (a) Confirm that the operator $\hat{U}_{nm}$ is unitary.
     (b) Show that the product of any two unitary operators of the form of eqn 5.57 gives a further operator of this form multiplied by a phase factor.

(5.36) Calculate $\Delta (x_A - x_B)^2$ and $\Delta (p_A + p_B)^2$ for the state in eqn 5.63.

(5.37) Alice wishes to teleport the motional state of a particle to Bob using the Gaussian state in eqn 5.63 If the motional state to be teleported is

$$\psi_a (x_a) = \left(\frac{b}{\pi} \right)^{1/4} \exp \left[-\frac{b (x_a - x_0)^2}{2} + i \frac{p_0 x_a}{\hbar} \right],$$

where x_0 and p_0 are real constants and b is a real positive constant, calculate the fidelity of the resulting teleported state.

Quantum information processing

6

We have seen how information can be encoded onto a quantum system by selecting the state in which it is prepared. Retrieving the information is achieved by performing a measurement, and the optimal measurement in any given situation is usually a generalized measurement. In between preparation and measurement, the information resides in the quantum state of the system, which evolves in a manner determined by the Hamiltonian. The associated unitary transformation may usefully be viewed as quantum information processing; if we can engineer an appropriate Hamiltonian then we can use the quantum evolution to assist in performing computational tasks.

Our objective in quantum information processing is to implement a desired unitary transformation. Typically this will mean coupling together a number, perhaps a large number, of qubits and thereby generating highly entangled states. It is fortunate, although by no means obvious, that we can realize any desired multiqubit unitary transformation as a product of a small selection of simple transformations and, moreover, that each of these need only act on a single qubit or on a pair of qubits. The situation is reminiscent of digital electronics, in which logic operations are decomposed into actions on a small number of bits. If we can realize and control a very large number of such operations in a single device then we have a computer. Similar control of a large number of qubits will constitute a quantum computer. It is the revolutionary potential of quantum computers, more than any other single factor, that has fuelled the recent explosion of interest in our subject. We shall examine the remarkable properties of quantum computers in the next chapter.

6.1 Digital electronics

In digital electronics, we represent bit values by voltages: the logical value 1 is a high voltage (typically $+5\,\mathrm{V}$) and 0 is the ground voltage ($0\,\mathrm{V}$). The voltage bits are coupled and manipulated by transistor-based devices, or gates. The simplest gates act on only one bit or combine two bits to generate a single new bit, the value of which is determined by the two input bits. For a single bit, with value 0 or 1, the only possible operations are the identity (which does not require a gate) and the bit

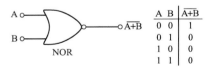

A	Ā
0	1
1	0

Fig. 6.1 The NOT gate and its truth table.

A	B	A·B
0	0	0
0	1	0
1	0	0
1	1	1

AND

A	B	A+B
0	0	0
0	1	1
1	0	1
1	1	1

OR

Fig. 6.2 The AND and OR gates and their truth tables.

Other notations   There are other notations in common use for the AND and OR operations. The AND operation $A \cdot B$ is also written as AB and as $A \wedge B$. The OR operation is also written as $A \vee B$.

A	B	$\overline{A \cdot B}$
0	0	1
0	1	1
1	0	1
1	1	0

NAND

A	B	$\overline{A+B}$
0	0	1
0	1	0
1	0	0
1	1	0

NOR

Fig. 6.3 The NAND and NOR gates and their truth tables.

flip. The latter is realized by the NOT gate, depicted in Fig. 6.1 together with its truth table, which presents all of the possible input values and the associated outputs. We represent the logical NOT operation by an overbar: $A \rightarrow \overline{A}$.

The simplest two-bit gates are the AND gate and the OR gate, depicted in Fig. 6.2. The AND gate outputs the value 1 if and only if both the inputs A *and* B have the value 1. For other inputs, the output value is 0. The AND operation is denoted by $A \cdot B$. The OR gate gives the value 1 if A *or* B has the value 1; it gives the value 1 unless A and B are both 0. The OR operation is denoted (somewhat confusingly) by $A + B$. There are two further two-bit gates which are in common use: the NAND (or NOT AND) and the NOR (or NOT OR) gates. The NAND gate has the effect of combining an AND gate and a following NOT gate and the NOR gate combines the NOT and OR operations. These gates, together with their associated truth tables, are presented in Fig. 6.3.

It is helpful to associate the bit values 1 and 0 with the logical statements TRUE and FALSE, respectively. An AND gate, for example, gives the output value 1, or TRUE, only if A *and* B are both TRUE. A NOR gate gives the output TRUE only if neither A *nor* B is TRUE, and so on. The 'digital algebra' governing gate operations is, in fact, that of true–false statements in logic. This was formulated by Boole in the middle of the nineteenth century and today is known as Boolean algebra. It was Shannon who first appreciated its power for information processing. Boole started with the NOT, AND, and OR functions and these led to a number of useful theorems. For a single Boolean variable A, we find

$$\begin{aligned} A \cdot 0 = 0, &\quad A + 0 = A, \quad \overline{\overline{A}} = A, \\ A \cdot 1 = A, &\quad A + 1 = 1, \\ A \cdot A = A, &\quad A + A = A, \\ A \cdot \overline{A} = 0, &\quad A + \overline{A} = 1. \end{aligned} \tag{6.1}$$

More generally, we find that Boolean algebra is commutative,

$$\begin{aligned} A + B = B + A, \\ A \cdot B = B \cdot A; \end{aligned} \tag{6.2}$$

it is associative,

$$\begin{aligned} A + (B + C) = (A + B) + C, \\ A \cdot (B \cdot C) = (A \cdot B) \cdot C; \end{aligned} \tag{6.3}$$

and it also obeys distribution rules,

$$\begin{aligned} A \cdot (B + C) = (A \cdot B) + (A \cdot C), \\ A + (B \cdot C) = (A + B) \cdot (A + C). \end{aligned} \tag{6.4}$$

In addition to these there are some unexpected, but very useful, rules which can be used to simplify logical functions. Two of these are the absorption rules, which state that

$$A + (A \cdot B) = A,$$
$$A \cdot (A + B) = A. \tag{6.5}$$

Yet more useful are De Morgan's theorems, the first of which states that a NOR gate is equivalent to an AND gate with NOT operations performed on each input:

$$\overline{A + B} = \overline{A} \cdot \overline{B}. \tag{6.6}$$

The second states that a NAND gate is equivalent to an OR gate with NOT operations performed on each input:

$$\overline{A \cdot B} = \overline{A} + \overline{B}. \tag{6.7}$$

These simple theorems were generalized by Shannon to cover more complicated Boolean expressions. Shannon's form of the theorems is, 'to obtain the inverse of any Boolean function, invert all variables ($A \to \overline{A}$, $B \to \overline{B}, \cdots$), and replace all OR gates by AND gates and all AND gates by OR gates.'

The basic gates can be combined to evaluate any possible Boolean function. A classic example is the half-adder, which is a primitive for performing binary addition. A half-adder circuit, together with its truth table, is depicted in Fig. 6.4. The circuit has two outputs, S and C, corresponding to the sum and carry of the binary digits A and B. The sum is the result of modulo addition of the two inputs:

$$S = \overline{A} \cdot B + A \cdot \overline{B}. \tag{6.8}$$

The carry is the value of the next column, the twos column, of the binary addition; the binary sum of the bits A and B is CS.

The half-adder is built from two different types of two-bit gates, namely three AND gates and one OR gate, together with two single-bit NOT gates. It is reasonable to ask whether or not we really need all of the five gates (NOT, AND, OR, NAND, and NOR) or if we need only a subset of these. An AND gate followed by a NOT gate performs the NAND operation and, indeed, a NAND gate followed by a NOT performs the AND operation. Similarly, an OR gate followed by a NOT gate is equivalent to a NOR gate. De Morgan's theorems, moreover, tell us that we can build a NOR gate out of NOT gates and an AND gate, or a NAND gate from NOT gates and an OR gate. It is clear that a combination of NOT gates and any one of the two-bit gates is universal, in that combinations of this type suffice to realize any Boolean function. In fact, we can go further and also dispense with the NOT gate if we have either NAND or NOR gates. If we set *both* inputs to A then the output is $\overline{A}$ (see Fig. 6.5).

For completeness, we note that there are two further two-bit gates: the XOR (or exclusive OR), denoted $A \oplus B$, and the XNOR, denoted $\overline{A \oplus B}$. The XOR operation gives as output the sum of A and B modulo

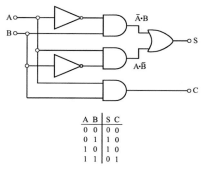

A B	S C
0 0	0 0
0 1	1 0
1 0	1 0
1 1	0 1

Fig. 6.4 The half-adder and its truth table.

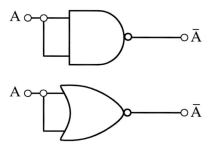

Fig. 6.5 The NAND and NOR versions of NOT.

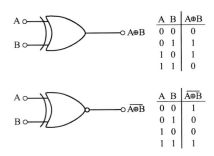

A B	A⊕B
0 0	0
0 1	1
1 0	1
1 1	0

A B	$\overline{A \oplus B}$
0 0	1
0 1	0
1 0	0
1 1	1

Fig. 6.6 The XOR and XNOR gates and their truth tables.

2. The XOR and XNOR gates and their truth tables are depicted in Fig. 6.6.

That the XOR and XNOR gates can be realized in terms of our other gates is clear from inspection of the half-adder in Fig. 6.4. The output S is precisely $A \oplus B$ and the addition of a further NOT gate gives the XNOR. There are, of course, very many other commonly used combinations of gates, including ones specially designed to act as memories, but further discussion of these would take us too far from our topic.

6.2 Quantum gates

In quantum information processing, the logical values 1 and 0 are replaced by the orthogonal qubit states $|1\rangle$ and $|0\rangle$. In place of the two possible values for each classical bit, we have a multitude of allowed states in the form of any superposition of $|0\rangle$ and $|1\rangle$. Quantum mechanics allows for the action of any unitary transformation on a qubit. We can envisage a device designed to enact a chosen unitary transformation on a qubit as a one-qubit gate. We often need to link quantum gates together, and to this end it is useful to have a diagrammatic representation of quantum gates. Figure 6.7 depicts a general one-qubit gate; the left-hand line represents the input state of the qubit ($|\psi\rangle$) and the right-hand line represents the output state as transformed by the gate ($\hat{U}|\psi\rangle$).

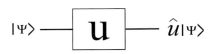

Fig. 6.7 A general one-qubit gate.

Among the multitude of possible one-qubit gates there are a few that occur sufficiently often to merit special symbols of their own. The most common of these, the Hadamard, Pauli-X, Pauli-Y, Pauli-Z, phase, and $\pi/8$ gates, together with their associated unitary operators, are depicted in Fig. 6.8. These gates are not independent, in that some of them can be realized as combinations of others. The Hadamard and $\pi/8$ gates are, in fact, universal in that a sufficiently long sequence of these can realize any desired unitary transformation to any required degree of precision. The choice of whether to use such a sequence of universal gates or a tailor-made gate U will depend on the physical system used to implement the gate.

The Pauli-X gate is sometimes referred to as the quantum NOT gate, as its action on a qubit state is to change $|0\rangle$ to $|1\rangle$ and $|1\rangle$ to $|0\rangle$. A universal NOT gate, however, would change any pure qubit state $|\psi\rangle$ into the orthogonal state $|\psi^{\perp}\rangle$. It is straightforward to show that such a universal NOT operation is impossible. To this end, let us suppose that there is a universal NOT operation associated with the operator $\hat{U}_{\text{NOT}}$. It would then follow that

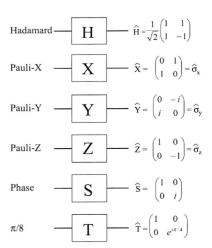

Fig. 6.8 Six common one-qubit gates and the associated unitary transformations.

$$\hat{U}_{\text{NOT}}|0\rangle = |1\rangle,$$
$$\hat{U}_{\text{NOT}}|1\rangle = |0\rangle. \tag{6.9}$$

The linearity of quantum mechanics then requires that the superposition state $2^{-1/2}(|0\rangle + |1\rangle)$ is left unchanged by the action of $\hat{U}_{\text{NOT}}$,

$$\hat{U}_{\text{NOT}}\frac{1}{\sqrt{2}}\left(|0\rangle + |1\rangle\right) = \frac{1}{\sqrt{2}}\left(|0\rangle + |1\rangle\right) \neq \frac{1}{\sqrt{2}}\left(|0\rangle - |1\rangle\right), \quad (6.10)$$

and so $\hat{U}_{\text{NOT}}$ does not perform the universal NOT operation. If we know the basis in which our qubit has been prepared then we can readily transform it into the orthogonal state. There is no operation, however, that will transform an unknown qubit state into the orthogonal state.

We might ask what is the nearest we can get to an ideal universal NOT. If we require that the quality of the NOT operation is the same for any pure qubit state then the optimal procedure is to implement an operation with the three effect operators

$$\hat{A}_1 = \frac{1}{\sqrt{3}}\hat{\sigma}_x, \quad \hat{A}_2 = \frac{1}{\sqrt{3}}\hat{\sigma}_y, \quad \hat{A}_3 = \frac{1}{\sqrt{3}}\hat{\sigma}_z. \quad (6.11)$$

We found in Section 2.4 that the density operator for any pure qubit state can be represented in the form

$$\hat{\rho} = \frac{1}{2}\left(\hat{I} + \vec{r}\cdot\hat{\vec{\sigma}}\right), \quad (6.12)$$

where $\vec{r}$ is a unit vector, the Bloch vector. The optimal universal NOT operation reverses the direction of the Bloch vector but also reduces its length:

$$\hat{\rho} \rightarrow \sum_i \hat{A}_i\hat{\rho}\hat{A}_i^\dagger = \frac{1}{2}\left(\hat{I} - \frac{1}{3}\vec{r}\cdot\hat{\vec{\sigma}}\right). \quad (6.13)$$

For a pure qubit state $|\psi\rangle$, our NOT operation produces a mixture of $|\psi\rangle$ and $|\psi^\perp\rangle$,

$$|\psi\rangle\langle\psi| = \frac{2}{3}|\psi^\perp\rangle\langle\psi^\perp| + \frac{1}{3}|\psi\rangle\langle\psi|, \quad (6.14)$$

corresponding to a success probability of 2/3 and a failure probability of 1/3. It is interesting to note that the optimal universal cloning operation, described in Appendix F, can also produce, at the same time, an optimal NOT operation.

Implementing a general unitary transformation on a set of qubits requires us to induce them to interact in a controlled way. At the simplest level this means realizing two-qubit unitary transformations, produced by associated two-qubit gates. More complicated multiqubit unitary transformations might then be constructed from combinations of one- and two-qubit gates in much the same way as general Boolean functions can be implemented using universal two-bit gates. There is one immediate difference between classical two-bit gates and quantum two-qubit gates and this is that the number of outputs from the quantum gate must equal the number of inputs: the gate changes a two-qubit state into another two-qubit state. The requirements of unitarity, moreover, mean that not all two-qubit transformations can be realized. Consider, for example, the transformation

$$|A\rangle \otimes |B\rangle \rightarrow |A\rangle \otimes |A \cdot B\rangle, \quad (6.15)$$

Universal NOT and Bell states
We can also prove the impossibility of the universal NOT operation by considering the effect such an operation would have on an entangled state. The Bell state $|\Psi^-\rangle$ is an eigenstate of $\hat{\sigma}_x \otimes \hat{\sigma}_x$, $\hat{\sigma}_y \otimes \hat{\sigma}_y$, and $\hat{\sigma}_z \otimes \hat{\sigma}_z$ with each of the eigenvalues being -1. A universal NOT operation performed on one of the two qubits would have to change all of these eigenvalues to $+1$, but the product of the three operators is $-\hat{I}$, so at least one of the eigenvalues *for any allowed state* must be -1.

Computational basis   The qubit basis states $|0\rangle$ and $|1\rangle$ are often referred to as the computational basis. This is because the output of a quantum information processor is, by convention, measured in this basis.

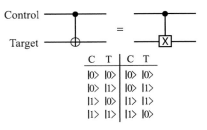

Fig. 6.9 The CNOT gate and its effect on the computational basis states.

where A and B denote the Boolean values 0 and 1 so that $|A\rangle$ and the other kets can be $|0\rangle$ or $|1\rangle$. We established in Section 2.3 that a unitary transformation conserves the overlap between pairs of states. The transformation in eqn 6.15 maps the two orthogonal states $|0\rangle \otimes |0\rangle$ and $|0\rangle \otimes |1\rangle$ onto the single state $|0\rangle \otimes |0\rangle$ and hence it is not unitary.

The most commonly encountered two-qubit gate is the controlled-NOT, or CNOT, gate. This acts on the state of two qubits, known as the control qubit, C, and the target qubit, T. The action of the CNOT gate is to change the state of the target qubit in a manner that depends on the state of the control qubit. If the control qubit is in the state $|0\rangle$ then the target qubit is left unchanged, but if it is in the state $|1\rangle$ then a Pauli-X gate is applied to the control qubit, changing $|0\rangle$ into $|1\rangle$ and $|1\rangle$ into $|0\rangle$. The CNOT appears sufficiently often to merit having its own symbol and this, together with is quantum truth table, is depicted in Fig. 6.9. It is important to identify correctly the control and target qubits; consider, for example the states that arise from the action of a CNOT on the states $|0\rangle_C |1\rangle_T$ and $|1\rangle_C |0\rangle_T$.

The CNOT operation has something in common with the classical XOR in that the final state may be expressed in terms of the Boolean XOR function:

$$|A\rangle \otimes |B\rangle \rightarrow |A\rangle \otimes |B \oplus A\rangle. \tag{6.16}$$

This transformation is a unitary one and is induced by the unitary operator

$$\hat{U}_{\text{CNOT}} = |0\rangle\langle 0| \otimes \hat{I} + |1\rangle\langle 1| \otimes \hat{\sigma}_x, \tag{6.17}$$

where the first operator in each product acts on the control qubit and the second on the target. If the control qubit is in the state $|0\rangle$ then the target qubit is unchanged, but if it is in the state $|1\rangle$ then the Bloch vector for the target qubit is rotated through π radians about the x-axis. It is often helpful to write quantum gate operations as matrices, and we do this in the natural basis $|0\rangle_C |0\rangle_T$, $|0\rangle_C |1\rangle_T$, $|1\rangle_C |0\rangle_T$, $|1\rangle_C |1\rangle_T$. The CNOT operator in eqn 6.17 then takes the form

$$\hat{U}_{\text{CNOT}} = \begin{pmatrix} 1 & 0 & 0 & 0 \\ 0 & 1 & 0 & 0 \\ 0 & 0 & 0 & 1 \\ 0 & 0 & 1 & 0 \end{pmatrix}. \tag{6.18}$$

When written in this way, it is obvious that $\hat{U}_{\text{CNOT}}$ is both Hermitian and its own inverse. It then follows immediately that it is also unitary.

The simple form of the transformation in eqn 6.16 might lead us to question whether the CNOT gate is performing an intrinsically quantum task. We could, after all, realize a Boolean map from A and B to A and $B \oplus A$ with a simple XOR gate. The quantum nature of the unitary CNOT operation is revealed, however, on considering superpositions of the states $|0\rangle$ and $|1\rangle$. If, for example, we were to prepare eigenstates of $\hat{\sigma}_x$,

$$|0'\rangle = \frac{1}{\sqrt{2}}\left(|0\rangle + |1\rangle\right),$$

$$|1'\rangle = \frac{1}{\sqrt{2}}\left(|0\rangle - |1\rangle\right), \tag{6.19}$$

hen we would find that the controlled-NOT operation produced

$$\hat{U}_{\mathrm{CNOT}}|A'\rangle_C|B'\rangle_T = |A' \oplus B'\rangle_C|B'\rangle_T. \tag{6.20}$$

We can see that in this basis the roles of the control and target have een interchanged; it is the state of the control qubit that is changed, while that of the target remains the same. There are no superpositions f bit values for classical bits and so there is no classical analogue of this asis-dependent behaviour.

A more compelling demonstration of the intrinsically quantum nature f the CNOT operation is its ability to create entangled states of a pair of qubits. Consider, for example, the action of $\hat{U}_{\mathrm{CNOT}}$ on the unentangled, r product, state $|0'\rangle_C|0\rangle_T$:

$$\begin{aligned}\hat{U}_{\mathrm{CNOT}}|0'\rangle_C|0\rangle_T &= \frac{1}{\sqrt{2}}\left(|0\rangle_C|0\rangle_T + |1\rangle_C|1\rangle_T\right)\\ &= |\Phi^+\rangle_{CT}. \tag{6.21}\end{aligned}$$

This is one of the Bell states given in eqn 2.108 and is, of course, a maximally entangled state of the control and target qubits. We saw n Chapter 5 that such states can exhibit non-local phenomena such as violation of a suitable Bell inequality. The ability to introduce such intrinsically quantum effects is a reliable indication of a quantum process nd we can be certain, therefore, that the CNOT gate has no classical equivalent. Many quantum information-processing protocols rely on he creation and manipulation of multiqubit entangled states and the mportance of the CNOT gate, therefore, should come as no surprise.

The CNOT is one of a large class of possible controlled-unitary gates, he general form of which is depicted in Fig. 6.10. The principle of these ates is that if the control qubit is in the state $|0\rangle$ then the target qubit s left unchanged, but if its state is $|1\rangle$ then the unitary operator $\hat{U}$ is pplied to the control qubit. The CNOT gate is clearly a simple example f a controlled-unitary (CU) gate, with the unitary operator in question eing $\hat{\sigma}_x$. Any of the single-qubit gates in Fig. 6.8, or indeed any other ingle-qubit unitary transformation, can appear in a controlled-unitary ate. The two most commonly encountered are the controlled-Z (CZ) nd the controlled-phase (C-phase) gate and these, together with their orresponding unitary matrices, are depicted in Figure 6.11. Like the CNOT, these gates can produce entanglement between the control and arget qubits and so are intrinsically quantum gates. Unlike the CNOT, owever, these gates are symmetrical with respect to the control and arget bits. We can show this explicitly by writing the unitary operators ssociated with the gates in the forms

$$\hat{U}_{\mathrm{CZ}} = \hat{I} \otimes \hat{I} - 2|1\rangle\langle1| \otimes |1\rangle\langle1|,$$

$$\hat{U}_{\mathrm{CS}} = \hat{I} \otimes \hat{I} - (1-i)|1\rangle\langle1| \otimes |1\rangle\langle1|, \tag{6.22}$$

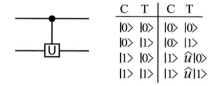

C	T	C	T
$|0\rangle$	$|0\rangle$	$|0\rangle$	$|0\rangle$
$|0\rangle$	$|1\rangle$	$|0\rangle$	$|1\rangle$
$|1\rangle$	$|0\rangle$	$|1\rangle$	$\hat{u}|0\rangle$
$|1\rangle$	$|1\rangle$	$|1\rangle$	$\hat{u}|1\rangle$

C	T	C	T
$|0\rangle$	$|0\rangle$	$|0\rangle$	$\hat{u}|0\rangle$
$|0\rangle$	$|1\rangle$	$|0\rangle$	$\hat{u}|1\rangle$
$|1\rangle$	$|0\rangle$	$|1\rangle$	$|0\rangle$
$|1\rangle$	$|1\rangle$	$|1\rangle$	$|1\rangle$

Fig. 6.10 The CU gate and its effect on the computational basis states.

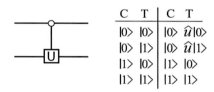

$$\hat{u}_{\mathrm{cz}} = \begin{pmatrix} 1 & 0 & 0 & 0 \\ 0 & 1 & 0 & 0 \\ 0 & 0 & 1 & 0 \\ 0 & 0 & 0 & -1 \end{pmatrix}$$

$$\hat{u}_{\mathrm{cs}} = \begin{pmatrix} 1 & 0 & 0 & 0 \\ 0 & 1 & 0 & 0 \\ 0 & 0 & 1 & 0 \\ 0 & 0 & 0 & i \end{pmatrix}$$

Fig. 6.11 The CZ and C-phase gates and their associated unitary matrices.

which are clearly invariant under interchange of the qubits.

As a final example of a two-qubit gate we consider the swap gate, the action of which is to interchange the states of the two qubits:

$$\hat{U}_{\text{swap}}|\alpha\rangle \otimes |\beta\rangle = |\beta\rangle \otimes |\alpha\rangle, \tag{6.23}$$

where $|\alpha\rangle$ and $|\beta\rangle$ are any qubit states. This means that the gate leaves the states $|0\rangle \otimes |0\rangle$ and $|1\rangle \otimes |1\rangle$ unchanged but interchanges the states $|0\rangle \otimes |1\rangle$ and $|1\rangle \otimes |0\rangle$. The swap gate is explicitly symmetric under exchange of the two qubits, and this symmetry is reflected in the symbol for this gate, depicted in Fig. 6.12. The swap gate does not, of course, introduce any entanglement between previously unentangled qubits.

There is a range of three-qubit and multiqubit gates which have been introduced. These operations can all be realized, however, in terms of a suitable set of one- and two-qubit gates. Combining quantum gates to produce more complicated unitary transformations, such as three-qubit operations, is the subject of the following section.

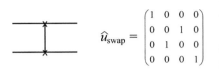

$$\hat{u}_{\text{swap}} = \begin{pmatrix} 1 & 0 & 0 & 0 \\ 0 & 0 & 1 & 0 \\ 0 & 1 & 0 & 0 \\ 0 & 0 & 0 & 1 \end{pmatrix}$$

Fig. 6.12 The swap gate and its associated unitary matrix.

6.3 Quantum circuits

The diagrammatic representation of gates comes into its own when we need to combine large numbers of one- and two-qubit operations in order to realize a multiqubit transformation. We can connect gates in a manner similar to the Boolean gates used in digital electronics, and this analogy leads us to refer to a collection of connected quantum gates as a quantum logic circuit or, simply, a quantum circuit. In a quantum circuit diagram, each qubit is represented by a single horizontal line, with gate operations depicted as in the preceding section. The sequence of gate operations is read from left to right. Consider, for example, the four-qubit circuit depicted in Fig. 6.13. The initial state of the four qubits is first transformed by a phase gate acting on qubit 1 and by a CNOT gate with qubit 3 acting as the control and qubit 4 as the target. The next transformation is a controlled-$\pi/8$ gate with qubit 4 acting as the control and qubit 2 as the target. Finally, qubit 4 is transformed by a Hadamard gate, and a CNOT gate is applied with qubits 3 and 1 being the control and target qubits, respectively.

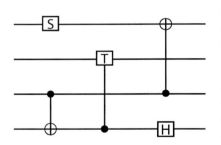

Fig. 6.13 A four-qubit circuit diagram.

It is remarkable to note that we can realize any multiqubit unitary transformation by combining single-qubit gates and a number of copies of a universal two-qubit gate. One such universal two-qubit gate is the CNOT gate. We present a proof of this important idea in Appendix L. Consider, as an example, the swap gate described in the previous section. We can summarize the action of this gate by its action on the Boolean state $|A\rangle \otimes |B\rangle$:

$$\hat{U}_{\text{swap}}|A\rangle \otimes |B\rangle = |B\rangle \otimes |A\rangle. \tag{6.24}$$

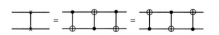

Fig. 6.14 A swap gate as three CNOT gates.

This swap operation may be implemented by a sequence of three CNOT gates as illustrated in Fig. 6.14. We can demonstrate this simply by calculating the effect of the gates on the state $|A\rangle \otimes |B\rangle$:

$$|A\rangle \otimes |B\rangle \rightarrow |A\rangle \otimes |B \oplus A\rangle$$
$$\rightarrow |A \oplus (B \oplus A)\rangle \otimes |B \oplus A\rangle = |B\rangle \otimes |B \oplus A\rangle$$
$$\rightarrow |B\rangle \otimes |(B \oplus A) \oplus B\rangle = |B\rangle \otimes |A\rangle. \qquad (6.25)$$

Any controlled-unitary gate can be realized using at most two CNOT gates and four single-qubit gates. We saw in Chapter 2 that the most general single-qubit unitary operator can be written in the form

$$\hat{U} = e^{i\alpha}\left(\cos\beta\hat{I} + i\sin\beta\vec{a}\cdot\hat{\vec{\sigma}}\right), \qquad (6.26)$$

where α and β are real and $\vec{a}$ is a unit vector. We can produce a controlled version of this unitary operator by the arrangement given in Fig. 6.15. In order to produce the controlled-unitary gate, we require the three V-gates to satisfy the conditions

$$\hat{V}_3\hat{V}_2\hat{V}_1 = \hat{I}$$
$$\hat{V}_3\hat{X}\hat{V}_2\hat{X}\hat{V}_1 e^{i\alpha} = \hat{U}. \qquad (6.27)$$

To see that this is always possible, we note that the first condition is automatically satisfied if $\hat{V}_2 = \hat{V}_3^\dagger\hat{V}_1^\dagger$, and it follows that the second will be satisfied if

$$\hat{V}_3\hat{\sigma}_x\hat{V}_3^\dagger\hat{V}_1^\dagger\hat{\sigma}_x\hat{V}_1 = \cos\beta\hat{I} + i\sin\beta\vec{a}\cdot\hat{\vec{\sigma}}. \qquad (6.28)$$

The two unitary operators $\hat{V}_3$ and $\hat{V}_1^\dagger$ produce a rotation on the Bloch sphere changing $\hat{\sigma}_x$ to $\vec{b}\cdot\hat{\vec{\sigma}}$ and $\vec{c}\cdot\hat{\vec{\sigma}}$, respectively, where $\vec{b}$ and $\vec{c}$ are any desired unit vectors. If we choose $\vec{b}$ and $\vec{c}$ (and hence $\hat{V}_3$ and $\hat{V}_1^\dagger$) such that

$$\vec{b}\cdot\vec{c} = \cos\beta, \qquad \vec{b}\times\vec{c} = \sin\beta\vec{a}, \qquad (6.29)$$

then the circuit depicted in Fig. 6.15 realizes the desired controlled-unitary gate.

The simplest three-qubit gate is the Toffoli or controlled-controlled-NOT gate, which performs a NOT operation on the third qubit if both the first and the second qubit are in the state $|1\rangle$:

$$\hat{U}_{\text{Toffoli}}|A\rangle \otimes |B\rangle \otimes |C\rangle = |A\rangle \otimes |B\rangle \otimes |C \oplus (A\cdot B)\rangle. \qquad (6.30)$$

The Toffoli gate and a quantum circuit to realize it based on two-qubit controlled gates are depicted in Fig. 6.16, where the W-gate produces the unitary transformation

$$\hat{W} = \frac{(1-i)}{2}\left(\hat{I} + i\hat{\sigma}_x\right), \qquad (6.31)$$

so that $\hat{W}^2 = \hat{\sigma}_x$. The fact that any controlled-unitary gate can be produced using CNOT gates and single-qubit gates then means that we can realize a Toffoli gate using only single-qubit and CNOT gates.

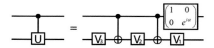

Fig. 6.15 A CU gate formed from CNOT gates and single-qubit gates.

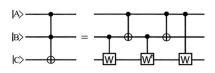

Fig. 6.16 The Toffoli gate.

A second commonly encountered three-qubit gate is the Fredkin or controlled-swap gate, which performs a swap operation on the second and third qubits if the first qubit is in the state $|1\rangle$, but leaves them unchanged if the first qubit is in the state $|0\rangle$. We can express this behaviour in terms of the effect on a Boolean state:

$$\hat{U}_{\text{Fredkin}}|A\rangle \otimes |B\rangle \otimes |C\rangle = |A\rangle \otimes |\overline{A} \cdot B + A \cdot C\rangle \otimes |\overline{A} \cdot C + A \cdot B\rangle. \quad (6.32)$$

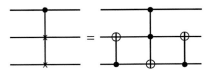

Fig. 6.17 The Fredkin gate.

The Fredkin gate can be realized by a circuit formed from a Toffoli gate and two CNOT gates, as depicted in Fig. 6.17. We can understand this very simply from the swap gate circuit given in Fig. 6.12 and the action of the Toffoli gate. If the first qubit is in the state $|0\rangle$ then the Toffoli gate has no effect on the state of the second and third qubits. In this case the effects of the two CNOT gates simply cancel out and the second and third qubits remain unchanged. If, however, the first qubit is in the state $|1\rangle$ then the second and third qubits are transformed by a sequence of three CNOT gates, which together comprise a swap gate.

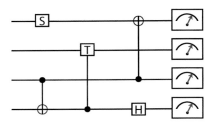

Fig. 6.18 A circuit diagram with meter symbols denoting a von Neumann measurement of the qubit in the computational basis.

The final step in any quantum information process is to perform a measurement on the qubits and thereby extract the processed information. The qubits, as transformed by the gates, can be measured in any basis or, indeed, any generalized measurement can be performed on them. We saw in Chapter 4 that we can realize a generalized measurement by means of a suitable unitary interaction between a set of qubits followed by a von Neumann measurement of each qubit. Any single-qubit von Neumann measurement, moreover, can be performed by means of a single-qubit unitary transformation followed by a measurement in the computational basis. It is convenient to think of the required transformations as part of the quantum circuit and the final measurement on each qubit as being in the computational basis. In Fig. 6.18, the meter symbol at the end of each qubit line denotes such a von Neumann measurement, so each qubit can provide up to one bit of information.

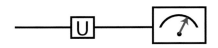

Fig. 6.19 Circuit realizing a general one-qubit von Neumann measurement.

Taken to its logical conclusion, we can view any quantum circuit, together with the measurements on the individual qubits, as a generalized measurement. At the simplest level, the circuit depicted in Fig. 6.19 represents a single-qubit von Neumann measurement in the basis $\hat{U}^\dagger|0\rangle$, $\hat{U}^\dagger|1\rangle$, with these states corresponding, respectively, to the measurement outcomes 0 and 1. The appearance of the conjugate of $\hat{U}$ might be unexpected but may be understood by considering the probabilities that the measuring device gives the results 0 and 1:

$$P(0) = \langle 0|\hat{U}\hat{\rho}\hat{U}^\dagger|0\rangle = \text{Tr}\left(\hat{\rho}\hat{U}^\dagger|0\rangle\langle 0|\hat{U}\right),$$

$$P(1) = \langle 1|\hat{U}\hat{\rho}\hat{U}^\dagger|1\rangle = \text{Tr}\left(\hat{\rho}\hat{U}^\dagger|1\rangle\langle 1|\hat{U}\right), \quad (6.33)$$

where $\hat{\rho}$ is the state of the qubit at the input to the circuit.

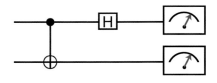

Fig. 6.20 A circuit realizing a Bell measurement.

A simple two-qubit circuit for realizing a measurement in the Bell-state basis is given in Fig. 6.20. The most straightforward way to determine the measurement realized by such a circuit is to start with

he states $|00\rangle$, $|01\rangle$, $|10\rangle$ and $|11\rangle$ corresponding to the possible mea-
·urement outcomes and to work *backwards* through the circuit to the
nput. Following this procedure, we find

$$|00\rangle \rightarrow \frac{1}{\sqrt{2}}\left(|00\rangle + |10\rangle\right) \rightarrow \frac{1}{\sqrt{2}}\left(|00\rangle + |11\rangle\right) = |\Phi^+\rangle,$$

$$|01\rangle \rightarrow \frac{1}{\sqrt{2}}\left(|01\rangle + |11\rangle\right) \rightarrow \frac{1}{\sqrt{2}}\left(|01\rangle + |10\rangle\right) = |\Psi^+\rangle,$$

$$|10\rangle \rightarrow \frac{1}{\sqrt{2}}\left(-|10\rangle + |00\rangle\right) \rightarrow \frac{1}{\sqrt{2}}\left(-|11\rangle + |00\rangle\right) = |\Phi^-\rangle,$$

$$|11\rangle \rightarrow \frac{1}{\sqrt{2}}\left(-|11\rangle + |01\rangle\right) \rightarrow \frac{1}{\sqrt{2}}\left(-|10\rangle + |01\rangle\right) = |\Psi^-\rangle, \quad (6.34)$$

o that the four possible results 00, 01, 10, and 11 correspond to mea-
urements of the two qubits in the Bell basis.

If the input qubits include ancillas, prepared in chosen states, then
he quantum circuit can perform a generalized measurement. Consider,
s an example, the set of probability operators

$$\hat{\pi}_0 = \frac{1}{2}|0\rangle\langle 0|, \qquad \hat{\pi}_1 = \frac{1}{2}|1\rangle\langle 1|,$$

$$\hat{\pi}_{0'} = \frac{1}{2}|0'\rangle\langle 0'|, \qquad \hat{\pi}_{1'} = \frac{1}{2}|1'\rangle\langle 1'|, \quad (6.35)$$

vhere the states $|0'\rangle$ and $|1'\rangle$ are the eigenstates of $\hat{\sigma}_x$ given in eqn 6.19.
Ve recall that the four states $|0\rangle$, $|1\rangle$, $|0'\rangle$, and $|1'\rangle$ are those used in the
3B84 protocol, described in Section 3.4, and either Bob or Eve might
nvisage using this generalized measurement. In the circuit depicted in
`ig. 6.21, the first (top) qubit is the one to be measured and the second,
·repared in the state $|0\rangle$, is the ancillary qubit. It is straightforward to
how that the probabilities for each of the possible measurement results
·re

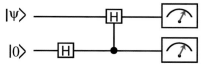

Fig. 6.21 Circuit realizing a generalized measurement for eavesdropping on
BB84.

$$P(0,0) = \frac{1}{2}|\langle\psi|0\rangle|^2 = \langle\psi|\hat{\pi}_0|\psi\rangle,$$

$$P(1,0) = \frac{1}{2}|\langle\psi|1\rangle|^2 = \langle\psi|\hat{\pi}_1|\psi\rangle,$$

$$P(0,1) = \frac{1}{2}|\langle\psi|0'\rangle|^2 = \langle\psi|\hat{\pi}_{0'}|\psi\rangle,$$

$$P(1,1) = \frac{1}{2}|\langle\psi|1'\rangle|^2 = \langle\psi|\hat{\pi}_{1'}|\psi\rangle, \quad (6.36)$$

orresponding to the desired generalized measurement.

We can realize quantum operations on a set of qubits by introducing
ncillary qubits but not measuring them. These qubits are prepared in
·elected states, induced to interact with the system of interest, and then
iscarded. This act of discarding qubits, as opposed to measuring them,
; represented in a quantum circuit diagram by a dustbin. As a simple
xample, consider the transformation of a single qubit of the form

$$\hat{\rho} \rightarrow (1-p)\hat{\rho} + p\hat{\sigma}_x\hat{\rho}\hat{\sigma}_x. \quad (6.37)$$

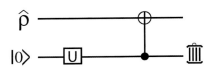

$$\hat{u} = \begin{pmatrix} \sqrt{1-p} & \sqrt{p} \\ -\sqrt{p} & \sqrt{1-p} \end{pmatrix}$$

Fig. 6.22 A quantum circuit in which a qubit is discarded rather than measured.

The operation is realized by the circuit given in Fig. 6.22. If we apply the two gates in turn to the input state then we generate the transformation

$$\hat{\rho} \otimes |0\rangle\langle 0| \rightarrow \hat{\rho} \otimes \left(\sqrt{1-p}|0\rangle + \sqrt{p}|1\rangle\right)\left(\sqrt{1-p}\langle 0| + \sqrt{p}\langle 1|\right)$$
$$\rightarrow (1-p)\hat{\rho} \otimes |0\rangle\langle 0| + \sqrt{p(1-p)}\left(\hat{\sigma}_x\hat{\rho} \otimes |1\rangle\langle 0| + \hat{\rho}\hat{\sigma}_x \otimes |0\rangle\langle 1|\right)$$
$$+ p\hat{\sigma}_x\hat{\rho}\hat{\sigma}_x \otimes |1\rangle\langle 1|. \tag{6.38}$$

Discarding the ancillary qubit amounts to taking the trace over its states to leave the reduced density operator for the first qubit. This procedure leaves only the terms $(1-p)\hat{\rho}$ and $p\hat{\sigma}_x\hat{\rho}\hat{\sigma}_x$, so giving the transformation in eqn 6.37. Naturally, we can also describe the effect of any circuit of this form as an operation on our qubit. Consider, for example, the three-qubit circuit in Fig. 6.23. Applying the gates to the input state $\hat{\rho} \otimes |0\rangle\langle 0| \otimes |0\rangle\langle 0|$ and then tracing out the second and third qubits produces the transformation

$$\hat{\rho} \rightarrow (1-p)(1-q)\hat{\rho} + p\hat{\sigma}_x\hat{\rho}\hat{\sigma}_x + q(1-p)\hat{\sigma}_z\hat{\rho}\hat{\sigma}_z. \tag{6.39}$$

This is an operation of the general form of eqn 4.74,

$$\hat{\rho} \rightarrow \sum_{i=0}^{2} \hat{A}_i\hat{\rho}\hat{A}_i^{\dagger}, \tag{6.40}$$

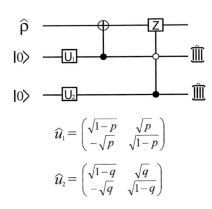

$$\hat{u}_1 = \begin{pmatrix} \sqrt{1-p} & \sqrt{p} \\ -\sqrt{p} & \sqrt{1-p} \end{pmatrix}$$

$$\hat{u}_2 = \begin{pmatrix} \sqrt{1-q} & \sqrt{q} \\ -\sqrt{q} & \sqrt{1-q} \end{pmatrix}$$

Fig. 6.23 A three-qubit circuit as a quantum operation on a single qubit.

with the three effect operators $\hat{A}_0 = [(1-p)(1-q)]^{1/2}\hat{I}$, $\hat{A}_1 = p^{1/2}\hat{\sigma}_x$, and $\hat{A}_2 = [q(1-p)]^{1/2}\hat{\sigma}_z$.

It is sometimes useful to represent a quantum operation or protocol as a quantum circuit. This situation is reminiscent of the equivalent circuits in electrical and electronic engineering, in which the operation of key aspects of a device is expressed in terms of a simpler circuit. As an example, we consider the teleportation of the state of a qubit as described in Section 5.5. The protocol is presented as a circuit in Fig. 6.24, in which the first two qubits are understood to be in Alice's domain and the last in Bob's. We recall that the input state can be written in the form (see eqn 5.45)

$$|\psi\rangle \otimes |\Psi^-\rangle = \frac{1}{2}\left[-|\Psi^-\rangle \otimes |\psi\rangle - |\Psi^+\rangle \otimes \hat{\sigma}_z|\psi\rangle\right.$$
$$\left. +|\Phi^-\rangle \otimes \hat{\sigma}_x|\psi\rangle - i|\Phi^+\rangle \otimes \hat{\sigma}_y|\psi\rangle\right], \tag{6.41}$$

so that a Bell-state measurement carried out on the first two qubits leaves the last in one of the states $|\psi\rangle$, $\hat{\sigma}_z|\psi\rangle$, $\hat{\sigma}_x|\psi\rangle$, and $\hat{\sigma}_y|\psi\rangle$. The CNOT and Hadamard gates followed by measurement in the computational basis constitutes, as we have seen, a Bell-state measurement. We associate the four possible measurement results 00, 01, 10, and 11 with the four Bell states as in eqn 6.34. This means that the state of Bob's qubit, given the measurement result, is

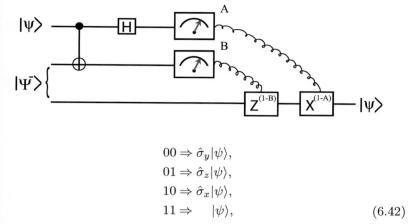

Fig. 6.24 A quantum circuit for teleportation.

$$00 \Rightarrow \hat{\sigma}_y|\psi\rangle,$$
$$01 \Rightarrow \hat{\sigma}_z|\psi\rangle,$$
$$10 \Rightarrow \hat{\sigma}_x|\psi\rangle,$$
$$11 \Rightarrow \quad |\psi\rangle, \tag{6.42}$$

where we have omitted the unimportant global phase factor. It only remains for Alice to send the measurement results to Bob so that he can apply the required unitary transformation and recover the state $|\psi\rangle$. In the circuit diagram, this classical communication is represented by a wire carrying a signal from Alice's measurements to the gates Z and X. These are applied if A and B, respectively, are 0. This is indicated on the circuit diagram by writing the gates as $Z^{(1-B)}$ and $X^{(1-A)}$. If both the Z and X gates operate then the result, of course, is the gate iY, which affects the required transformation if the Bell measurement gives the result 00, or $|\Phi^+\rangle$.

Whether or not the Z and X gates act, in quantum teleportation, is determined by the results of measurements performed on Alice's two qubits and hence by the pre-measurement state of the qubits. It follows, therefore, that the same outcome can be achieved by the use of controlled-Z and CNOT gates as represented in Fig. 6.25. Here the controlled gates act on the last qubit only if the associated control qubit is in the state $|0\rangle$. The measurements performed on the first and second qubits serve only to reveal the transformation applied to the final qubit. The preparation of the third qubit in the state $|\psi\rangle$ does not require the measurements to be performed on the first two qubits and the outcome would be the same if the meter symbols were replaced by dust-bins. A pair of quantum circuits may be equivalent in this sense but, naturally, one version may be easier to implement or be more practical. The teleportation circuit in Fig. 6.24, for example, requires only local operations and classical communications rather than the two controlled gates required for the circuit in Fig. 6.25.

6.4 Quantum error correction

One of the central ideas in our subject is that all information-processing devices are imperfect and their operation leads, inevitably, to errors. Rather than trying to eradicate all possible sources of error, Shannon's noisy-channel coding theorem tells us how to combat errors in a classical

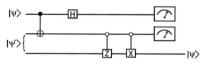

Fig. 6.25 The controlled-gate version of the teleportation circuit.

Recycling qubits If the physical qubits are a valuable resource then it is desirable to be able to reuse them rather than discarding them after use. We can achieve this simply by measuring them in the computational basis and applying an X gate if they are found to be in the state $|1\rangle$. This procedure resets the qubits to the state $|0\rangle$.

Errors in quantum information processing   A simple qualitative calculation serves to illustrate the importance of error correction. Let us suppose that single-qubit errors occur at a rate R and that the time required for one gate operation is T. The probability that no such error occurs is simply $\exp(-RT)$. If our processor has n qubits then the probability that no error occurs in time T is simply $\exp(-nRT)$. A general algorithm might require every qubit to interact with each of the others, and this may involve $\mathrm{O}(n^2)$ steps, to give a zero-error probability of the order of $\exp(-n^3RT)$. By optimizing the arrangement of the gates using, for example, ideas from fast Fourier transforms, we might reduce the $\mathrm{O}(n^2)$ to $\mathrm{O}(n\log n)$ to give a zero-error probability of the order of $\exp[-(n^2\log n)RT]$. Clearly, as n grows, this probability rapidly tends to zero.

communication channel by introducing redundancy. In quantum information, errors are more of a problem than they are in classical systems. The first reason for this is that quantum systems tend to be significantly smaller than their classical counterparts and are far more susceptible to environmental influences. Secondly, a classical bit has only two states, associated with the logical values 0 and 1, and so there is only one type of error, that which flips the bit value, $0 \leftrightarrow 1$. The state of a qubit, however, can be modified in a wide variety of ways by interaction with its environment. This environment is unmonitored and, to a large extent, uncontrolled and it follows that each qubit can be subjected to a variety of possible error-inducing operations. Finally, there are infinitely many possible quantum states (any qubit state of the form $\alpha|0\rangle + \beta|1\rangle$ is allowed), so how are we to ensure that precisely the desired state is restored by any error-correcting protocol?

It is far from obvious that anything can be done to detect and correct quantum errors. To start with, we can only detect any errors by performing measurements, but the act of making these tends to result in a change of state, something we are trying to avoid. The no-cloning theorem proven in Section 3.2, moreover, establishes that we cannot make copies of the unknown state of a qubit. When these problems are combined with the wide variety of possible changes that can be induced and the continuous range of possible quantum states, the situation might appear hopeless. That it is not was established independently by Shor and Steane, who each proposed multiqubit states, or quantum codewords, for which, remarkably, arbitrary single-qubit errors can be detected and corrected.

Shor's and Steane's error-correcting protocols use nine or seven qubits, respectively, to protect a single logical qubit. There also exists a five-qubit codeword, and this is the smallest that allows us to correct an arbitrary single-qubit error. Before describing these larger codewords, it is instructive to consider a simpler, if less effective, protocol based on three qubits. We start by representing the logical qubit states $|0_{\ell 3}\rangle$ and $|1_{\ell 3}\rangle$ by the three-qubit states

$$|0_{\ell 3}\rangle = |000\rangle, \qquad |1_{\ell 3}\rangle = |111\rangle. \qquad (6.43)$$

If interaction with the environment causes one of the qubits to be flipped, $|0\rangle \leftrightarrow |1\rangle$, then we might use majority voting to detect this and to make the necessary correction by applying an X gate to the qubit. Superposition states of our logical qubit become

$$\alpha|0_{\ell 3}\rangle + \beta|1_{\ell 3}\rangle = \alpha|000\rangle + \beta|111\rangle. \qquad (6.44)$$

Were we to simply check for errors by measuring $\hat{\sigma}_z$ on each qubit, then the superposition would be destroyed. We can protect the state and also detect the presence of a bit flip by performing collective measurements on the qubits of the collective observables corresponding to the operators

$$\mathrm{ZZI} = \hat{\sigma}_z \otimes \hat{\sigma}_z \otimes \hat{\mathrm{I}},$$
$$\mathrm{IZZ} = \hat{\mathrm{I}} \otimes \hat{\sigma}_z \otimes \hat{\sigma}_z. \qquad (6.45)$$

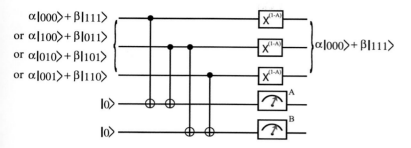

Fig. 6.26 A circuit for three-qubit error correction.

Here we have introduced a condensed, and hopefully self-explanatory, notation for such multiqubit operations. This will be especially useful when we consider states of larger numbers of qubits. Any desired state in the form of 6.44 is an eigenstate of both ZZI and IZZ with eigenvalues +1. If we flip any one of the three qubits then we change one or both of these eigenvalues. There are three possible states that can be produced by such an error, and these satisfy the eigenvalue equations

$$ZZI\,(\alpha|100\rangle + \beta|011\rangle) = -\,(\alpha|100\rangle + \beta|011\rangle),$$
$$IZZ\,(\alpha|100\rangle + \beta|011\rangle) = +\,(\alpha|100\rangle + \beta|011\rangle),$$
$$ZZI\,(\alpha|010\rangle + \beta|101\rangle) = -\,(\alpha|010\rangle + \beta|101\rangle),$$
$$IZZ\,(\alpha|010\rangle + \beta|101\rangle) = -\,(\alpha|010\rangle + \beta|101\rangle),$$
$$ZZI\,(\alpha|001\rangle + \beta|110\rangle) = +\,(\alpha|001\rangle + \beta|110\rangle).$$
$$IZZ\,(\alpha|001\rangle + \beta|110\rangle) = -\,(\alpha|001\rangle + \beta|110\rangle). \tag{6.46}$$

If we measure the two observables ZZI and IZZ and find the result +1 in both cases then we can be sure that no (simple bit-flip) error has occurred. If either or both of the results is −1 then we can correct the error by applying an X gate to the indicated qubit. For example, if we find that both measurements give the result −1 then eqn 6.46 implies that the second qubit has been flipped, so that the state has become $|010\rangle + \beta|101\rangle$. Applying an X gate to the second qubit, corresponding to acting with the operator $\hat{I} \otimes \hat{\sigma}_x \otimes \hat{I}$, restores the original state given in eqn 6.44.

This system of error correction can be represented, of course, as a quantum circuit, and this is given in Fig. 6.26. Five qubits are necessary, three to carry the logical state $\alpha|0_{\ell 3}\rangle + \beta|1_{\ell 3}\rangle$ and two to provide the outcomes of the required measurements. The error is corrected by acting on the identified qubit with an X gate. This is indicated in the circuit diagram by an exponent: $\hat{X}^0 = \hat{I}$, $\hat{X}^1 = \hat{X}$.

The three-qubit quantum codewords do not allow us to detect and correct the most general single-qubit errors. Suppose, for example, that the phase of the first qubit is shifted so that $|0\rangle \to |0\rangle$ and $|1\rangle \to -|1\rangle$. The resulting effect on our state in eqn 6.44 is to change it to $\alpha|000\rangle - \beta|111\rangle$, but this error is not detected in our error-correcting protocol. It is highly desirable, even essential, to be able to correct arbitrary errors. This can be achieved for single-qubit errors using five-, seven-, or nine-qubit codewords. We shall treat only the seven-qubit Steane code in any detail. A brief presentation of the nine- and five-qubit protocols is given in Appendix M.

The Steane code represents the logical qubit states $|0_{\ell 7}\rangle$ and $|1_{\ell 7}\rangle$ by entangled states of seven qubits in the form

$$
\begin{aligned}
|0_{\ell 7}\rangle = 2^{-3/2} \, (&|0000000\rangle + |1010101\rangle + |0110011\rangle \\
&+ |1100110\rangle + |0001111\rangle + |1011010\rangle \\
&+ |0111100\rangle + |1101001\rangle) \, , \\
|1_{\ell 7}\rangle = 2^{-3/2} \, (&|1111111\rangle + |0101010\rangle + |1001100\rangle \\
&+ |0011001\rangle + |1110000\rangle + |0100101\rangle \\
&+ |1000011\rangle + |0010110\rangle) \, .
\end{aligned}
\tag{6.47}
$$

We note that the 16 kets superposed to form these two quantum codewords embody bit sequences that differ from each other at not less than three places. As such, they correspond to the Hamming [7,4] code. The classical Hamming codes were, in fact, the inspiration for the Steane code. Our ability to detect single-qubit errors relies on the fact that the logical qubit states $|0_{\ell 7}\rangle$ and $|1_{\ell 7}\rangle$ are both eigenstates of each of the six mutually commuting multiqubit operators

$$
\begin{aligned}
\mathbf{J}_1 &= \text{IIIXXXX}, & \mathbf{K}_1 &= \text{IIIZZZZ}, \\
\mathbf{J}_2 &= \text{IXXIIXX}, & \mathbf{K}_2 &= \text{IZZIIZZ}, \\
\mathbf{J}_3 &= \text{XIXIXIX}, & \mathbf{K}_3 &= \text{ZIZIZIZ},
\end{aligned}
\tag{6.48}
$$

with eigenvalue $+1$. We can check this by direct calculation, but it is more elegant to rewrite the states using these six operators, to which we add

$$
\mathbf{I} = \text{IIIIIII}, \qquad \mathbf{X} = \text{XXXXXXX}.
\tag{6.49}
$$

When expressed in terms of these eight compatible operators, our logical states are

$$
\begin{aligned}
|0_{\ell 7}\rangle &= 2^{-3/2}(\mathbf{I} + \mathbf{J}_1)(\mathbf{I} + \mathbf{J}_2)(\mathbf{I} + \mathbf{J}_3)|0000000\rangle, \\
|1_{\ell 7}\rangle &= 2^{-3/2}(\mathbf{I} + \mathbf{J}_1)(\mathbf{I} + \mathbf{J}_2)(\mathbf{I} + \mathbf{J}_3)|1111111\rangle \\
&= 2^{-3/2}(\mathbf{I} + \mathbf{J}_1)(\mathbf{I} + \mathbf{J}_2)(\mathbf{I} + \mathbf{J}_3)\mathbf{X}|0000000\rangle \\
&= \quad \mathbf{X}|0_{\ell 7}\rangle.
\end{aligned}
\tag{6.50}
$$

The square of each of the $\mathbf{J}$ operators is $\mathbf{I}$, and hence $\mathbf{J}_i(\mathbf{I}+\mathbf{J}_i) = (\mathbf{I}+\mathbf{J}_i)$ $(i = 1, 2, 3)$. It follows immediately that

$$
\mathbf{J}_{1,2,3} \left(\alpha|0_{\ell 7}\rangle + \beta|1_{\ell 7}\rangle \right) = \left(\alpha|0_{\ell 7}\rangle + \beta|1_{\ell 7}\rangle \right).
\tag{6.51}
$$

Similarly, because the state $|0000000\rangle$ is an eigenstate of each of the three $\mathbf{K}$ operators with eigenvalue $+1$, we also have the eigenvalue equation

$$
\begin{aligned}
\mathbf{K}_{1,2,3} \left(\alpha|0_{\ell 7}\rangle + \beta|1_{\ell 7}\rangle \right) &= \mathbf{K}_{1,2,3} \left(\alpha\mathbf{I} + \beta\mathbf{X} \right)|0_{\ell 7}\rangle \\
&= \left(\alpha|0_{\ell 7}\rangle + \beta|1_{\ell 7}\rangle \right).
\end{aligned}
\tag{6.52}
$$

Any single-qubit error will change at least one of the six measurement results to -1, and we can use the pattern of these to diagnose and then correct the error. Consider, for example, the effect of an undesired single-qubit unitary transformation on one of the qubits comprising our codeword. We showed in Section 2.4 that such a general single-qubit unitary operator can be written in the form

$$\hat{U} = \exp\left(i\gamma\hat{I} + i\delta\vec{a}\cdot\hat{\vec{\sigma}}\right)$$
$$= e^{i\gamma}\left(\cos\delta\,\hat{I} + i\sin\delta\,\vec{a}\cdot\hat{\vec{\sigma}}\right), \qquad (6.53)$$

where γ and δ are real and $\vec{a}$ is a unit vector. Let us suppose that a transformation of this form acts on the third of our qubits so that the logical state $|\psi_{\ell 7}\rangle = \alpha|0_{\ell 7}\rangle + \beta|1_{\ell 7}\rangle$ becomes

$$\hat{U}_3|\psi_{\ell 7}\rangle = e^{i\gamma}\left[\cos\delta\,\hat{I}_3 + i\sin\delta\left(a_x\hat{X}_3 + a_y\hat{Y}_3 + a_z\hat{Z}_3\right)\right]|\psi_{\ell 7}\rangle. \quad (6.54)$$

The states $|\psi_{\ell 7}\rangle$, $\hat{X}_3|\psi_{\ell 7}\rangle$, $\hat{Y}_3|\psi_{\ell 7}\rangle$ and $\hat{Z}_3|\psi_{\ell 7}\rangle$ are all eigenstates of the six operators in eqn 6.48. We can see this very straightforwardly by first recalling that $|\psi_{\ell 7}\rangle$ is an eigenstate of each of these operators with eigenvalue $+1$ and by noting that the operators $\hat{X}_3$, $\hat{Y}_3$, and $\hat{Z}_3$ commute or anticommute with each of the six $\mathbf{J}$ and $\mathbf{K}$ operators. For example, $\hat{X}_3$ commutes with the operators $\mathbf{J}_1$, $\mathbf{J}_2$, $\mathbf{J}_3$, and $\mathbf{K}_1$, but anticommutes with $\mathbf{K}_2$ and $\mathbf{K}_3$. It follows that $\hat{X}_3|\psi_{\ell 7}\rangle$ is an eigenstate of $\mathbf{J}_1$, $\mathbf{J}_2$, $\mathbf{J}_3$ and $\mathbf{K}_1$ with eigenvalue $+1$ but that it is an eigenstate of $\mathbf{K}_2$ and $\mathbf{K}_3$ with eigenvalue -1:

$$\mathbf{J}_1\hat{X}_3|\psi_{\ell 7}\rangle = \hat{X}_3\mathbf{J}_1|\psi_{\ell 7}\rangle = +\hat{X}_3|\psi_{\ell 7}\rangle,$$
$$\mathbf{J}_2\hat{X}_3|\psi_{\ell 7}\rangle = \hat{X}_3\mathbf{J}_2|\psi_{\ell 7}\rangle = +\hat{X}_3|\psi_{\ell 7}\rangle,$$
$$\mathbf{J}_3\hat{X}_3|\psi_{\ell 7}\rangle = \hat{X}_3\mathbf{J}_3|\psi_{\ell 7}\rangle = +\hat{X}_3|\psi_{\ell 7}\rangle,$$
$$\mathbf{K}_1\hat{X}_3|\psi_{\ell 7}\rangle = \hat{X}_3\mathbf{K}_1|\psi_{\ell 7}\rangle = +\hat{X}_3|\psi_{\ell 7}\rangle,$$
$$\mathbf{K}_2\hat{X}_3|\psi_{\ell 7}\rangle = -\hat{X}_3\mathbf{K}_2|\psi_{\ell 7}\rangle = -\hat{X}_3|\psi_{\ell 7}\rangle,$$
$$\mathbf{K}_3\hat{X}_3|\psi_{\ell 7}\rangle = -\hat{X}_3\mathbf{K}_3|\psi_{\ell 7}\rangle = -\hat{X}_3|\psi_{\ell 7}\rangle. \quad (6.55)$$

This pattern of measurement outcomes unambiguously determines that the state has been transformed by action of the operator $\hat{X}_3$. The error has been detected, and can be corrected by acting on the third qubit with an X gate. A transformation of the form of eqn 6.54 will generate no detected errors with probability $\cos^2\delta$ and an error will be generated by $\hat{X}_3$, $\hat{Y}_3$, or $\hat{Z}_3$ with respective probabilities $a_x^2\sin^2\delta$, $a_y^2\sin^2\delta$, and $a_z^2\sin^2\delta$. Each of these possibilities has its own unique signature in the error-detection protocol:

$$(\mathbf{J}_1,\mathbf{J}_2,\mathbf{J}_3,\mathbf{K}_1,\mathbf{K}_2,\mathbf{K}_3) = (+1,+1,+1,+1,+1,+1) \Rightarrow |\psi_{\ell 7}\rangle,$$
$$(\mathbf{J}_1,\mathbf{J}_2,\mathbf{J}_3,\mathbf{K}_1,\mathbf{K}_2,\mathbf{K}_3) = (+1,+1,+1,+1,-1,-1) \Rightarrow \hat{X}_3|\psi_{\ell 7}\rangle,$$
$$(\mathbf{J}_1,\mathbf{J}_2,\mathbf{J}_3,\mathbf{K}_1,\mathbf{K}_2,\mathbf{K}_3) = (+1,-1,-1,+1,-1,-1) \Rightarrow \hat{Y}_3|\psi_{\ell 7}\rangle,$$
$$(\mathbf{J}_1,\mathbf{J}_2,\mathbf{J}_3,\mathbf{K}_1,\mathbf{K}_2,\mathbf{K}_3) = (+1,-1,-1,+1,+1,+1) \Rightarrow \hat{Z}_3|\psi_{\ell 7}\rangle.$$
$$(6.56)$$

In each case, the pattern of measurement results reveals a unique and readily correctable error. All single-qubit errors are detectable and correctable in this way. This is true also for more general single-qubit operations on the initial state.

Naturally, a quantum circuit can be designed to perform error detection and correction for a logical qubit protected using the Steane code. Six measurements need to be performed, each giving two possible outcomes; it follows that a suitable circuit requires six ancillary qubits in addition to the seven comprising the logical qubit. The state of each of the ancillary qubits needs to be modified according to the properties of the seven qubits forming the quantum codeword. This is readily achieved by a suitable arrangement of CNOT and controlled-Z gates.

This short discussion of quantum error correction does not adequately represent the full sophistication of the topic but has been written, rather, to illustrate the important ideas. Two remaining issues deserve to be mentioned: these are how to account for errors in the error-detection and correction process itself, and what level of errors can be tolerated in the operation of a quantum information processor. Addressing these will be an important step in the development of any practical device.

6.5   Cluster states

The model for information processing presented so far is one in which a required multiqubit unitary transformation is constructed, step by step, using gates. We start with a set of qubits, often prepared in the product state $|0\rangle \otimes \cdots \otimes |0\rangle$, and the gates act to produce an entangled state. The final step is information extraction by measurement of the individual qubits. A dramatically different approach, proposed by Raussendorf and Briegel, is based on the idea that we might start with an initially entangled state of a specific form (a cluster state) and coerce a subset of the qubits into the required state by single-qubit measurements followed by single-qubit unitary transformations. The idea is best appreciated by examples.

We start by constructing the cluster states; to do so, it is convenient to picture the qubits as being arranged on a grid. A cluster state is prepared by the action of a number of controlled-Z gates acting on selected nearest neighbours. Each of the qubits is first prepared in the state $|0'\rangle = 2^{-1/2}(|0\rangle + |1\rangle)$, and then a controlled-Z gate is symmetrical with respect to exchange of the control and target qubits and it is not necessary to distinguish between them. Figure 6.27 represents a selection of some of the simplest cluster states. Qubits that are coupled by the action of a controlled-Z gate are joined by a straight line. It is straightforward to show, following this algorithm, that the states corresponding to the first four diagrams in Fig. 6.27 are

Graph states   A more general arrangement of qubits and controlled-Z gates is also possible.   If we allow any of the qubits to be connected by controlled-Z gates (and not just nearest neighbours) then the result is known as a graph state.

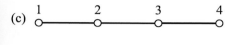 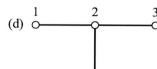

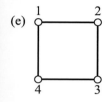

Fig. 6.27 The five simplest cluster states.

$$|\mathcal{C}_2\rangle = \frac{1}{\sqrt{2}}\left(|0\rangle_1|0'\rangle_2 + |1\rangle_1|1'\rangle_2\right),$$

$$|\mathcal{C}_3\rangle = \frac{1}{\sqrt{2}}\left(|0'\rangle_1|0\rangle_2|0'\rangle_3 + |1'\rangle_1|1\rangle_2|1'\rangle_3\right),$$

$$|\mathcal{C}_4^-\rangle = \frac{1}{2}\left(|0'\rangle_1|0\rangle_2|0'\rangle_3|0\rangle_4 + |1'\rangle_1|1\rangle_2|1'\rangle_3|1\rangle_4\right.$$
$$\left.+|1'\rangle_1|1\rangle_2|1'\rangle_3|0\rangle_4 + |1'\rangle_1|1\rangle_2|0'\rangle_3|1\rangle_4\right),$$

$$|\mathcal{C}_4^\top\rangle = \frac{1}{\sqrt{2}}\left(|0'\rangle_1|0\rangle_2|0'\rangle_3|0'\rangle_4 + |1'\rangle_1|1\rangle_2|1'\rangle_3|1'\rangle_4\right). \quad (6.57)$$

These are all highly entangled states. A simple, but by no means complete, indication of this is the fact that for each qubit in each of these pure states the reduced density operator is $\hat{I}/2$.

Given a sufficiently complicated cluster state, we can produce any desired multiqubit state. We shall not attempt to prove this but, rather, shall simply demonstrate the principle. The simplest cluster state, $|\mathcal{C}_2\rangle$, can be used to prepare any desired single-qubit pure state $|\psi\rangle$. One way is to measure the second qubit in the basis $|0'\rangle$, $|1'\rangle$ and then apply the indicated unitary operator to the first qubit:

$$|0'\rangle_2 \Rightarrow \hat{U} = |\psi\rangle_{11}\langle 0| + |\psi^\perp\rangle_{11}\langle 1|,$$
$$|1'\rangle_2 \Rightarrow \hat{U} = |\psi^\perp\rangle_{11}\langle 0| + |\psi\rangle_{11}\langle 1|. \quad (6.58)$$

As a second example, let us suppose that we wish to prepare the state $|\Phi^+\rangle_{12}$ given the cluster state $|\mathcal{C}_3\rangle$. Measuring the third qubit in the computational basis will leave the remaining two qubits in one of two different maximally entangled states

$$|0\rangle_3 = \frac{1}{\sqrt{2}}\left(|0'\rangle_1|0\rangle_2 + |1'\rangle_1|1\rangle_2\right),$$
$$|1\rangle_3 = \frac{1}{\sqrt{2}}\left(|0'\rangle_1|0\rangle_2 - |1'\rangle_1|1\rangle_2\right). \quad (6.59)$$

In the first case we can recover the required state by applying a Hadamard gate to the first qubit. For the second outcome we can again apply a Hadamard gate to the first qubit, followed by a Pauli-Z gate.

Cluster-state quantum information processing produces the required state by a sequence of single-qubit measurements and subsequent single-qubit unitary transformations. The state of the unmeasured qubits, following a suitably defined sequence of such operations, will be equivalent to the output of a properly constructed quantum circuit. It then only remains to perform measurements on each qubit to complete the quantum information-processing task.

Suggestions for further reading

Ekert, A., Hayden, P., Inamori, H., and Oi, D. K. L. (2001). What is quantum computation? *International Journal of Modern Physics A* **16**, 3335.

Kaye, P., Laflamme, R., and Mosca, M. (2007). *An introduction to quantum computing*. Oxford University Press, Oxford.

Lo, H.-K., Popescu, S., and Spiller, T. (eds) (1998). *Introduction to quantum computation and information*. World Scientific, Singapore.

Macchiavello, C., Palma, G. M., and Zeilinger, A. (eds) (2000). *Quantum computation and quantum information theory*. World Scientific, Singapore.

Mermin, N. D. (2007). *Quantum computer science: an introduction*. Cambridge University Press, Cambridge.

Nielsen, M. A. and Chuang, I. L. (2000). *Quantum computation and quantum information*. Cambridge University Press, Cambridge.

Raussendorf, R., Browne, D. E., and Briegel, H. J. (2003). Measurement- based quantum computation on cluster states *Physical Review A* **68**, 022312.

Smith, R. J. (1983). *Circuits, devices and systems* (4th edn). Wiley, New York.

Steane, A. M. (1998). Quantum computing. *Reports on Progress in Physics* **61**, 117.

Stenholm, S. and Suominen, K.-A. (2005). *Quantum approach to informatics*. Wiley, Hoboken, NJ.

Vedral, V. (2006). *Introduction to quantum information science*. Oxford University Press, Oxford.

Exercises

6.1) Prove the following theorems of Boolean algebra:

    (a) $A + A = A$;

    (b) $A + \overline{A} = 1$;

    (c) $A \cdot A = A$;

    (d) $A \cdot \overline{A} = 0$.

6.2) Use the properties of Boolean algebra to simplify the following expressions:

    (a) $A \cdot B \cdot C + A \cdot B \cdot \overline{C}$;

    (b) $A \cdot (\overline{A} + B)$;

    (c) $(A + C) \cdot (A + D) \cdot (B + C) \cdot (B + D)$;

    (d) $(A + B) \cdot (\overline{A} + C)$.

6.3) Prove the two absorption rules given in eqn 6.5.

6.4) Prove De Morgan's two theorems given in eqns 6.6 and 6.7. In each case, draw the equivalent patterns of gates.

6.5) Construct an OR gate using only NAND gates.

6.6) Draw logic circuits to realize the following Boolean functions:

    (a) $\overline{A + B} + A \cdot B + \overline{A + B}$;

    (b) $(\overline{A \cdot B} + A \cdot B) \cdot \overline{A + B}$;

    (c) $\overline{A + B} \cdot \overline{A \cdot B} + A \cdot B$.

6.7) Construct a sequence of H and T gates to produce each of the other one-qubit transformations in Fig. 6.8.

6.8) If the Pauli-X gate is the quantum NOT gate then is there a quantum 'square root of NOT gate'?

6.9) Confirm that the three operators in eqn 6.11 form an acceptable operation and confirm that this operation transforms the density operator in eqn 6.12 into that in eqn 6.13.

6.10) Calculate the density operator for the state produced by the action of the optimal universal NOT operation on *one* of the two qubits prepared in the Bell state $|\Psi^-\rangle$. What is the resulting state if the universal NOT operation is applied to both qubits?

6.11) Which of the following two-qubit transformations are unitary?

    (a) $|A\rangle \otimes |B\rangle \rightarrow |B\rangle \otimes |\overline{A}\rangle$;

    (b) $|A\rangle \otimes |B\rangle \rightarrow |\overline{A}\rangle \otimes |A\rangle$;

    (c) $|A\rangle \otimes |B\rangle \rightarrow |A \oplus B\rangle \otimes |A\rangle$;

    (d) $|A\rangle \otimes |B\rangle \rightarrow 2^{-1/2}\left(|A\rangle \otimes |B\rangle + |\overline{A}\rangle \otimes |\overline{B}\rangle\right)$;

    (e) $|A\rangle \otimes |B\rangle \rightarrow (-1)^{A+B}|A\rangle \otimes |B\rangle$.

6.12) Is there an operation which implements the transformation in eqn 6.15? If there is, write down a suitable set of effect operators.

6.13) In the figure depicting the half-adder (Fig. 6.4), each (voltage) bit is passed in parallel into more than one gate. Is a similar operation possible for qubits?

6.14) This question is about the design of a quantum circuit that realizes the half-adder operation. We require a quantum circuit such that two of the output qubits give the sum and carry for the first two qubits.

    (a) Show that no two-qubit circuit can realize the required operation.

    (b) Design a three-qubit circuit half-adder.

6.15) Confirm the form of the transformation performed by the CNOT gate on eigenstates of $\hat{\sigma}_x$ as given in eqn 6.20. What is the form of the resulting state if a CNOT gate acts on a pair of qubits both of which have been prepared in eigenstates of $\hat{\sigma}_y$?

6.16) Calculate the state generated by a CNOT gate acting on each of the states in which the control qubit is an eigenstate of $\hat{\sigma}_x$ and the target qubit is an eigenstate of $\hat{\sigma}_z$.

6.17) Determine the most general form of an initially unentangled state of a control and a target qubit such that a CNOT gate generates a maximally entangled state.

6.18) Find the most general form of the controlled-unitary operation for which the resulting state is independent of which qubit is labelled as the control and which as the target.

6.19) Consider the quantum circuit diagram in Fig. 6.13. Calculate the state produced by the gates for the input states

    (a) $|0000\rangle$;

    (b) $\hat{H} \otimes \hat{I} \otimes \hat{H} \otimes \hat{I}|0000\rangle$;

    (c) $\hat{H} \otimes \hat{H} \otimes \hat{H} \otimes \hat{H}|0000\rangle$.

6.20) What is the most general controlled-unitary gate that can be produced using single-qubit gates and just one CNOT gate?

(6.21) Construct the simplest possible realization of a controlled-unitary gate in which the unitary operator is $e^{i\alpha}\hat{I}$.

(6.22) Design a quantum circuit using only single-qubit gates and CNOT gates that realizes a Toffoli gate.

(6.23) Design a quantum circuit, using only CNOT gates and single-qubit gates, to realize the root-swap transformation

$$\hat{U}_{swap}^{1/2}|00\rangle = |00\rangle,$$

$$\hat{U}_{swap}^{1/2}|01\rangle = \frac{1}{\sqrt{2}}\left(|01\rangle + |10\rangle\right),$$

$$\hat{U}_{swap}^{1/2}|10\rangle = \frac{1}{\sqrt{2}}\left(|01\rangle - |10\rangle\right),$$

$$\hat{U}_{swap}^{1/2}|11\rangle = |11\rangle.$$

(6.24) Which of the following sets of gates is universal?

   (a) Single-qubit gates and CS gates.
   (b) Single-qubit gates and swap gates.
   (c) Single-qubit gates and CZ gates.

   [*Hint:* you might try using these combinations of gates to construct a CNOT gate.]

(6.25) Design a three-qubit circuit, with measurements in the computational basis, for discriminating between the eight three-qubit states

$$\frac{1}{\sqrt{2}}\left(|0, A, B\rangle \pm |1, \overline{A}, \overline{B}\rangle\right),$$

   where A and B take the values 0 and 1.

(6.26) The quantum circuit depicted in Fig. 6.21 includes a controlled-Hadamard gate. Is it possible to construct such a gate using only one CNOT gate and the single-qubit gates given in Fig. 6.8?

(6.27) Design a two-qubit circuit to implement optimal unambiguous discrimination between the two equiprobable single-qubit states

$$|\psi_1\rangle = \cos\theta|0\rangle + \sin\theta|1\rangle,$$
$$|\psi_2\rangle = \cos\theta|0\rangle - \sin\theta|1\rangle.$$

   Note that the circuit will have four possible measurement outcomes, but we require only three: 'the state was $|\psi_1\rangle$', 'the state was $|\psi_2\rangle$', and 'I don't know'. This means that one of the four possible measurement results for the circuit will be redundant.

(6.28) Construct a circuit equivalent to that in Fig. 6.23 using only CNOT gates and single-qubit gates.

(6.29) Construct a quantum circuit realizing the optimal universal NOT operation given by the transformation in eqn 6.13.

(6.30) In Figure 6.28 (below) $\hat{U} = \vec{a}\cdot\hat{\vec{\sigma}}$ is the operator corresponding to any component of spin, determined by the unit vector $\vec{a}$. What is the function of the circuit?

(6.31) Show that the measurements in any quantum circuit can always be moved to the end of the circuit by introducing suitable controlled gates.

(6.32) A qubit state is protected by using an n-qubit quantum codeword. Let the probability that any one of the component qubits produces an error be p. Calculate the probability that *at most* one error occurs. Show that this probability differs from unity by terms of order p^2.

(6.33) A quantum information-processing device admits only single-qubit errors of the (rather contrived) form: $|0\rangle \rightarrow |1\rangle$, $|1\rangle \rightarrow -|0\rangle$. Construct suitable three-qubit quantum codewords and a set of suitable observables with which to detect errors.

(6.34) Prepare a circuit diagram equivalent to that given in Fig. 6.26 for three-qubit error correction but using additional controlled gates in place of measurements on the two ancillary qubits.

(6.35) Construct a table like eqn 6.56 but extended to detect the states $|\psi_{\ell7}\rangle$, $\hat{X}_i|\psi_{\ell7}\rangle$, $\hat{Y}_i|\psi_{\ell7}\rangle$, and $\hat{Z}_i|\psi_{\ell7}\rangle$ $(i = 1, \cdots, 7)$. Hence show that the results of measuring the six $\mathbf{J}$ and $\mathbf{K}$ uniquely discriminate between these 22 states.

(6.36) Suppose that our quantum state $|\psi_{\ell7}\rangle$ is subjected to a dramatic interaction such that all information concerning its sixth qubit is lost. This corresponds to changing the state to

$$|\psi_{\ell7}\rangle\langle\psi_{\ell7}| = \hat{\rho}_{\ell7}$$
$$\rightarrow \frac{1}{4}\left(\hat{\rho}_{\ell7} + \hat{X}_6\hat{\rho}_{\ell7}\hat{X}_6\right.$$
$$\left. + \hat{Y}_6\hat{\rho}_{\ell7}\hat{Y}_6 + \hat{Z}_6\hat{\rho}_{\ell7}\hat{Z}_6\right).$$

   Show that even this dramatic single-qubit error can be corrected.

(6.37) Design a quantum circuit for implementing error detection and correction for the Steane code.

(6.38) Calculate the state $|\mathcal{C}_4^{\square}\rangle$. What is the effect on the state of the remaining qubits of measuring on any one of the qubits the observable corresponding to (a) $\hat{\sigma}_z$, (b) $\hat{\sigma}_z$, or (c) $\hat{\sigma}_y$?

(6.39) Show how the state $(\sqrt{3}|0\rangle_1|0\rangle_2 + |1\rangle_1|1\rangle_2)/2$ can be prepared from the cluster state $|\mathcal{C}_3\rangle$.

(6.40) Show how the state $|\text{GHZ}\rangle$ can be constructed by measuring any one of the qubits in the cluster state $|\mathcal{C}_4^\top\rangle$.

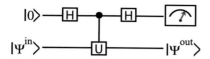

Fig. 6.28 Figure for Exercise (6.29).

Quantum computation

In the preceding chapter we established that a suitable set of quantum gates, complemented by quantum error correction, allows us to produce a desired multiqubit unitary transformation. This transformation is one of the three steps in a quantum computation; the others, of course, are the preparation of the qubits in their initial state and the measurement of them after the transformation has been implemented.

A quantum computation is designed to solve a problem or class of problems. The power of quantum computers is that they can do this, at least for some problems, very much more efficiently and quickly than any conventional computer based on classical logic operations. If we can build a quantum computer then a number of important problems which are currently intractable will become solvable.

The potential for greatly enhanced computational power is, in itself, reason enough to study quantum computers, but there is another. Moore's law is the observation that the number of transistors on a chip doubles roughly every eighteen months. A simple corollary is that computer performance also doubles on the same timescale. Associated with this exponential improvement is a dramatic reduction in the size of individual components. If the pace is to be kept up then it is inevitable that quantum effects will become increasingly important and ultimately will limit the operation of the computer. In these circumstances it is sensible to consider the possibility of harnessing quantum effects to realize quantum information processors and computers.

7.1 Elements of computer science

We start with a brief introduction to the theory of computer science, the principles of which underlie the operation of what we shall refer to as classical computers. These include all existing machines and any based on the manipulation of classical bits.

The development of computer science owes much to Turing, who devised a simple but powerful model of a computing device: the Turing machine. It its most elementary form, this consists of four elements. (i) A *tape* for data storage, which acts as a memory. This tape has a sequence of spaces, each of which has on it one of a finite set of symbols. (ii) A *processor*, which controls the operations of the machine. The processor is characterized by a finite number of internal states. (iii) A finite *instruction set*, which determines the action of the processor, depending on the tape symbol and on the internal state of the processor. (iv) A

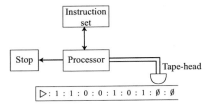

Fig. 7.1 Schematic representation of a Turing machine.

tape head, which can read a symbol on the tape and, if instructed to do so by the processor, erase the symbol and replace it with another one. Also included is the ability to move the head along the tape, either to the right or left, if instructed to do so. A schematic representation of these four elements and their arrangement is given in Fig. 7.1. The tape includes a start marker on one of its spaces, and the processor has an initial, or start, configuration and also a stop configuration. When this is reached, the computation is complete and the processor reports this fact to the user. At this stage, the required value should be on (a section of) the tape.

A simple example will serve to illustrate the operation of a Turing machine. We let each of the spaces on the tape carry one of the four symbols $\{\triangleright, 0, 1, \emptyset\}$; the symbol $\triangleright$ is the start marker, the digits 0 and 1 comprise the input data, and the symbol $\emptyset$ denotes a blank space. Our processor has four possible states $\{S, I, II, F\}$, of which S is the initial configuration and F the final or stop state. The model is completed by an instruction set, which we select to be

$$
\begin{aligned}
(S, \triangleright) &\Rightarrow (I, \triangleright), \\
(I, 0) &\Rightarrow (II, 0), \\
(I, 1) &\Rightarrow (II, 1), \\
(II, 0) &\Rightarrow (F, 1), \\
(II, 1) &\Rightarrow (II, 0), \\
(II, \emptyset) &\Rightarrow (I, 1), \\
(I, \emptyset) &\Rightarrow (F, \emptyset).
\end{aligned}
\tag{7.1}
$$

All of the instructions also include a move to the next space (to the right) on the tape.

The purpose of this simple program is revealed by working through some examples. Suppose, first, that the tape is prepared in the state

$$
\boxed{\triangleright \ : \ 1 \ : \ 0 \ : \ 1 \ : \ \emptyset \ : \ \emptyset \ : \ \cdots}.
$$

The steps in the program, remembering to move on one square after each instruction, are

$$
(S, \triangleright) \Rightarrow (I, 1), \quad (I, 1) \Rightarrow (II, 1), \quad (II, 0) \Rightarrow (F, 1),
\tag{7.2}
$$

at which point the tape configuration is

$$
\boxed{\triangleright \ : \ 1 \ : \ 1 \ : \ 1 \ : \ \emptyset \ : \ \emptyset \ : \ \cdots}.
$$

As a second example, let us suppose that the tape is prepared in the state

$$
\boxed{\triangleright \ : \ 0 \ : \ 1 \ : \ 1 \ : \ \emptyset \ : \ \emptyset \ : \ \cdots}.
$$

In this case the steps in the program are

$$(S, \triangleright) \Rightarrow (I, \triangleright),$$
$$(I, 0) \Rightarrow (II, 0),$$
$$(II, 1) \Rightarrow (II, 0),$$
$$(II, 1) \Rightarrow (II, 0),$$
$$(II, \emptyset) \Rightarrow (I, 1),$$
$$(I, \emptyset) \Rightarrow (F, \emptyset), \tag{7.3}$$

t which point the tape shows

| $\triangleright$ | : | 0 | : | 0 | : | 0 | : | 1 | : | $\emptyset$ | : | $\cdots$ |

f we read the tape data from the right then our programme executes he mapping

$$101 \Rightarrow 111,$$
$$110 \Rightarrow 1000, \tag{7.4}$$

which corresponds to adding two (or 10) to the number on the tape.

The theoretical significance of Turing machines derives from the fact hat, despite their simplicity, they encapsulate the idea of an algorithm a set of procedures for evaluating a function). Indeed, it was established y Turing and by Church that the functions which can be evaluated by Turing machine correspond precisely to those that can be calculated y a (classical) computer. This means that studying Turing machines llows us to make strong statements about the operation of all possible omputers. By providing a systematic but general description of an algorithm, moreover, the Turing machine has also had an impact on the evelopment of mathematics. The most celebrated example is Hilbert's *Intscheidungsproblem*: Is there an algorithm with which we can determine whether or not any given mathematical statement is true? Turing urned this into a computing problem by asking if we can find a function which can be evaluated on a Turing machine, the value of which tells s whether or not any given program will terminate. The fact that no uch function exists is the so-called halting problem and established that here is, in general, no algorithm of the type sought by Hilbert.

For practical purposes, it is at least as important to determine whether roblems can be solved efficiently as it is to know whether they can be olved in principle. The key idea in addressing this question is the way a which the resources required to solve any given problem scale with the ze of the input data. It is reasonable to expect that adding together a air of ten-bit numbers will be quicker and take less memory (or space) han adding hundred-bit numbers. An efficient algorithm will be one hat requires resources which increase only slowly with the size of the nput data; an inefficient one will require resources which grow rapidly ith the size of the input data. Suppose, for example, that we have two roblems and have designed algorithms to tackle them. For the first we nd that the time required to run the algorithm with an n-bit input tring is, for large n,

$$T_1(n) = a_1 n^2, \tag{7.5}$$

Probabilistic algorithms Theoretical computer scientists recognize the possibility of Turing machines modified by the addition of a random element, such as coin tossing. There exist problems which can be solved efficiently on such machines but for which no corresponding efficient and deterministic algorithm is known.

and for the second,

$$T_2(n) = a_1 2^n, \tag{7.6}$$

The time taken will, in general, be a complicated function of n. In writing these equations, we are writing upper bounds on the time taken for strings with n bits. A more quantitative discussion of this is given in Appendix N.

where a_1 and a_2 are constants. These constants may be of very different magnitude, but for sufficiently large values of n the second algorithm will always require the greater time to perform and is, in this sense, less efficient. Let us suppose that our problem is at the very limit of what is possible using our computer but that we need to solve the problem for an $n + 1$-bit input. The additional time required for the first algorithm is

$$T_1(n + 1) - T_1(n) \approx 2a_1 n, \tag{7.7}$$

which, for large n, is very small compared with $T_1(n)$. For our second problem, however, adding one more bit *doubles* the time required. This exponential dependence on the number of bits to be processed means that with increasing n, the algorithm rapidly becomes impractical.

It is clear that problems requiring a time that is polynomial in n are, in the sense of scaling with large n, *easier* than those for which the time depends exponentially on n. We use this idea to define the polynomial class of problems **P** as the set of problems for which the required time, for large n, is proportional to n^k for some integer k and so is polynomial in n. (Typically, k is a small integer, 1, 2 or 3.) Simple examples include addition and multiplication of two binary numbers. If these numbers both have n bits then addition requires time $T(n) \propto n$ and multiplication, using the most familiar algorithm learnt at school, requires $T(n) \propto n^2$.

Most efficient algorithms   The most obvious algorithm is not always the most efficient. It is important, of course, to find, where possible, the fastest algorithm. Only then can we make a decisive statement about the difficulty of a mathematical problem. We see in Appendix N, for example, that it is possible to perform multiplication in time $T(n) \propto n^{\log 3}$ or even better.

A more general and difficult class of problems is the non-deterministic polynomial, or **NP**, class. These are problems for which no efficient (class **P**) algorithm is known but for which the solution, once found, can be verified as correct in polynomial time. A useful analogy is the search for a needle in a stack of hay; the search is hard but, once the needle has been found, it is easy to discriminate between the needle and a blade of hay. A simple example of great importance is finding the two prime factors of a large product, the difficulty of which, as we saw in Section 3.1, underlies the security of the RSA cryptosystem. The obvious way to find the two prime factors of an n-bit number N is to try dividing it by each of the numbers up to $\sqrt{N}$, which is a number of divisions that is exponential in n ($\approx 2^{n-1}$). Once the solution has been found, of course, a single division is all that is required to verfiy that the solution found is, indeed, a factor of N. It is intriguing to note, however, that there is no proof that $\mathbf{P} \neq \mathbf{NP}$ or, to put it another way, it is possible that problems such as factoring can be solved in polynomial time but that we have not yet found an efficient algorithm.

NP   Algorithms for the efficient solution of **NP** problems have been proposed but require the ability to follow, in parallel, very large numbers of logical paths. It is not possible to run these on conventional classical computers. Theoretical computer scientists recognize several hundred complexity classes in addition to **P** and **NP**. These include classes that depend not just on computing time but also on space, and which address a wide variety of types of problem.

An important problem for our subject is the efficient simulation of quantum systems. Let us suppose that we wish to study the evolution of a system with a basis of 2^n states, such as a collection of n interacting qubits. Keeping track of the state vector requires us to store 2^n probability amplitudes or $2(2^n - 1)$ independent real numbers. A general Hamiltonian will have the form of a $2^n \times 2^n$ Hermitian matrix, with 2^{2n}

ndependent real parameters. Storing the state vector requires a space
hat is exponential in n and calculating the time evolution requires, in
he absence of any helpful symmetry or other simplification, an expo-
ential number of computational steps and so a time that is exponential
1 n. For such problems, it clearly makes sense to use a quantum sys-
em, which naturally evolves linearly in time, to simulate the system of
nterest. The natural way to model complex quantum systems is to use
 quantum computer.

7.2 Principles of quantum computation

 quantum information processor, or quantum computer, has a number
f elements in common with a Turing machine and also some important
ifferences. In place of the tape, we have a string of qubits; these can
e prepared in any initial state, usually an unentangled pure state, and
ogether comprise the input data to our processor. The processor acts
n the qubits following an instruction set encoded in the arrangement
f its quantum gates. Finally, the information is extracted at the end
f the process by measuring the state of each qubit in a suitable basis,
sually the computational basis of $|0\rangle$ and $|1\rangle$. The processor can induce
ny desired unitary transformation on the qubit string. Our quantum
omputer differs from a Turing machine in two important ways. First,
ne input qubits can be prepared not only in the computational basis,
ut also in any superposition state. Secondly, a Turing machine proceeds
y a deterministic sequence of classical operations; this means that we
ould, at any stage, stop its operation, examine it, and then make it
ontinue its task. For a quantum information processor, of course, any
ich intervention would modify the state of our qubits and it would then
ot be possible to resume the processing operation.

Our quantum computation is composed of three parts. (i) A finite
ollection, or string, of *qubits*, the initial state of which encodes the
nput data. Each qubit has only two orthogonal states, so there is no
ossibility of including start or blank symbols. (ii) An arrangement
f *quantum gates*, designed so as to perform a preselected multiqubit
nitary transformation on the input state of the qubits. (iii) Finally,
e perform a *measurement* on the output state of the individual qubits,
sually in the computational basis. This measurement process should
eveal, at least with a sufficiently large probability, the required result.

It is convenient to use a shortened notation for the states of our qubit
tring. We do this by first labelling the basis states, in the computational
asis, by the associated binary string. For example, the five-qubit state
$|0\rangle \otimes |1\rangle \otimes |1\rangle \otimes |0\rangle \otimes |0\rangle$ is denoted $|01100\rangle$ and is used to represent the
inary number 1100, or twelve. We simplify this further by denoting the
tate by the binary number encoded, so that the six-qubit state $|000100\rangle$
 written as $|100\rangle$. This means, of course, that there is no longer an
xplicit reference in the state label to the number of qubits. Where this
night cause confusion, we shall use a more explicit representation of

Note that this means that our string is
labelled in the opposite sense, or order
of bits, to that used for the tape in the
Turing machine of the previous section.

the state. In particular, the state $|0\rangle$ might denote a single qubit or a string of n qubits, each prepared in the single-qubit state $|0\rangle$. Where an explicit representation of the n-qubit state is required, we write

$$|0\rangle^{\otimes n} = \underbrace{|0\rangle \otimes \cdots \otimes |0\rangle}_{n \text{ terms}}. \tag{7.8}$$

Ideally, our quantum information processor would enact the unitary transformation

$$|a\rangle \rightarrow \hat{U}|a\rangle = |f(a)\rangle, \tag{7.9}$$

where a is any desired binary number (up to $2^n - 1$, where n is the number of qubits in the input string) and $f(a)$ is any Boolean function of a (where $0 \leq f(a) \leq 2^n - 1$). Measuring each qubit would then reveal the desired value $f(a)$. We saw in Section 3.2, however, that this process is not allowed for all possible functions. The reason is that a unitary transformation necessarily preserves the overlap between any pair of states. If our function has the same value for two distinct strings a_1 and a_2 then

$$|\langle f(a_2)|f(a_1)\rangle| = 1, \tag{7.10}$$

but the states $|a_1\rangle$ and $|a_2\rangle$ correspond to distinct binary numbers and so are orthogonal:

$$\langle a_2|a_1\rangle = 0 \Rightarrow \left(\langle a_2|\hat{U}^\dagger\right)\left(\hat{U}|a_1\rangle\right) = 0. \tag{7.11}$$

It follows that

$$\hat{U}|a\rangle \neq |f(a)\rangle, \tag{7.12}$$

for at least some values of a.

In order to be able to compute *any* function using a quantum processor, we introduce a second qubit string, prepared in a state $|b\rangle$. The general arrangement is depicted in Fig. 7.2. Our quantum processor is then designed so as to perform the unitary transformation

$$|a\rangle \otimes |b\rangle \rightarrow \hat{U}_f|a\rangle \otimes |b\rangle = |a\rangle \otimes |b \oplus f(a)\rangle, \tag{7.13}$$

where $b \oplus f(a)$ represents a string each bit of which is determined by the modulo 2 addition of the corresponding bits in the strings b and $f(a)$. The states $|a_1\rangle \otimes |b \oplus f(a_1)\rangle$ and $|a_2\rangle \otimes |b \oplus f(a_2)\rangle$ are now orthogonal even if $f(a_1) = f(a_2)$. If we choose $b = 0$, of course, then measuring the final state of the second string of qubits in the computational basis reveals, directly, the required function $f(a)$. The second string need not have the same number of qubits as the first, and so we can accommodate problems for which $f(a)$ and a are strings with different numbers of bits.

The power of a quantum computer derives largely from the fact that we can input not just a state corresponding to a single number a, but a *superposition* of many. Indeed, we can prepare an equally weighted superposition of all of these states by starting with each qubit in the

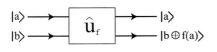

Fig. 7.2 Schematic representation of a quantum information processor.

tate $|0\rangle$ and applying to each of these a Hadamard gate so that

$$|0\rangle^{\otimes n} \rightarrow \hat{H}^{\otimes n}|0\rangle^{\otimes n}$$
$$= 2^{-n/2}\left(|0\rangle + |1\rangle\right) \otimes \left(|0\rangle + |1\rangle\right) \otimes \cdots \otimes \left(|0\rangle + |1\rangle\right)$$
$$= 2^{-n/2} \sum_{a=0}^{2^n-1} |a\rangle. \tag{7.14}$$

The single quantum processor then calculates simultaneously the values of $f(a)$ for all a in the sense that states corresponding to all of these values are present in the transformed state. For n qubits in the first string we generate, as depicted in Fig. 7.3, a highly entangled superposition of n product states:

$$2^{-n/2} \sum_{a=0}^{2^n-1} |a\rangle \otimes |0\rangle \rightarrow 2^{-n/2} \sum_{a=0}^{2^n-1} |a\rangle \otimes |f(a)\rangle. \tag{7.15}$$

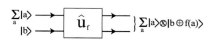

Fig. 7.3 A quantum information processor with an entangled output.

The form of this state indicates the origin of the increased computing speed possible with a quantum computer. Performing the required unitary transformation with an array of quantum gates requires a time that is polynomial in n. The prepared state, however, contains a superposition of 2^n computed values, so our processor has performed an exponential (in n) number of calculations in a polynomial time. This suggests that we might be able to tackle problems in the **NP** class. We can expect to find, at least for some problems, an exponential decrease in computing time when using a quantum computer instead of a classical device.

The first explicit demonstration of a computational task which could be performed faster on a quantum computer than on any classical machine was Deutsch's algorithm. The problem to be solved may seem contrived, but the problem itself is not the purpose of the algorithm. It was devised solely to illustrate the potential efficiency gains provided by a quantum computer. Consider the four possible one-bit functions which map the values $\{0,1\}$ onto $\{0,1\}$. There are two constant functions,

$$f(0) = 0, \qquad f(1) = 0,$$
$$\text{and} \quad f(0) = 1, \qquad f(1) = 1, \tag{7.16}$$

and also two 'balanced' functions (balanced in the sense that the calculated values, 0 and 1, occur equally often),

$$f(0) = 0, \qquad f(1) = 1,$$
$$\text{and} \quad f(0) = 1, \qquad f(1) = 0. \tag{7.17}$$

Let us suppose that we have a 'black box' or 'oracle', the operation of which is to calculate one of these functions. The internal workings of the oracle are hidden from us and we can only run the algorithm by inputting a 0 or a 1. Our task is to determine whether the function is constant or balanced.

It is obvious that a classical computation can answer this question only by addressing the oracle twice, so as to evaluate both $f(0)$ and $f(1)$. If we have only $f(0)$ or $f(1)$ then we have no information about whether the function is constant or balanced. The constant or balanced nature of the function is intrinsically a property of both $f(0)$ *and* $f(1)$. If our oracle is a quantum processor, with qubit inputs, then we can solve this problem in just a *single* run.

For this simple problem, each of the qubit strings in Fig. 7.2 is just a single qubit. If we input the state $|A\rangle \otimes |B\rangle$ (with A and B taking the values 0 and 1) then the oracle performs the transformation

$$|A\rangle \otimes |B\rangle \rightarrow |A\rangle \otimes |B \oplus f(A)\rangle. \tag{7.18}$$

If our second qubit is prepared in the superposition state $2^{-1/2}(|0\rangle - |1\rangle)$ then our oracle leaves the input state unchanged apart from a global change of sign:

$$|A\rangle \otimes \frac{1}{\sqrt{2}}(|0\rangle - |1\rangle) \rightarrow (-1)^{f(A)}|A\rangle \otimes \frac{1}{\sqrt{2}}(|0\rangle - |1\rangle). \tag{7.19}$$

The overall sign is unobservable for these states, but we can make use of it by preparing a superposition state for the first qubit so that

$$\frac{1}{2}(|0\rangle + |1\rangle) \otimes (|0\rangle - |1\rangle) \rightarrow \frac{1}{2}\left[(-1)^{f(0)}|0\rangle + (-1)^{f(1)}|1\rangle\right] \otimes (|0\rangle - |1\rangle). \tag{7.20}$$

The state of the second qubit remains unchanged but that of the first contains the answer to our question. If f is a constant function then we find $\pm 2^{-1/2}(|0\rangle + |1\rangle)$, but if it is balanced then it is left in the orthogonal state $\pm 2^{-1/2}(|0\rangle - |1\rangle)$. We can now determine readily whether the function is constant or balanced by measuring $\hat{\sigma}_x$ for our first qubit or, equivalently, by applying to it a Hadamard gate and then making a measurement in the computational basis.

The power of this quantum algorithm becomes apparent if we consider an extension of it, the Deutsch–Jozsa algorithm, to a function of n bits. In this case our input is an n-bit number a and our function is again a single bit, either 0 or 1. The function is either constant, $f(a) = 0$ or $f(a) = 1$, or it is balanced, in that it returns the value 0 for exactly half of the 2^n input strings and 1 for the remaining inputs. Our task is to design an algorithm which determines *with certainty* whether f is constant or balanced. A classical solution to this problem can only proceed by giving the oracle a sequence of different numbers $a_1, a_2, \cdots, a_m$ to process, and in each case it will return the value $f(a_i)$. As soon as we have two different output values then the process can stop, as we know that the function is balanced. If, however, we find that our output values are all the same then we do not know whether the function is constant or balanced. As the sequence of similar values increases in length we become more confident that the function is constant, but to know for certain, we must try more than half of the possible input values, or $2^{n-1} + 1$. Only then does a sequence of output values, all of which are

he same, imply that the function is balanced. Thus our test requires,
n the worst-case scenario, a number of trials that is exponential in n.

A quantum processor allows us to solve the Deutsch–Jozsa problem in
single run. The algorithm is a simple generalization of that for a single
qubit. We prepare our first string in an equally weighted superposition
of all the states $|a\rangle$ and our second, single-qubit, string in the state
$2^{-1/2}(|0\rangle - |1\rangle)$:

$$2^{-(n+1)/2}(|0\rangle + |1\rangle)^{\otimes n} \otimes (|0\rangle - |1\rangle) = 2^{-(n+1)/2}\sum_{a=0}^{2^n-1}|a\rangle \otimes (|0\rangle - |1\rangle).$$

(7.21)

he action of the quantum processor transforms this into the state

$$\hat{U}_f 2^{-(n+1)/2}\sum_{a=0}^{2^n-1}|a\rangle \otimes (|0\rangle - |1\rangle) = 2^{-(n+1)/2}\sum_{a=0}^{2^n-1}(-1)^{f(a)}|a\rangle \otimes (|0\rangle - |1\rangle).$$

(7.22)

f the function is constant then the first n qubits remain in the state

$$2^{-n/2}\sum_{a=0}^{2^n-1}|a\rangle = \hat{H}^{\otimes n}|0\rangle,$$

(7.23)

ut if the function is balanced then the state will be orthogonal to this
ne. Measuring $\hat{\sigma}_x$ for each of the first n qubits, or applying a Hadamard
ate to each and then measuring in the computational basis, then suffices
o determine whether the function is constant or balanced.

In solving the problem in a single shot, the Deutsch–Jozsa algorithm
rovides a clear example of exponential speed-up. We have a prob-
em for which the classical algorithm is exponential in n (taking up to
$2^{n-1} + 1$ trials) but for which the quantum algorithm takes only a sin-
le trial, whatever the value of n. The Deutsch–Jozsa algorithm works
ecause we can encode the required information in the phases of a set of
quantum states. This information becomes readable by again exploiting
he superposition principle, or quantum interference. A further exam-
le of such an algorithm, due to Bernstein and Vazirani, is described in
Appendix O.

We conclude this section by describing Simon's algorithm, a method
or determining the period of an unknown function in far fewer steps
han can be achieved classically. Suppose that we have an oracle which
alculates a function $f(a)$ of an n-qubit input a. This function has the
roperty that its value is periodic under bitwise modulo 2 addition so
hat

$$f(a \oplus b) = f(a)$$

(7.24)

or all a, and that the values of f for all other a are different. Our
ask is to determine the value of the non-zero n-bit number b using the
inimum number of queries of the oracle.

To find b on a classical computer, all we can do is to provide a sequence
f values of a, list the computed values of $f(a)$, and keep going until we

We can state these conditions more formally as

$$f(a_1) = f(a_2)$$

if and only if $a_1 = a_2$ or $a_1 = a_2 \oplus b$.

find two identical computed values. If these are $f(a_1)$ and $f(a_2)$ then it is a simple matter to find b:

$$b = a_1 \oplus a_2. \tag{7.25}$$

Picking, at random, values of a to input requires a number of trials proportional to $2^{n/2}$ in order for there to be a significant probability of finding a pair of numbers with the same computed value of f. We can do a little better than this, but the required number of trials remains exponential in n.

Simon's algorithm requires a number of computations that is only linear in n. We start, in the same manner as for the Deutsch–Jozsa algorithm, by preparing a first n-qubit register in a superposition of all the states $|a\rangle$. We add to this a second register in the form of a string of qubits each of which is prepared, in this case, in the state $|0\rangle$. Our oracle then performs the transformation

$$2^{-n/2} \sum_{a=0}^{2^n-1} |a\rangle \otimes |0\rangle \rightarrow 2^{-n/2} \sum_{a=0}^{2^n-1} |a\rangle \otimes |f(a)\rangle. \tag{7.26}$$

The states $|f(a)\rangle$ are all mutually orthogonal apart, that is, from those for which the values of a are related by eqn 7.25. We can use this property to rewrite the output state given by eqn 7.26 in the form

$$2^{-n/2} \sum_{a=0}^{2^n-1} |a\rangle \otimes |f(a)\rangle = 2^{-(n-1)/2} \sum_{f(a)} \frac{1}{\sqrt{2}} \left(|a\rangle + |a \oplus b\rangle \right) \otimes |f(a)\rangle,$$
$$\tag{7.27}$$

The right-hand side of eqn 7.27 is, of course, the Schmidt decomposition of the output state.

where the final sum runs over the 2^{n-1} distinct values of $f(a)$. If we make a measurement of the second register in the computational basis then we will find, with equal probability, any one of the values of $f(a)$. Let us suppose that such a measurement has been performed and has given the answer $f(a_1)$. It would then follow that the first register would be left in the superposition state

$$2^{-n/2} \sum_{a=0}^{2^n-1} |a\rangle \otimes |f(a)\rangle \rightarrow \frac{1}{\sqrt{2}} \left(|a_1\rangle + |a_1 \oplus b\rangle \right). \tag{7.28}$$

Performing a measurement in the computational basis on these qubits will give *one* of the two random numbers a_1 or $a_1 \oplus b$ and so does not help us. If we give up any information about a_1, however, then we can determine something about b, and we do this by first applying a Hadamard gate to each qubit and only then measuring each in the computational basis. The transformed state is

$$\hat{H}^{\otimes n} \frac{1}{\sqrt{2}} \left(|a_1\rangle + |a_1 \oplus b\rangle \right) = 2^{-(n+1)/2} \sum_{c=0}^{2^n-1} \left[(-1)^{a_1 \cdot c} + (-1)^{(a_1 \oplus b) \cdot c} \right] |c\rangle$$

$$= 2^{-(n+1)/2} \sum_{c=0}^{2^n-1} (-1)^{a_1 \cdot c} \left[1 + (-1)^{b \cdot c} \right] |c\rangle$$

$$= 2^{-(n-1)/2} \sum_{b \cdot c=0} (-1)^{a_1 \cdot c} |c\rangle, \tag{7.29}$$

where the final sum runs over all of the 2^{n-1} values of c for which $b \cdot c = 0$ mod 2. If we now perform a measurement in the computational basis then we find a single value of c for which $b \cdot c = 0$ mod 2 and this, unless $= 0$, provides some information about b. The number of computations needed to determine b in this way cannot be less than n, as each gives only a single bit of information and b is n bits in length. It may take more than n computations, as the information provided by the latest computed value of c may not be independent of the information already obtained from previously computed values. It may be shown, however, that the probability of obtaining the value of b in $n + m$ trials is greater than $1 - 2^{-(m+1)}$, which tends to unity exponentially in m. Only a small number of additional trials (independent of the value of n) will find b with a very high probability.

To see that our computation gives only one bit, we need only note that half of all the 2^n possible n-bit numbers b satisfy the equation $b \cdot c = 0$ mod 2. (The other half satisfy the equation $b \cdot c = 1$ mod 2.) It follows that learning a value of c eliminates half of the possible values of b and so provides one bit of information.

.3 The quantum Fourier transform

We have seen that we can represent the set of integers $0, 1, \cdots, N - 1$ by a quantum state $|a\rangle$, in which the binary digits forming the number are associated with the computational-basis states of a string of qubits. N is less than or equal to 2^n, then n qubits suffice for this purpose. is useful to define a basis which is conjugate to that formed by the ates $\{|a\rangle\}$. The easiest and most natural way to do this is by means of the quantum Fourier transform, in which the new basis states (labelled ith a tilde) are related to the computational basis by a discrete Fourier ansform:

$$|\tilde{b}\rangle = \frac{1}{\sqrt{N}} \sum_{a=0}^{N-1} \exp\left(i\frac{2\pi ab}{N}\right) |a\rangle. \quad (7.30)$$

We summarize the main properties of the discrete Fourier transform in ppendix P. The two bases are related by means of a unitary transformation $\hat{U}_{\text{QFT}}$, with matrix elements

$$\langle a|\hat{U}_{\text{QFT}}|b\rangle = \frac{1}{\sqrt{N}} \exp\left(i\frac{2\pi ab}{N}\right), \quad (7.31)$$

that

$$|\tilde{b}\rangle = \hat{U}_{\text{QFT}}|b\rangle. \quad (7.32)$$

is straightforward to verify that the operator $\hat{U}_{\text{QFT}}$ is indeed unitary, id it follows that we can invert the quantum Fourier transform by leans of the unitary operator $\hat{U}_{\text{QFT}}^{-1} = \hat{U}_{\text{QFT}}^{\dagger}$:

$$|b\rangle = \hat{U}_{\text{QFT}}^{\dagger}|\tilde{b}\rangle. \quad (7.33)$$

The quantum Fourier transform is a unitary transformation and it llows that an n-qubit quantum circuit can be designed to implement for any value of N, provided, of course, that $2^n \geq N$. The circuit is mplest, however, when $N = 2^n$ and the Fourier transform acts on the

Conjugate bases We can define a pair of canonically conjugate bases by imposing two requirements: (i) each of the basis states in one basis is an equally weighted (apart from phases) superposition of all of the states in the conjugate basis, and (ii) if we form an observable from the basis states, so that

$$\hat{A} = \sum_{a} a|a\rangle\langle a|,$$

then this operator generates a shift in the conjugate basis. This second condition means that for some constant κ,

$$\exp\left(i\kappa\hat{A}n\right)|\tilde{b}\rangle = |\widetilde{b + n} \bmod N\rangle.$$

The quantum Fourier transform naturally produces an appropriate conjugate basis. We note that this idea is important in quantum optics and in quantum mechanics, where it is used to introduce operators for optical or harmonic-oscillator phase and for the azimuthal angular coordinate conjugate to the z-component of orbital angular momentum.

entire state space of the n qubits:

$$|\tilde{b}\rangle = \hat{U}_{\text{QFT}}|b\rangle = 2^{-n/2} \sum_{a=0}^{2^n-1} \exp\left(i\frac{2\pi ab}{2^n}\right)|a\rangle. \qquad (7.34)$$

It is helpful to be able to write this state in terms of the n individual qubits, and in order to do this we note that if a is the number corresponding to the bit string $A_n A_{n-1} \cdots A_1$ then

$$\frac{a}{2^n} = \frac{A_n}{2} + \frac{A_{n-1}}{2^2} + \frac{A_{n-2}}{2^3} + \cdots + \frac{A_1}{2^n}. \qquad (7.35)$$

It follows that the exponential in our quantum Fourier transform given in eqn 7.34 can be written as a product of factors, one for each qubit:

$$\exp\left(i\frac{2\pi ab}{2^n}\right) = \exp\left(i\frac{2\pi A_n b}{2}\right) \exp\left(i\frac{2\pi A_{n-1} b}{2^2}\right) \cdots \exp\left(i\frac{2\pi A_1 b}{2^n}\right)$$

$$= \prod_{\ell=1}^{n} \exp\left(i\frac{\pi A_\ell b}{2^{n-\ell}}\right). \qquad (7.36)$$

The transformed states $|\tilde{b}\rangle$ are, therefore, product states of the form

$$|\tilde{b}\rangle = \frac{1}{2^{n/2}}\left[|0\rangle + \exp\left(i\pi b\right)|1\rangle\right] \otimes \left[|0\rangle + \exp\left(i\frac{\pi b}{2}\right)|1\rangle\right] \otimes$$

$$\cdots \otimes \left[|0\rangle + \exp\left(i\frac{\pi b}{2^{n-1}}\right)|1\rangle\right]. \qquad (7.37)$$

There is a simplification which can be made; we can use the fact that $e^{i2\pi m} = 1$ for all integers m to write

You will sometimes find a *binary point* used as the natural analogue of the more familiar decimal point. For example,

$$B_3 \cdot B_2 B_1 = \frac{B_3 B_2 B_1}{2^2}$$
$$= B_3 2^0 + B_2 2^{-1} + B_1 2^{-2}.$$

$$\exp\left(i\pi b\right) = \exp\left(i\pi B_1\right),$$

$$\exp\left(i\pi\frac{b}{2}\right) = \exp\left(i\pi\frac{B_2 B_1}{2}\right),$$

$$\exp\left(i\pi\frac{b}{2^2}\right) = \exp\left(i\pi\frac{B_3 B_2 B_1}{2}\right),$$

$$\vdots \qquad\qquad \vdots$$

$$\exp\left(i\pi\frac{b}{2^{n-1}}\right) = \exp\left(i\pi\frac{B_n B_{n-1}\cdots B_1}{2^{n-1}}\right), \qquad (7.38)$$

where the number b corresponds to the bit string $B_n B_{n-1}\cdots B_1$. It might seem that we need only to perform a suitable set of single-qubit transformations in order to perform the quantum Fourier transform, but this is not correct, for the reason that the phases in the superposition for each qubit depend on the value b encoded in the whole string of qubits.

The n-qubit quantum Fourier transform can be implemented using a sequence of Hadamard gates and controlled-phase gates, generalized to the required phases for each qubit. We denote these phase gates as R_k and define them by the unitary transformation

$$\hat{R}_k|0\rangle = |0\rangle,$$
$$\hat{R}_k|1\rangle = e^{i\pi/2^{k-1}}|1\rangle, \qquad (7.39)$$

$$|B_2\rangle -\boxed{H}-\boxed{R_2}-\times \quad \frac{1}{\sqrt{2}}\left(|0\rangle + e^{i\pi B_1}|1\rangle\right)$$

$$|B_1\rangle -\bullet-\boxed{H}-\times \quad \frac{1}{\sqrt{2}}\left(|0\rangle + e^{i\pi B_2/2}|2\rangle\right)$$

Fig. 7.4 An $N = 4$ quantum Fourier transform circuit.

r, in matrix form,

$$R_k = \begin{pmatrix} 1 & 0 \\ 0 & e^{i\pi/2^{k-1}} \end{pmatrix}. \tag{7.40}$$

Note that for $k = 1, 2, 3$ this gate becomes, respectively, the Pauli-Z, phase, and $\pi/8$ gates described in Section 6.2.

Figure 7.4 depicts a simple two-qubit circuit with which to implement the $N = 4$ quantum Fourier transform. It is instructive to consider the actions, on the input state $|B_2 B_1\rangle$, of each of the four gates in turn. The first Hadamard gate produces the transformation

$$|B_2\rangle \otimes |B_1\rangle \rightarrow \frac{1}{\sqrt{2}}\left(|0\rangle + (-1)^{B_2}|1\rangle\right) \otimes |B_1\rangle$$

$$= \frac{1}{\sqrt{2}}\left(|0\rangle + e^{i\pi B_2}|1\rangle\right) \otimes |B_1\rangle. \tag{7.41}$$

The controlled-phase gate corrects the phase of the superposition state of the first qubit depending on the state of the second:

$$\frac{1}{\sqrt{2}}\left(|0\rangle + e^{i\pi B_2}|1\rangle\right) \otimes |B_1\rangle \rightarrow \frac{1}{\sqrt{2}}\left(|0\rangle + e^{i\pi B_2}e^{i\pi B_1/2}|1\rangle\right) \otimes |B_1\rangle$$

$$= \frac{1}{\sqrt{2}}\left(|0\rangle + e^{i\pi B_2 B_1/2}|1\rangle\right) \otimes |B_1\rangle. \tag{7.42}$$

The second Hadamard gate prepares the second qubit in the required state:

$$\frac{1}{\sqrt{2}}\left(|0\rangle + e^{i\pi B_2 B_1/2}|1\rangle\right) \otimes |B_1\rangle \rightarrow$$

$$\frac{1}{2}\left(|0\rangle + e^{i\pi B_2 B_1/2}|1\rangle\right) \otimes \left(|0\rangle + e^{i\pi B_1}|1\rangle\right). \tag{7.43}$$

Finally, a swap gate puts the qubits in the correct order:

$$\frac{1}{2}\left(|0\rangle + e^{i\pi B_2 B_1/2}|1\rangle\right) \otimes \left(|0\rangle + e^{i\pi B_1}|1\rangle\right) \rightarrow$$

$$\left(|0\rangle + e^{i\pi B_1}|1\rangle\right) \frac{1}{2}\left(|0\rangle + e^{i\pi B_2 B_1/2}|1\rangle\right). \tag{7.44}$$

The last swap gate will not be necessary if we can perform the further transformations and measurements intended for qubit 1 on qubit 2 and those for qubit 2 on qubit 1.

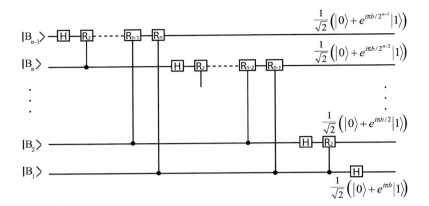

Fig. 7.5 A circuit diagram for the $N = 2^n$ quantum Fourier transform.

Extending the quantum Fourier transform circuit to n qubits and $N = 2^n$ is straightforward. Figure 7.5 depicts the required circuit, with the final set of swap gates omitted. The phase of the superposition for the final qubit in the transformed state given in eqn 7.37 depends on the logical values associated with all of the qubits and it follows, therefore, that this n-bit phase requires the action of single-qubit gates (the initial Hadamard gate) and $n - 1$ controlled-phase gates.

The most obvious feature of the quantum Fourier transform is that it performs the transformation between the computational basis $\{|a\rangle\}$ and the conjugate basis $\{|\tilde{a}\rangle\}$. This means that we can use the quantum Fourier transform to prepare states in the conjugate basis from a string of qubits prepared in the computational basis. We can also make measurements in the conjugate basis by performing an inverse quantum Fourier transform and then measuring each qubit in the computational basis. As an illustration, suppose that we are presented with a string of qubits prepared in the state

$$|\psi(\varphi)\rangle = \frac{1}{\sqrt{N}} \sum_{a=0}^{N-1} e^{ia\varphi} |a\rangle \qquad (7.45)$$

and are asked to determine, by means of a suitable measurement, the parameter φ. The first thing to notice is that the state $|\psi(\varphi)\rangle$ is periodic in φ with period 2π so that $|\psi(\varphi + 2\pi)\rangle = |\psi(\varphi)\rangle$. This means that any measurement can only determine the value of φ modulo 2π. The natural way to proceed is to make an effective measurement in the conjugate basis by performing an inverse quantum Fourier transform and then measuring each qubit in the computational basis. The probability that this gives the result b is

$$P(b) = \left| \langle \tilde{b} | \psi(\varphi) \rangle \right|^2$$
$$= \left| \langle b | \hat{U}_{\text{QFT}}^{\dagger} | \psi(\varphi) \rangle \right|^2$$
$$= \frac{1}{N^2} \left| \sum_{a=0}^{N-1} \exp\left[ia\left(\varphi - \frac{2\pi b}{N} \right) \right] \right|^2$$

$$= \frac{1}{N^2} \frac{\sin^2\left(\frac{1}{2}N\varphi\right)}{\sin^2\left[\frac{1}{2}\left(\varphi - 2\pi b/N\right)\right]}. \tag{7.46}$$

This takes the value unity for one value of b if φ is an integer multiple of $2\pi/N$. If this is not the case then the measurement will give one of the two integer multiples of $2\pi/N$ nearest to φ with a probability in excess of 0.81.

A Fourier transform provides the frequency components forming a given signal or function. An important and natural application, therefore, is determining the frequency of a periodic function. Let us suppose that we have a function of our input $a = 0, 1, \cdots, N-1$ which is periodic with period r. This means that

$$f(a + r \bmod N) = f(a) \tag{7.47}$$

for all a. Note that this condition can only hold if r is a factor of N. We shall also assume that f does not take any given value more than once in any period. If we attempt to solve this problem classically then we might proceed, in the absence of any further information about the function, to calculate values of f for different inputs a, until we find two similar values. If we start by calculating $f(0)$ and then $f(1), f(2), \cdots$, then this process will require r calculations. We might be lucky and find the period in fewer steps by guessing inputs, but on average we shall do no better than r, or $O(N)$, trials.

A quantum computer, as described in Section 7.2, can calculate the values of the function for all N values of a in parallel:

$$\frac{1}{\sqrt{N}} \sum_{a=0}^{N-1} |a\rangle \otimes |0\rangle \rightarrow \frac{1}{\sqrt{N}} \sum_{a=0}^{N-1} |a\rangle \otimes |f(a)\rangle. \tag{7.48}$$

As this state has encoded within it all of the values of $f(a)$, correlated with the associated inputs a, the required periodicity r is present in the state. If we measure the second register in the computational basis then we shall find, at random, any one of the allowed values of $f(a)$. If, for example, we find the value $f(a_0)$, then the first register is left in the superposition state

$$|\psi\rangle = \sqrt{\frac{r}{N}} \sum_{m=0}^{N/r-1} |a_0 + mr\rangle. \tag{7.49}$$

The value of a_0, which lies in the range 0 to $r-1$, has been generated at random by our measurement of the second register. It is clear, therefore, that measuring the state of the first register in the computational basis will produce only a random number in the range 0 to $N-1$ and so tells us nothing about the periodicity of f. We can find the required value of r by measuring the first register in the conjugate basis or, equivalently, by performing a quantum Fourier transform and then measuring the register in the computational basis. Applying the unitary quantum

Fourier transform produces the state

$$\hat{U}_{\mathrm{QFT}}|\psi\rangle = \sqrt{\frac{r}{N}} \sum_{m=0}^{N/r-1} \frac{1}{\sqrt{N}} \sum_{a=0}^{N-1} \exp\left(i2\pi \frac{a(a_0 + mr)}{N}\right)|a\rangle$$

$$= \frac{1}{\sqrt{r}} \sum_{\ell=0}^{r-1} \exp\left(i2\pi \frac{a_0 \ell}{r}\right) \left|\frac{\ell N}{r}\right\rangle, \qquad (7.50)$$

where we have used the summation in eqn P.12 from Appendix P. If we make a measurement in the computational basis then we shall obtain one of the values

$$b = \frac{\ell N}{r}, \qquad \ell = 0, 1, \cdots, r-1. \qquad (7.51)$$

Our measurement has given the value of b and we know, from the formulation of the problem, the value of N. Hence we can rearrange eqn 7.51 in the form

$$\frac{\ell}{r} = \frac{b}{N}. \qquad (7.52)$$

We can cancel the common factors in b and N and, if ℓ and r are relatively prime, this simplified fraction will be ℓ/r and hence give us the required periodicity r.

It is important, for practical applications, to be able to assess the efficiency of the quantum Fourier transform and to compare it with competing classical algorithms. The discrete Fourier transform, described in Appendix P, is most simply expressed as a multiplication by an $N \times N$ matrix in which each element is (apart from a factor $N^{-1/2}$) one of the Nth roots of unity. Performing the necessary N^2 multiplications and N summations suggests that the time required will scale like N^2 or 2^{2n}, where N is an n-bit number. In terms of computational complexity, described in Appendix N, this means that the time to perform a direct discrete Fourier transform is

$$T_{\mathrm{DFT}}(n) = O(N^2) = O\left(2^{2n}\right). \qquad (7.53)$$

This time is clearly exponential in the number of bits forming the input. A faster classical algorithm, which is in common use, is the fast Fourier transform, or FFT. It achieves its increased speed by breaking a required discrete Fourier transform of size $N = N_1 N_2$ into a number of smaller transforms, of sizes N_1 and N_2, together with $O(N)$ multiplications by complex roots of unity. This has the effect of reducing the time required to be proportional to $N \log N$ or $n2^n$ for large N. In the language of computational complexity, the required time is

$$T_{\mathrm{FFT}}(n) = O(N \log N) = O\left(2^{n+\log n}\right). \qquad (7.54)$$

This is dramatically faster than the time required for a more conventional Fourier transform, but it is still exponential in n. The quantum Fourier transform, as depicted in Fig. 7.5, requires a number of gates that scales as n^2 and it follows that the time taken is

$$T_{\mathrm{QFT}}(n) = O(n^2) = O\left(\log^2 N\right). \qquad (7.55)$$

For large r, the number of primes less than or equal to r tends to $r/\ln r$. The value of ℓ generated, therefore, will be prime with probability $(1/r) \times r/\ln r = 1/\ln r$ and will be relatively prime to r with at least this probability. It follows that $O(\log N)$ trials will suffice to obtain, unambiguously, the value of r.

This time grows only *polynomially* in n and so places the problem into the class **P**. If we have access to a suitably large quantum processor then it will be possible to perform Fourier transforms efficiently.

As an illustration of the power of the quantum Fourier transform, we note that it can find the period of a function in time $O(\log^3 N)$, made up of $O(\log^2 N)$ for the quantum Fourier transform and $O(\log N)$ trials. This should be compared with the $O(N)$ trials required for trial and error using a classical computer.

7.4 Shor's factoring algorithm

Undoubtedly the most famous quantum algorithm is that devised by Shor for factoring. Given an integer N, the task is to find a number m which divides it exactly. We recall that the RSA public-key cryptosystem, described in Section 3.1, relies for its security on the difficulty of performing precisely this task. Shor's algorithm, in providing an efficient method for factoring, presents a significant threat to public-key cryptosystems and it is this that has, perhaps more than anything else, sparked widespread interest in quantum computation and quantum information. Before discussing Shor's quantum algorithm, we should note that the best known classical algorithm is the general number field sieve. We shall not describe how this works, but need only note that factoring a large n-bit number N takes a time which scales as $\exp[c(\log N)^{1/3}(\log \log N)^{2/3}]$ for some constant c, so that

$$T_{\text{GNFS}}(n) = \exp\left[\Theta\left(n^{1/3}\log^{2/3}n\right)\right], \qquad (7.56)$$

which is clearly superpolynomial in n; it grows faster, for large numbers, than any power of n.

Shor's algorithm derives its efficiency from that of the quantum Fourier transform as a method for determining the period of a function. Before presenting the algorithm, however, we need to introduce some results from number theory. We start with a randomly selected integer, y, which is relatively prime (or coprime) to N. We can readily check that y is relatively prime to N using the Euclidean algorithm, described in Appendix E. If, by chance, y is not relatively prime to N then we shall have found a factor of N and performed the required task! Having selected a value for y, we form the function

$$f(a) = y^a \bmod N. \qquad (7.57)$$

For $a = 0$, of course, the function takes the value unity, and we seek the smallest subsequent value for which it is again unity:

$$f(r) = y^r \bmod N = 1. \qquad (7.58)$$

The value of r is the period of the function, after which the sequence of values repeats itself. Let us assume that we have determined the value

An indication of the significance of this is that Euler's theorem (eqn E.13) shows that

$$a^{\varphi(N)} \bmod N = 1,$$

where $\varphi(N)$ is Euler's φ-function. We recall that $\varphi(N)$ is used to prepare RSA private and public keys.

of r and see how this information allows us to determine a factor of N. We start by rewriting eqn 7.58 in the form

$$(y^r - 1) \bmod N = 0, \tag{7.59}$$

which tells us that $y^r - 1$ is an integer multiple of N. Next we factor $y^r - 1$ as the difference of two squares, to give

$$\left(y^{r/2} + 1\right)\left(y^{r/2} - 1\right) \bmod N = 0. \tag{7.60}$$

It follows that $y^{r/2} + 1$ and $y^{r/2} - 1$ are factors of λN:

$$\left(y^{r/2} + 1\right)\left(y^{r/2} - 1\right) = \lambda N, \tag{7.61}$$

for some integer λ. This result might not be useful to us if r is odd, as in this case our factors will probably not be integers. It will also not be useful if $y^{r/2} + 1$ or $y^{r/2} - 1$ is an integer multiple of N, as then neither $y^{r/2} + 1$ nor $y^{r/2} - 1$ provides any new information. Thankfully, it can be shown that the probability for either of these two unhelpful outcomes to occur is less than $1/2$. If either does occur then we need to choose a different value for y and start again. In all other cases, both $y^{r/2} + 1$ and $y^{r/2} - 1$ will have a non-trivial common divisor with N and we can find this efficiently using the Euclidean algorithm.

As an illustration, let us suppose that our number N is the product of precisely two primes p and q, so that

$$N = pq. \tag{7.62}$$

If $y^{r/2} + 1$ and $y^{r/2} - 1$ are both integers and neither one is an integer multiple of N, then it necessarily follows that one is an integer multiple of p and the other is an integer multiple of q:

$$\begin{aligned} y^{r/2} + 1 &= \lambda_p p, \\ y^{r/2} - 1 &= \lambda_q q. \end{aligned} \tag{7.63}$$

A search for the greatest common divisor of $y^{r/2} + 1$ and N will give the prime factor p, while a similar search using $y^{r/2} - 1$ yields q.

A very simple example will serve to demonstrate the technique. Let us attempt to factor 15 using the relatively prime values 2 and 11. For $y = 2$ we find the function

$$\begin{aligned} 2^0 \bmod 15 &= 1, \\ 2^1 \bmod 15 &= 2, \\ 2^2 \bmod 15 &= 4, \\ 2^3 \bmod 15 &= 8, \\ 2^4 \bmod 15 &= 1, \\ \vdots \quad \vdots \end{aligned} \tag{7.64}$$

o that the required period is $r = 4$. This is even and we find, in this ase,

$$y^{r/2} + 1 = 2^2 + 1 = 5,$$
$$y^{r/2} - 1 = 2^2 - 1 = 3, \tag{7.65}$$

which are the required factors. For $y = 11$ we find the function

$$11^0 \bmod 15 = 1,$$
$$11^1 \bmod 15 = 11,$$
$$11^2 \bmod 15 = 1,$$
$$\vdots \qquad \vdots \tag{7.66}$$

o that the period is $r = 2$. In this case we find

$$y^{r/2} + 1 = 11^1 + 1 = 12,$$
$$y^{r/2} - 1 = 11^1 - 1 = 10. \tag{7.67}$$

The greatest common divisor of 12 and 15 is 3, and $\gcd(10, 15) = 5$. Once again, the algorithm provides the required factors.

Each of the steps in the above factoring algorithm is simple and can be performed efficiently on an existing computer, with the single exception of finding the period r. It is the use of a quantum processor to perform his task that offers the prospect of efficient factoring and the associated hreat to the RSA cryptosystem.

The quantum Fourier transform is, as we saw in the preceding sec-on, very well suited to determining the period of a function. It is no surprise, therefore, that it plays a central role in Shor's algorithm. We cart by preparing a first register of qubits and transform these into the superposition state

$$|\psi(L)\rangle = \frac{1}{\sqrt{L}} \sum_{a=0}^{L-1} |a\rangle, \tag{7.68}$$

orresponding to an equally weighted superposition of the integers $0, 1,$
$\cdots, L - 1$. We leave unspecified, for the moment, the value of L, but ote that it should at least exceed N. We shall determine an appropriate alue for this towards the end of this section. To this register we add a econd of at least $\log N$ qubits and prepare these in a state corresponding o the bit string $y^a \bmod N$, so that our two-register state becomes

$$|\psi(L)\rangle \otimes |0\rangle \rightarrow \frac{1}{\sqrt{L}} \sum_{a=0}^{L-1} |a\rangle \otimes |y^a \bmod N\rangle. \tag{7.69}$$

This state depends on all of the values of $y^a \bmod N$ and it follows that contains, encoded within it, the required periodicity r. This can be extracted purely by operations on the first register, but it is easier to ee what is happening if we start by making a measurement, in the omputational basis, on the second register. In doing so, we note that the smallest non-zero value of a for which $y^a \bmod N = 1$ is r and that no value of $y^a \bmod N$ appears more than once in any period. Let us denote the measurement result by $y^{a_0} \bmod N$, where $0 \leq a_0 < r$, corresponding to the input numbers $a_0, a_0 + r, a_0 + 2r, \cdots, a_0 + Qr$, where $Q \leq (L - 1 - a_0)/r$ so that we are restricted to values of a in range 0 to $L - 1$. Performing this measurement on the second register leaves the first in the state

$$|\varphi(a_0, r)\rangle = \frac{1}{\sqrt{Q+1}} \sum_{m=0}^{Q} |a_0 + mr\rangle, \qquad (7.70)$$

which is of the same form as eqn 7.49, encountered in our discussion of period-finding. If $Q + 1$ has r as a factor then, as demonstrated in the preceding section, performing a quantum Fourier transform will lead us to the value of r. Here, however, the value of Q depends on that chosen for L and it is most unlikely that r will be a factor of $Q+1$, so we cannot rely on the period-finding algorithm as it was described.

Let us proceed by performing a quantum Fourier transform on the state in eqn 7.70, choosing L as the size of the transform matrix so that

$$\hat{U}_{\text{QFT}}|a\rangle = \frac{1}{\sqrt{L}} \sum_{b=0}^{L-1} \exp\left(i\frac{2\pi ab}{L}\right) |b\rangle. \qquad (7.71)$$

This produces the state

$$\hat{U}_{\text{QFT}}|\varphi(a_0, r)\rangle = \frac{1}{\sqrt{(Q+1)L}} \sum_{m=0}^{Q} \sum_{b=0}^{L-1} \exp\left(i2\pi\frac{(a_0 + mr)b}{L}\right) |b\rangle. \qquad (7.72)$$

If we perform a measurement in the computational basis then the probability that this gives the value b is

$$P(b) = \frac{1}{L(Q+1)} \left| \sum_{m=0}^{Q} \exp\left(i2\pi\frac{(a_0 + mr)b}{L}\right) \right|^2$$

$$= \frac{1}{L(Q+1)} \left| \sum_{m=0}^{Q} \exp\left(i2\pi\frac{m(rb \bmod L)}{L}\right) \right|^2. \qquad (7.73)$$

We note that the probabilities $P(b)$ are independent of the value a_0 associated with the result of the measurement carried out on the second register. This is indicative of the fact that, as stated, no measurement on the second register is necessary.

This probability will take a significant value if the terms in the summation are almost in phase. We can identify the values of $rb \bmod L$ for which this is true by recalling that $Q < L/r$. This means, in particular, that each term in the summation in eqn 7.73 will have an imaginary part with the same sign (so that their phases all lie within a range of π) if

$$-\frac{r}{2} \leq rb \bmod L \leq \frac{r}{2}. \qquad (7.74)$$

It is straightforward to show that there are r values of b for which this inequality is satisfied.

Our measurement of b will give one of the likely values with a probability in excess of 0.40 and we can reasonably base our determination of on getting one of these outcomes. We can see how to do this by noting that a value of b satisfying eqn 7.74 also satisfies the inequality

$$|rb - \kappa L| \leq \frac{r}{2} \tag{7.75}$$

or an integer κ, where $0 \leq \kappa \leq r - 1$. Dividing this by rL gives

$$\left| \frac{b}{L} - \frac{\kappa}{r} \right| \leq \frac{1}{2L}. \tag{7.76}$$

Here b is our measurement result and L was preselected by our choice of the input state given in eqn 7.68. If L is sufficiently large then we can use this inequality to determine the ratio κ/r and hence r itself. It may be shown, in particular, that if $L \geq N^2$ then there is exactly one fraction, κ/r, satisfying eqn 7.76 and this, together with the required value of κ, can be determined efficiently by means of a continued fraction. It is sensible to choose $L = 2^n$, making $N^2 < L < 2N^2$, as then we know how to construct an efficient circuit with which to perform the required quantum Fourier transform.

The quantum Fourier transform can be performed in polynomial time and it follows, therefore, that factoring using a quantum processor can also be performed in polynomial time. We have already mentioned the threat posed by Shor's algorithm to the RSA cryptosystem, but should note that related algorithms can also evaluate discrete logarithms efficiently and so challenge Diffie–Hellman key exchange. Further variations present a similar threat to other public-key cryptosystems. These features are, in themselves, sufficient to motivate interest in Shor's algorithm, but there is also the more fundamental point that factoring on a classical computer appears to be a class **NP** problem. In showing that factoring using a quantum processor is a class **P** problem we may also have learnt something fundamental about computational complexity and, indeed, about mathematics.

There are r integer multiples of L in the range from 0 to $r(L-1)$. There are also $L-1$ multiples of r in this range and these are spaced, of course, by r. It follows that for each of the r multiples of L there must be one multiple of r within a distance $r/2$.

7.5 Grover's search algorithm

Not all quantum algorithms that have been devised exhibit the dramatic decrease in computing time associated with the quantum Fourier transform. Some, such as Grover's search algorithm, provide a more modest increase in efficiency but this can represent, nevertheless, the difference between a practical solution and an impractical one. The problem addressed is how to search for a desired entry in an unstructured database. If there are N entries in the database then, classically, there is no better strategy than to look at the entries in turn until we find the one we require. On average, this will take $N/2$ trials and, in terms of computational complexity, we can say that the required time is $O(N)$. Grover's algorithm finds the required item more quickly, needing only $O(\sqrt{N})$ queries of the database.

Let us suppose that each element in our database is labelled by an integer a, ranging from 0 to $N-1$, so that we can represent each element

Unstructured database As an illustration of structured and unstructured databases, let us consider a telephone directory: a book in which names are listed in alphabetical order together with their telephone numbers. Finding the number for a given individual is easy because the database is ordered alphabetically. Finding the person with a given telephone number, however, is difficult as the numbers are not ordered.

by a string of n bits where $2^n \geq N$. We can think of our problem as a computation of a one-bit function $f(a)$, where $f(a) = 1$ if a is the required element and $f(a) = 0$ otherwise. If we can arrange for this function to be computed by a quantum black box or oracle, then we have the means to implement the unitary transformation

$$|a\rangle \otimes |B\rangle \rightarrow |a\rangle \otimes |B \oplus f(a)\rangle, \tag{7.77}$$

where $|B\rangle$ is a second register of just a single qubit. This transformation is reminiscent of that in eqn 7.18, encountered in our discussion of Deutsch's algorithm. Here, as there, it is useful to prepare the second register in the state $2^{-1/2}(|0\rangle - |1\rangle)$, so that $f(a)$ appears as a phase factor:

$$|a\rangle \otimes \frac{1}{\sqrt{2}}(|0\rangle - |1\rangle) \rightarrow (-1)^{f(a)}|a\rangle \otimes \frac{1}{\sqrt{2}}(|0\rangle - |1\rangle). \tag{7.78}$$

In this way, the oracle labels the required element by a π phase shift. This phase is not an observable property of the transformed state given in eqn 7.78, but we can access it by the now familiar process of preparing our first register in the superposition state

$$|\psi\rangle = \frac{1}{\sqrt{N}}\sum_{a=0}^{N-1}|a\rangle, \tag{7.79}$$

so that the oracle unitary-transformation produces the state

$$\hat{U}_{\text{oracle}}|\psi\rangle \otimes \frac{1}{\sqrt{2}}(|0\rangle - |1\rangle) = \frac{1}{\sqrt{N}}\sum_{a=0}^{N-1}(-1)^{f(a)}|a\rangle \otimes \frac{1}{\sqrt{2}}(|0\rangle - |1\rangle). \tag{7.80}$$

The state of the second register qubit is unchanged by the oracle transformation and so acts like a catalyst in a chemical reaction. You will often find that explicit reference to this qubit is omitted in discussions of Grover's algorithm.

A measurement performed on the first register, in the computational basis, is no more likely to select the required entry than any of the others. We can change this, however, by performing, on the first register, the so-called diffusion unitary transformation defined by the n-bit unitary operator

$$\hat{D} = 2|\psi\rangle\langle\psi| - \hat{I}^{\otimes n}, \tag{7.81}$$

where $|\psi\rangle$ is the superposition state in eqn 7.79 and $\hat{I}^{\otimes n}$ is the n-qubit identity operator. The action of this operator produces the state

$$\left(\hat{D} \otimes \hat{I}\right)\hat{U}_{\text{oracle}}|\psi\rangle \otimes \frac{1}{\sqrt{2}}(|0\rangle - |1\rangle)$$
$$= \left[\left(1 - \frac{4}{N}\right)|\psi\rangle + \frac{2}{\sqrt{N}}|a_0\rangle\right] \otimes \frac{1}{\sqrt{2}}(|0\rangle - |1\rangle), \tag{7.82}$$

where $|a_0\rangle$ is the state corresponding to the database element we are trying to find. The amplitude for this state has been increased from $N^{-1/2}$ to $N^{-1/2}(3 - 4/N)$. Hence a measurement of the first register, in the computational basis, on this state will produce the desired result, o, with about nine times the probability that it produces any of the other possible results.

Before continuing with our analysis of Grover's algorithm, we pause to consider how the diffusion transformation in eqn 7.81 can be implemented. We start by recalling that the state $|\psi\rangle$ can be generated from the zero state by applying a Hadamard gate to each qubit:

$$|\psi\rangle = \hat{H}^{\otimes n}|0\rangle. \tag{7.83}$$

It follows that we can write our diffusion operator in the form

$$\hat{D} = \hat{H}^{\otimes n}\left(2|0\rangle\langle 0| - \hat{I}^{\otimes n}\right)\hat{H}^{\otimes n}, \tag{7.84}$$

where $|0\rangle\langle 0|$ is the projector onto the n-qubit state $|0\rangle \otimes |0\rangle \otimes \cdots \otimes |0\rangle$. Thus we can realize the unitary diffusion operation by performing a Hadamard transformation on each qubit, followed by a conditional phase shift with all the n-qubit states except $|0\rangle$ acquiring a minus sign, and finally a further n-qubit Hadamard transformation. The conditional phase shift requires $O(n)$ gates and there are, in addition, $2n$ Hadamard gates so that the total number of one- and two-qubit gate operations required to implement $\hat{D}$ is $O(n)$.

The combination of the operation of the oracle and of the diffusion transformation significantly increases the amplitude for the desired state and, with it, the probability that a measurement will give the result a_0. If we can access the oracle a second time then the same sequence of operations will further amplify this amplitude:

> Alternatively, of course, we could change only the sign of the amplitude for the state $|0\rangle$.

$$\hat{G}^2|\psi\rangle \otimes \frac{1}{\sqrt{2}}\left(|0\rangle - |1\rangle\right)$$

$$= \left[\left(1 - \frac{12}{N} + \frac{16}{N^2}\right)|\psi\rangle + \frac{4}{\sqrt{N}}\left(1 - \frac{2}{N}\right)|a_0\rangle\right]$$

$$\otimes \frac{1}{\sqrt{2}}\left(|0\rangle - |1\rangle\right), \tag{7.85}$$

where $\hat{G} = (\hat{D} \otimes \hat{I})\hat{U}_{\text{oracle}}$. The second iteration has further amplified the amplitude of the state $|a_0\rangle$ to $N^{-1/2}[5 - (20/N) + (16/N^2)]$, which, for large N, is about five times greater than it was initially.

At this stage we can reasonably ask whether further iterations, corresponding to further action of the operator $\hat{G}$, can increase the amplitude so that the probability that a measurement gives the value a_0 approaches unity. In order to address this question, we note that the state remains a superposition of just two orthonormal states, $|a_0\rangle$ and

$$|a_0^\perp\rangle = \frac{1}{\sqrt{N-1}}\sum_{a \neq a_0}|a\rangle. \tag{7.86}$$

The initial state can now be written as

$$|\psi\rangle \otimes \frac{1}{\sqrt{2}}\left(|0\rangle - |1\rangle\right) = \left(\frac{1}{\sqrt{N}}|a_0\rangle + \sqrt{\frac{N-1}{N}}|a_0^\perp\rangle\right) \otimes \frac{1}{\sqrt{2}}\left(|0\rangle - |1\rangle\right), \tag{7.87}$$

or as the column vector

$$|\psi\rangle \otimes \frac{1}{\sqrt{2}}(|0\rangle - |1\rangle) = \frac{1}{\sqrt{N}}\begin{pmatrix} 1 \\ \sqrt{N-1} \end{pmatrix}. \qquad (7.88)$$

In this representation, our operator $\hat{G}$ has the simple matrix form

$$\hat{G} = \frac{1}{N}\begin{pmatrix} N-2 & 2\sqrt{N-1} \\ -2\sqrt{N-1} & N-2 \end{pmatrix}. \qquad (7.89)$$

This operator induces a rotation in the effective two-dimensional state space, and we can see this directly by introducing an angle θ defined so that

$$\hat{G} = \begin{pmatrix} \cos\theta & \sin\theta \\ -\sin\theta & \cos\theta \end{pmatrix}, \qquad (7.90)$$

where $\sin\theta = 2(N-1)^{1/2}/N$ so that, for large N, $\theta \approx 2N^{-1/2}$. In terms of θ, the initial state has the simple form

$$|\psi\rangle = \begin{pmatrix} \sin(\theta/2) \\ \cos(\theta/2) \end{pmatrix}. \qquad (7.91)$$

The advantage of this representation is that successive interactions correspond to further rotations through the same angle so that ℓ such processes correspond to the action of the operator

$$\hat{G}^\ell = \begin{pmatrix} \cos(\ell\theta) & \sin(\ell\theta) \\ -\sin(\ell\theta) & \cos(\ell\theta) \end{pmatrix}. \qquad (7.92)$$

After ℓ iterations, therefore, our initial state will have been transformed into

$$\hat{G}^\ell \frac{1}{\sqrt{N}}\begin{pmatrix} 1 \\ \sqrt{N-1} \end{pmatrix} = \begin{pmatrix} \sin\left[\left(\ell+\frac{1}{2}\right)\theta\right] \\ \cos\left[\left(\ell+\frac{1}{2}\right)\theta\right] \end{pmatrix}. \qquad (7.93)$$

It only remains to choose a value for ℓ that is not too large, but is such that $\sin\left[\left(\ell+\frac{1}{2}\right)\theta\right]$ is close to unity so that a measurement in the computational basis is likely to reveal the desired result. The natural way to achieve this is to select the value of ℓ such that $\left(\ell+\frac{1}{2}\right)$ is as close as possible to $\pi/2$:

$$\left(\ell+\frac{1}{2}\right)\theta \approx \frac{\pi}{2}. \qquad (7.94)$$

We can readily find the required value of ℓ for any given N, but an approximate solution suffices to determine the way that ℓ scales with N. We have already noted that, for large N, $\theta \approx 2N^{-1/2}$, and this means that

$$\ell \approx \frac{\pi}{4}\sqrt{N}, \qquad (7.95)$$

so that the required number of iterations is about $\sqrt{N}$. This means that we shall find the required element of our database, with high probability, using only $O(\sqrt{N})$ queries of our quantum oracle. This contrasts with the $O(N)$ queries required to find the desired element in an unstructured classical list.

We conclude our analysis of Grover's algorithm by addressing the question of how it might be employed to give a real advantage over classical searches. The unstructured list or database used to motivate our discussion is, in truth, unlikely to be a sensible application. The reason for this is that the N elements of the list will each have to be associated with a quantum state $|a\rangle$, and this encoding step will require a time which scales like $O(N)$ and will provide the rate-limiting step. Unless the list is already encoded as a set of quantum states, Grover's algorithm will, for this reason, fare no better (in terms of computational complexity) than a classical search. Where a quantum search algorithm may pay dividends, however, will be in any situation in which we can readily create an input state representing the possibilities. A simple example would be if we needed to find a number in the range 0 to $N - 1$ with a desired numerical property. This suggests, in particular, using an algorithm, based on Grover's, to speed up the time taken to solve a problem in the **NP** class. Such a problem may take a time that is exponential in the number of bits in the input, but the possibility of halving the exponent ($2^{n/2}$ instead of 2^n) might make solvable a previously intractable problem.

One intriguing possibility would be to use a quantum search algorithm to speed up the analysis of secret communications encoded using short keys, for example DES (Data Encryption Standard) or AES (Advanced Encryption Standard). We would proceed by preparing a superposition of all possible keys and searching among the decrypted signals for any (hopefully one) that makes linguistic sense.

7.6 Physical requirements

Having seen some of the remarkable things that can be done with a quantum processor, it is only natural to ask how we might build one. It is probably no surprise that many teams around the world are actively pursuing precisely this goal. Progress has been impressive but, at the time of writing, no satisfactory large-scale processor has been demonstrated. There have, however, been a number of important demonstration experiments, which include the implementation of quantum algorithms for small numbers of qubits. Practical quantum processors and computers that can compete with and even outperform the best classical devices seem a long way off.

It is not even possible, yet, to identify a winning technology, the physical system upon which the first practical quantum processors will be based. For this reason, perhaps, there is a bewildering array of rival ideas under active investigation. Among the most prominent are the following.

Trapped ions Arrays of single atomic ions, frozen into their motional ground state, form the basis for the ion-trap quantum computer. The qubit states are realized as stable or metastable electronic levels in the ions. Transformations are induced by applying laser pulses and interactions between the qubits are mediated by exciting collective vibrations of the ions.

Nuclear magnetic resonance   Nuclear spins have a magnetic moment and so can be oriented by an applied magnetic field. The states of a qubit might then be represented by a spin-1/2 nucleus aligned parallel or antiparallel to this field. Transformations are induced by applying a resonant radio frequency field. Interactions between qubits occur as each nucleus evolves in the magnetic field generated by its neighbours.

Trapped neutral atoms   It is possible to trap and cool neutral atoms, either in optical lattices or by magnetic fields generated by current-carrying wires. The atoms can be manipulated by applying external fields and made to interact by inducing controlled collisions.

Cavity quantum electrodynamics   Very high-quality optical cavities can trap light, even single photons, for a useful period of time. A single atom, with a qubit encoded in a pair of electronic energy levels, can interact resonantly with a mode of such a cavity and emit, reversibly, a photon into it. This process can be used to induce coherent single-photon exchange between pairs of atoms.

Single-photon linear optics   A qubit can be encoded in the polarization of a single photon, as described in Section 3.3 or, indeed, by its path through a set of optical components. Single-qubit transformations can be performed using readily available components, and interactions can be induced two-photon interference of the kind demonstrated by Hong, Ou, and Mandel and described in Appendix G.

Coupled quantum dots   Existing computers are based on semiconductor physics, and the advanced technological level of this makes it an attractive area for quantum information processing. Individual quantum dots can be used, with charges or electron spins used as qubits. The coupling between the dots can be controlled by selectively applying voltages to local electrodes.

Superconducting Josephson junctions   SQUIDs (superconducting quantum interference devices) enclose within them a quantized magnetic flux. This can be used to embody a qubit, as can the charge and phase difference across the junction. Flux linkage between SQUIDs allows the qubits to interact.

We shall not attempt to describe any of these in detail, for two reasons. First, we may have our favourites, but it is difficult to know which, if any, of these will turn out to be important. Second, the field is developing rapidly and any review will soon become outdated. It is worth mentioning, however, the conditions required to realize a quantum information processor. There are five recognized criteria, formulated by DiVincenzo, and these provide the means by which we can compare developments in competing systems and measure progress towards a quantum computer.

The five DiVincenzo criteria are the following.

. **Well-defined state space**   The system should be scalable and have well-defined qubits. It is essential that we can identify precisely our qubits and access the state space of each of these, so that we can realize any desired unitary transformation. We need to be able to upgrade the processor by being able to add further qubits.

. **Initialization**   We need to be able to initialize the system of qubits in a unique pure state. If we can prepare our n qubits in a product state and let this represent $|0\rangle \otimes |0\rangle \otimes \cdots \otimes |0\rangle$, then the controlled application of selected quantum gates will allow us to generate any desired multiqubit state.

. **Long coherence times**   Interaction of our qubits with their environment can rapidly and uncontrollably modify our quantum state and with it ruin the desired transformation. Even with efficient quantum error correction, we need decoherence times that are very long compared with gate operation times. This is a great challenge, as we would like our qubits to interact *strongly* with controlling external influences and with each other, but *weakly* with everything else.

. **Universal set of quantum gates**   We need to be able to apply the desired sequence of one- and two-qubit unitary transformations. This requires us to be able to interact coherently with single qubits and with specified pairs of qubits without affecting the states of their neighbours.

. **Qubit-specific measurements**   In order to obtain our readout, we need to be able to perform projective von Neumann measurements of each of the qubits.

None of the existing systems perform entirely satisfactorily against all of these criteria. The DiVincenzo criteria are helpful, at this stage of development, in providing a sensible method for comparing rival technologies, and will continue to do so as the field develops.

Suggestions for further reading

Bouwmeester, D., Ekert, A., and Zeilinger, A. (eds) (2000). *The physics of quantum information*. Springer-Verlag, Berlin.

Bruß, D. and Leuchs, G. (eds) (2007). *Lectures on quantum information theory*. Wiley-VCH, Weinheim.

Chen, G., Church, D. A., Englert, B.-G., Henkel, C., Rohwedder, B., Scully, M. O., and Zubairy, M. S. (2007). *Quantum computing devices: principles, designs, analysis*. Chapman and Hall/CRC, Boca Raton, FL.

Copeland, B. J. (2004). *The essential Turing*. Oxford University Press, Oxford.

Ekert, A. and Jozsa, R. (1996). Quantum computation and Shor's factoring algorithm. *Reviews of Modern Physics* **68**, 733.

Jaeger, G. (2007). *Quantum information: an overview.* Springer, New York.

Kaye, P., Laflamme, R., and Mosca, M. (2007). *An introduction to quantum computing.* Oxford University Press, Oxford.

Le Bellac, M. (2006). *A short introduction to quantum information and quantum computation.* Cambridge University Press, Cambridge.

Lo, H.-K., Popescu, S., and Spiller, T. (eds) (1998). *Introduction to quantum computation and information.* World Scientific, Singapore.

Macchiavello, C., Palma, G. M., and Zeilinger, A. (eds) (2000). *Quantum computation and quantum information theory.* World Scientific, Singapore.

Mermin, N. D. (2007). *Quantum computer science: an introduction.* Cambridge University Press, Cambridge.

Mertens, S. and Moore, C. (in preparation). *The nature of computation.*

Nielsen, M. A. and Chuang, I. L. (2000). *Quantum computation and quantum information.* Cambridge University Press, Cambridge.

Schleich, W. P. and Walther, H. (eds) (2007). *Elements of quantum information.* Wiley-VCH, Weinheim.

Stenholm, S. and Suominen, K.-A. (2005). *Quantum approach to informatics.* Wiley, Hoboken, NJ.

Vedral, V. (2006). *Introduction to quantum information science.* Oxford University Press, Oxford.

Exercises

(7.1) A Turing machine has the four tape symbols $\{\triangleright, 0, 1, \emptyset\}$ and the four processor states $\{S, I, II, F\}$. The instruction set, or program, is

$$(S, \triangleright) \Rightarrow (I, \triangleright),$$
$$(I, 0) \Rightarrow (I, 0),$$
$$(I, 1) \Rightarrow (II, 0),$$
$$(II, 0) \Rightarrow (I, 1),$$
$$(II, 1) \Rightarrow (II, 1),$$
$$(II, \emptyset) \Rightarrow (I, 1),$$
$$(I, \emptyset) \Rightarrow (F, \emptyset),$$

where, as in the text, each instruction is followed by moving the tape head one place to the right.

(a) Calculate the effect of this program on tapes in the following two initial configurations:

$$\boxed{\triangleright \ : \ 0 \ : \ 1 \ : \ \emptyset \ : \ \emptyset \ : \ \cdots},$$

$$\boxed{\triangleright \ : \ 1 \ : \ 0 \ : \ 1 \ : \ \emptyset \ : \ \cdots}.$$

(b) What simple mathematical operation does the program perform?

(7.2) Devise an instruction set for a Turing machine, the action of which calculates the parity of a bit string printed on the tape. (The parity is 1 if there is an odd number of 1s on the tape and is 0 otherwise.)

(7.3) Consider three algorithms for which the required computing times are

(a) $T_1(n) = b_1 \tau n$,
(b) $T_2(n) = b_2 \tau n^k$,
(c) $T_3(n) = b_3 \tau 2^{kn}$,

where τ is the typical processor time for a single operation. In each case we need to add one bit $(n \to n+1)$ but keep the time taken unchanged. By how much do we need to decrease τ in each case?

7.4) The time taken to run an algorithm is $T(n) = 3 \log n + 4$. Show that $T(n) = O(n^{\ell})$ for all $\ell > 0$.

7.5) Solve, by recurrence, eqns N.8 and N.11.

7.6) A list of n elements is to be sorted by applying a sequence of pairwise comparisons and, if appropriate, swap operations. Show that after ℓ such operations, at most 2^{ℓ} of the possible $n!$ initial orderings have been arranged in the correct order. Hence show that $T(n) = \Omega(n \log n)$ for this process.

7.7) It may have occurred to you that factoring a product of primes N by trying all integers less than or equal to $\sqrt{N}$ is far from optimal, as we need try only prime numbers in this range. Show that this simplification does not change the complexity class from **NP** to **P**.

[*Hint:* you might find it useful to refer to the discussion in Appendix E.]

7.8) How many real numbers are required to specify a general *mixed* state of n qubits? How does this compare with the number required to specify a product state, that is, one in which the density operator is simply a tensor product of n single-qubit density operators?

7.9) Write the unitary operator required for the transformation in eqn 7.13 in terms of the states $|a\rangle$ and $|f(a)\rangle$ and confirm that it is unitary.

7.10) A quantum processor is designed to implement the copying transformation

$$|a\rangle \otimes |0\rangle \rightarrow |a\rangle \otimes |a\rangle$$

where $|a\rangle$ encodes the bit string a.

(a) Does this not conflict with the no-cloning theorem?

(b) Design a simple quantum circuit for implementing this transformation.

7.11) Quantify the entanglement between the two qubit registers for the state in eqn 7.15 by calculating $\text{Tr}(\hat{\rho}_1^2)$, where $\hat{\rho}_1$ is the reduced density operator for the first register, if:

(a) The functions $f(a)$ are all different, so that $f(a) \neq f(b)$ unless $a = b$.

(b) There are m pairs of values for which f has the same value, that is,

$$f(a_i) = f(a_{i+m}), \quad i = 1, \cdots, m,$$

and all the other values are distinct.

Under what conditions will the two registers be unentangled?

7.12) Construct quantum circuits, the actions of which calculate each of the four one-bit constant and balanced functions and so form possible oracles for testing using Deutsch's algorithm.

7.13) A classical algorithm is to solve the Deutsch–Jozsa problem by a number of trials to determine, with certainty, whether an oracle computes a constant or a balanced function. If the function is equally likely to be constant or balanced, how many trials are required, on average, to determine the required nature of the function?

7.14) Let us suppose, in the preceding problem, that we tolerate a probability for getting a wrong answer of ε. What is the maximum number of trials required in order to satisfy these conditions?

7.15) (a) Show that the state in eqn 7.22, if the function is constant, is orthogonal to all of the possible states generated by balanced functions.

(b) Are the states generated by different balanced functions mutually orthogonal?

7.16) The Hadamard gates used to prepare the input of n qubits for the Deutsch–Jozsa algorithm have a systematic error which means that they perform the transformation

$$|0\rangle \rightarrow \cos\theta|0\rangle + \sin\theta|1\rangle,$$
$$|1\rangle \rightarrow \cos\theta|1\rangle - \sin\theta|0\rangle,$$

where $\theta \approx \pi/4$.

(a) Calculate the probability that the algorithm correctly identifies a balanced function.

(b) Show that we require

$$\theta - \frac{\pi}{4} = \frac{x}{\sqrt{n}}$$

for some constant x as $n \rightarrow \infty$, if this probability is to tend to a non-zero constant.

7.17) A quantum computer running the Bernstein–Vazirani algorithm (see Appendix O) suffers from the same systematic error as that described in the preceding question. Calculate the probabilities (for large n) that the output string has

(a) no errors;

(b) precisely one error;

(c) precisely ℓ errors, where $\ell \ll n$.

(7.18) We attempt to solve Simon's problem classically by selecting, at random, different input strings a, and we require that the process succeeds with a fixed probability p.

  (a) Show that the number of trials required for large n is proportional to $2^{n/2}$.
  (b) In terms of computational complexity, is the required number of trials for the *optimal* classical solution of this problem $O(2^{n/2})$, $\Omega(2^{n/2})$, or $\Theta(2^{n/2})$?

(7.19) If $|a\rangle$ is the n-qubit state encoding the n-bit string a, show that

$$\hat{H}^{\otimes n}|a\rangle = 2^{-n/2}\sum_{c=0}^{2^n-1}(-1)^{a\cdot c}|c\rangle.$$

(7.20) Show that two n-bit strings b and c satisfy the equation $b\cdot c = 0 \bmod 2$ if and only if they have the same parity.

(7.21) Calculate the reduced density operator for the first register from eqn 7.27. Hence show that it is not necessary to perform a measurement of the second register in order to run Simon's algorithm.

(7.22) Confirm the unitarity of $\hat{U}_{\mathrm{QFT}}$ by showing that

$$\langle \tilde{b}|\tilde{c}\rangle = \delta_{bc}.$$

(7.23) Calculate the discrete Fourier transforms of the following functions:

  (a) $x_a = \gamma^a$, where γ is a constant.
  (b) the binomial function

$$x_a = \frac{(N-1)!}{a!(N-1-a)!}p^a q^{N-1-a},$$

  where p and q are constants
  (c) the rectangular 'top-hat' function

$$x_a = \frac{1}{\sqrt{T+1}}, \qquad a = a_0, a_0+1,\cdots, a_0+T,$$

  for some a_0, and $x_a = 0$ for other values of a.

(7.24) (a) Determine the effect of the operator $\hat{U}_{\mathrm{QFT}}^2$.
  (b) Hence, or otherwise, show that $\hat{U}_{\mathrm{QFT}}^4 = \hat{I}$.

(7.25) Design a two-qubit circuit which realizes the quantum Fourier transform for $N = 3$.
  [*Hint:* you might find it helpful to refer to the discussion in Appendix L.]

(7.26) Repeat the analysis leading to the state in eqn 7.44 for a three-qubit circuit designed to implement the $N = 8$ quantum Fourier transform.

(7.27) (a) How many Hadamard gates and how many controlled-phase gates are required in order to perform an n-qubit, $N = 2^n$ quantum Fourier transform?

  (b) How many swap gates are needed if the ordering of the qubits is important?

(7.28) Design a circuit to perform the inverse quantum Fourier transform for $N = 2^n$.

(7.29) (a) Confirm that the probabilities $P(b)$ given in eqn 7.46, sum to unity as they should.

  (b) Show that a measurement will give a value of b corresponding to one of the two values of $2\pi b/N$ nearest to φ with a probability greater than $8/\pi^2$.

(7.30) What happens if we attempt to use a quantum Fourier transform to find the period of a periodic function in which a value of f appears twice in a single period?

(7.31) Compare the discrete Fourier transform, fast Fourier transformation and quantum Fourier transform by evaluating 2^{2n}, $n2^n$, and n^2 for $n = 10$, $n = 100$, and $n = 1000$.

(7.32) (a) If $N = pq$, where p and q are primes, what is the probability that a number y selected randomly in the range $1 < y < N$ will be relatively prime with N?

  (b) Show that the probability that two large numbers selected at random are coprime is approximately $6/\pi^2$.

(7.33) Find the period of each of the following functions and, if appropriate, use this information to factor the designated number:

  (a) $11^a \bmod 133$, to factor 133.

  (b) $2^a \bmod 221$, to factor 221.

(7.34) Show that, for large L, the probabilities in eqn 7.73, where b satisfies eqn 7.74, are each greater than $4/(\pi^2 r)$. (It follows that the probability that the measurement gives one of these r likely values exceeds $4/\pi^2$.)

.35) There is a prize for factoring the products of large prime numbers. The latest challenge is

$$RSA\text{-}704 = 74037563479561712828046796$$
$$09742957314259318888923128$$
$$90849362326389727650340282$$
$$66276891996419625117843995$$
$$89433050212758537011896809$$
$$82867331732731089309005525$$
$$05116877063299072396380786$$
$$71008609696253793465056379$$
$$6359.$$

How many qubits would be needed to tackle this problem using Shor's algorithm?

.36) (a) Confirm that the diffusion operator $\hat{D}$ defined in in eqn 7.81 is indeed unitary.

(b) Show that the action of this operator on an arbitrary state is

$$\hat{D} \sum_{a=0}^{N-1} c_a |a\rangle = \sum_{a=0}^{N-1} (2\bar{c} - c_a)|a\rangle,$$

where $\bar{c}$ is the average value of the amplitudes c_a:

$$\bar{c} = \frac{1}{N} \sum_{a=0}^{N-1} c_a.$$

The diffusion operator, because of this formula, is also referred to as inversion about the mean.

(7.37) Design a quantum circuit, using one- and two-qubit gates, to perform the unitary diffusion transformation.

(7.38) Find the optimal number of iterations for Grover's search algorithm by solving eqn 7.94 for $N = 100$ and for $N = 1000$. Compare these exact results with the approximate expression in eqn 7.95.

(7.39) Suppose that we have an unstructured database of N elements and that we need to find any one of M of these, with $M \ll N$. We can adapt Grover's algorithm by preparing an oracle that labels all M possible solutions with a change of sign. Show that finding one of the desired elements takes $O(\sqrt{N/M})$ iterations.

(7.40) Would a quantum search algorithm be of any assistance in deciphering a message encrypted using the Vernam cipher?

Quantum information theory

The astute reader might have formed the impression that quantum information science is a rather qualitative discipline because we have not, as yet, explained how to quantify quantum information. There are three good reasons for leaving this important question until the final chapter. Firstly, quantum information theory is technically demanding and to treat it at an earlier stage might have suggested that our subject was more complicated than it is. Secondly, there is the fact that many of the ideas in the field, such as teleportation and quantum circuits, are unfamiliar and it was important to present these as simply as possible. Finally, and most importantly, the theory of quantum information is not yet fully developed. It has not yet reached, in particular, the level of completeness of its classical counterpart. For this reason we can answer only some of the many questions we would like a quantum theory of information to address. Having said this, we can say that however, there are beautiful and useful mathematical results and it seems certain that these will continue to form an important part of the theory as it develops.

We noted in the introduction to Chapter 1 that 'quantum mechanics a probabilistic theory and so it was inevitable that a quantum information theory would be developed'. A presentation of at least the beginnings of a quantitative theory is the objective of this final chapter.

8.1 The von Neumann entropy

The entropy or information derived from a given probability distribution , as we have seen, a convenient measure of the uncertainty associated with the distribution. If many of the probabilities are large, so that many of the possible events are comparably likely, then the entropy will be large. If one probability is close to unity, however, then the entropy will be small. It is convenient to introduce entropy in quantum mechanics as a measure of the uncertainty, or lack of knowledge, of the form of the state vector. If we know that our system is in a particular pure state then the associated uncertainty or entropy should be zero. For mixed states, however, it will take a non-zero value. The most natural way to define this entropy is to adopt von Neumann's form,

$$S(\hat{\rho}) = -\text{Tr}(\hat{\rho}\log\hat{\rho}),\qquad(8.1)$$

where $\hat{\rho}$ is the density operator for the system. Here, as elsewhere in the book, we use 'log' to denote a logarithm in base 2, so that our entropy is expressed in bits. It is sometimes more convenient to work with natural logarithms, in which case our entropy is

$$S_e\left(\hat{\rho}\right) = -\text{Tr}\left(\hat{\rho}\ln\hat{\rho}\right). \tag{8.2}$$

If we multiply $S_e\left(\hat{\rho}\right)$ by Boltzmann's constant then we have the thermodynamic entropy. Many of the properties of classical information described in Section 1.3 apply also to the von Neumann entropy, while others are subtly different.

The density operator is a positive Hermitian operator and can be written in the diagonal form

$$\hat{\rho} = \sum_m \rho_m |\rho_m\rangle\langle\rho_m|, \tag{8.3}$$

where the states $|\rho_m\rangle$ are the orthonormal eigenvectors of $\hat{\rho}$, and the ρ_m are the associated (non-negative) eigenvalues, which sum to unity. When written in this form, our von Neumann entropy becomes

$$S\left(\hat{\rho}\right) = -\sum_m \rho_m \log \rho_m, \tag{8.4}$$

which has the same form as the Shannon entropy for a distribution of probabilities with the values $\{\rho_m\}$. The von Neumann entropy takes its minimum value of zero if and only if one of the eigenvalues is unity, so that the others are zero. The system in this case will be in a pure state:

$$\hat{\rho} = |\psi\rangle\langle\psi| \quad \Leftrightarrow \quad S\left(\hat{\rho}\right) = 0. \tag{8.5}$$

The von Neumann entropy takes its maximum value of $\log d$, where d is the dimension of the state space, only if all of the ρ_m are equal and take the value $1/d$. The density operator in this case takes its most mixed form and is proportional to the identity operator:

$$\hat{\rho} = \frac{1}{d}\hat{I} \quad \Leftrightarrow \quad S\left(\hat{\rho}\right) = \log d. \tag{8.6}$$

The von Neumann entropy is invariant under a unitary transformation:

$$S\left(\hat{U}\hat{\rho}\hat{U}^\dagger\right) = S\left(\hat{\rho}\right). \tag{8.7}$$

To see this, we need only note that the entropy is a function only of the eigenvalues of $\hat{\rho}$ and that these, unlike the associated eigenvectors, are unchanged by a unitary transformation. It follows that the von Neumann entropy for an isolated quantum system is unchanged by its natural evolution.

We can write our density operator in any basis and, unless we use the eigenstates of $\hat{\rho}$, there will be off-diagonal elements $\rho_{nm} = \langle\lambda_n|\hat{\rho}|\lambda_m\rangle$. If we suppress these then the result is a more mixed state with a greater entropy. This means that

$$-\sum_n \rho_{nn} \log \rho_{nn} \geq S\left(\hat{\rho}\right). \tag{8.8}$$

The proof of this is a straightforward consequence of the fact that equal-
izing the probabilities, as in eqn 1.39, can only increase the information
or leave it unchanged. Here the probabilities ρ_m appearing in $S(\hat{\rho})$ are
replaced by $\sum_m \rho_m |\langle \lambda_n | \rho_m \rangle|^2$, so that the positive quantities $|\langle \lambda_n | \rho_m \rangle|^2$
play the role of the λ_{ij} in eqn 1.39. As an illustration of this, consider
a qubit with density operator

$$\hat{\rho} = \frac{1}{2}\left(\hat{I} + \vec{r} \cdot \hat{\vec{\sigma}}\right), \qquad (8.9)$$

where $\vec{r} = (u, v, w)$ is the Bloch vector. We found in Section 2.4 that
the eigenvalues of this density operator are $\frac{1}{2}(1+r)$ and $\frac{1}{2}(1-r)$, where
$r = |\vec{r}|$. It follows that

$$S(\hat{\rho}) = -\frac{1}{2}(1+r)\log\frac{1}{2}(1+r) - \frac{1}{2}(1-r)\log\frac{1}{2}(1-r). \qquad (8.10)$$

If we retain only the diagonal components in the computational basis,
however, we find

$$-\sum_{i=0,1} \langle i | \hat{\rho} | i \rangle \log \langle i | \hat{\rho} | i \rangle = -\frac{1}{2}(1+|w|)\log\frac{1}{2}(1+|w|)$$

$$-\frac{1}{2}(1-|w|)\log\frac{1}{2}(1-|w|), \qquad (8.11)$$

which is greater than $S(\hat{\rho})$ as $|w|$ is less than r, so that the two prob-
abilities appearing in eqn 8.11 are each closer to the maximum entropy
of $1/2$ than those in eqn 8.10.

If we form a linear combination of two density operators then the
result also tends to be a more mixed state than that associated with
either of the component density operators. The precise statement of
this property, known as concavity, is as follows. If $\hat{\rho} = p_1 \hat{\rho}_1 + p_2 \hat{\rho}_2$,
where $\hat{\rho}_1$ and $\hat{\rho}_2$ are density operators and p_1 and p_2 are probabilities
$p_1 + p_2 = 1$), then

$$S(\hat{\rho}) \geq p_1 S(\hat{\rho}_1) + p_2 S(\hat{\rho}_2), \qquad (8.12)$$

with the equality holding only if p_1 or p_2 is zero or if $\hat{\rho}_1 = \hat{\rho}_2$. It is worth
taking time to prove this important inequality. We start by making use
of the diagonal representation of $\hat{\rho}$ given in eqn 8.3 to write $S(\hat{\rho})$ in the
form

$$S(\hat{\rho}) = -\sum_m \rho_m \log \rho_m = \sum_m s\left(\langle \rho_m | \hat{\rho} | \rho_m \rangle\right), \qquad (8.13)$$

where we have introduced the function $s(x) = -x \log x$. The function
$s(x)$ is concave and hence

$$s\left(\langle \rho_m | \hat{\rho} | \rho_m \rangle\right) \geq p_1 s\left(\langle \rho_m | \hat{\rho}_1 | \rho_m \rangle\right) + p_2 s\left(\langle \rho_m | \hat{\rho}_2 | \rho_m \rangle\right), \qquad (8.14)$$

so that

$$S(\hat{\rho}) \geq p_1 \sum_m s\left(\langle \rho_m | \hat{\rho}_1 | \rho_m \rangle\right) + p_2 \sum_m s\left(\langle \rho_m | \hat{\rho}_2 | \rho_m \rangle\right)$$

$$\geq p_1 \sum_m \langle \rho_m | s(\hat{\rho}_1) | \rho_m \rangle + p_2 \sum_m \langle \rho_m | s(\hat{\rho}_2) | \rho_m \rangle$$

$$= p_1 S(\hat{\rho}_1) + p_2 S(\hat{\rho}_2), \qquad (8.15)$$

Projective measurements   This
means that the operation of performing
a von Neumann measurement, as in eqn
4.9, cannot decrease the entropy.

where the second inequality is a consequence of eqn 8.8. This concavity condition generalizes to more than two component density operators in the form of the inequality

$$S\left(\sum_i p_i \hat{\rho}_i\right) \geq \sum_i p_i S\left(\hat{\rho}_i\right). \tag{8.16}$$

The possibility of preparing different but non-orthogonal states means that the Shannon entropy and the von Neumann entropy can take different values for the same set of preparation probabilities. Let the event A be the selection and preparation of one of a set of pure states. If A takes the value a_i, then we prepare the state $|\psi_i\rangle$. We found in Section 1.3 that the Shannon information, or entropy, for this is

$$H(A) = -\sum_i P(a_i) \log P(a_i). \tag{8.17}$$

The von Neumann entropy, however, is that associated with the a priori density operator,

$$S\left(\hat{\rho}\right) = -\text{Tr}\left(\hat{\rho} \log \hat{\rho}\right), \tag{8.18}$$

where

$$\hat{\rho} = \sum_i P(a_i)|\psi_i\rangle\langle\psi_i|. \tag{8.19}$$

The Shannon information is strictly greater than or equal to the von Neumann entropy:

$$H(A) \geq S\left(\hat{\rho}\right), \tag{8.20}$$

with the equality holding only if the states $\{|\psi_i\rangle\}$ are all mutually orthogonal. We shall see in Section 8.5 that this inequality represents the natural redundancy associated with encoding using non-orthogonal states. It also reflects the difficulty in discriminating between non-orthogonal states. The inequality in eqn 8.20 can be extended to apply to a selection of mixed states:

$$S\left(\sum_i P(a_i)\hat{\rho}_i\right) \leq H(A) + \sum_i P(a_i)S\left(\hat{\rho}_i\right). \tag{8.21}$$

A derivation of this useful inequality is given in Appendix Q. The combination of eqns 8.21 and 8.16 means that we can place both upper and lower bounds on $S\left(\hat{\rho}\right)$:

$$\sum_i p_i S\left(\hat{\rho}_i\right) \leq S\left(\sum_i p_i \hat{\rho}_i\right) \leq \sum_i p_i S\left(\hat{\rho}_i\right) - \sum_i p_i \log p_i. \tag{8.22}$$

It is useful to define a quantum relative entropy. For two density operators $\hat{\rho}$ and $\hat{\sigma}$, the quantum relative entropy is defined to be

$$S(\hat{\sigma}\|\hat{\rho}) = \text{Tr}\left[\hat{\sigma}\left(\log \hat{\sigma} - \log \hat{\rho}\right)\right]. \tag{8.23}$$

This quantity is the natural analogue of the relative entropy $H(P\|Q)$ introduced in Section 1.3. Like its classical counterpart, the quantum relative entropy is greater than or equal to zero:

$$S(\hat{\sigma}\|\hat{\rho}) \geq 0, \tag{8.24}$$

with the equality holding if and only if $\hat{\sigma} = \hat{\rho}$. We prove this important inequality in Appendix R. There is no upper bound on the value of $S(\hat{\sigma}\|\hat{\rho})$ and, in particular, it will take an infinite value if one of the non-zero-eigenvalue eigenstates of $\hat{\sigma}$ is also an eigenstate of $\hat{\rho}$ with eigenvalue zero.

The quantum relative entropy and the associated inequality given in eqn 8.24 can be used to establish a number of important results in quantum information theory and statistical mechanics. As a simple example of this, we can show that the state with the greatest von Neumann entropy for a given mean energy is the Boltzmann thermal state with density operator

$$\hat{\rho}_\beta = \frac{\exp(-\beta\hat{H})}{\text{Tr}\left[\exp(-\beta\hat{H})\right]}, \tag{8.25}$$

where $\hat{H}$ is the Hamiltonian for the system and $\beta = (k_B T)^{-1}$ is the inverse temperature. The mean energy $\bar{E}$ is, of course, the expectation value of the Hamiltonian:

$$\bar{E} = \text{Tr}\left(\hat{\rho}_\beta\hat{H}\right). \tag{8.26}$$

Let $\hat{\sigma}$ be the density operator for a different state with the same mean energy as for $\hat{\rho}_\beta$ so that

$$\bar{E} = \text{Tr}\left(\hat{\sigma}\hat{H}\right). \tag{8.27}$$

We start by evaluating the two quantities

$$\text{Tr}\left(\hat{\sigma}\log\hat{\rho}_\beta\right) = \text{Tr}\left[\hat{\sigma}\log\left(e^{-\beta\hat{H}}\right)\right] - \log\left[\text{Tr}\left(e^{-\beta\hat{H}}\right)\right]$$

$$= -\frac{\beta}{\ln 2}\bar{E} - \log\left[\text{Tr}\left(e^{-\beta\hat{H}}\right)\right],$$

$$\text{Tr}\left(\hat{\rho}_\beta\log\hat{\rho}_\beta\right) = -\frac{\beta}{\ln 2}\bar{E} - \log\left[\text{Tr}\left(e^{-\beta\hat{H}}\right)\right], \tag{8.28}$$

so that $\text{Tr}\left(\hat{\sigma}\log\hat{\rho}_\beta\right) = \text{Tr}\left(\hat{\rho}_\beta\log\hat{\rho}_\beta\right)$. It then follows from the inequality in eqn 8.24 that

$$S\left(\hat{\sigma}\right) = -\text{Tr}\left(\hat{\sigma}\log\hat{\sigma}\right)$$

$$\leq -\text{Tr}\left(\hat{\sigma}\log\hat{\rho}_\beta\right)$$

$$= S\left(\hat{\rho}_\beta\right). \tag{8.29}$$

We note that $S(\hat{\sigma}\|\hat{\rho}) = 0$ if and only if $\hat{\sigma} = \hat{\rho}$ (or $\text{Tr}\left(\hat{\sigma}\log\hat{\sigma}\right) = \text{Tr}\left(\hat{\sigma}\log\hat{\rho}\right)$). There is no such requirement, of course, for the equality of $\text{Tr}\left(\hat{\sigma}\log\hat{\rho}\right)$ and $\text{Tr}\left(\hat{\rho}\log\hat{\rho}\right)$.

follows that $\hat{\rho}_\beta$ is the density operator with maximum von Neumann entropy. It is also possible to derive the form of the state with maximum von Neumann entropy using Lagrange's method of undetermined multipliers, as shown in Appendix B.

8.2   Composite systems

Our study of classical information in Chapter 1 demonstrated the significance of the joint probability distribution $P(a_i, b_j)$ for two events A and B and of the associated entropy $H(A, B)$. We also encountered the mutual information $H(A : B)$ as a measure of correlation between the events A and B, and the conditional entropy $H(B|A)$. It is natural, in quantum information theory, to define analogous properties based on the von Neumann entropy for the state of two quantum systems, A and B, which we denote $S(A, B)$:

$$S(A, B) = S(\hat{\rho}_{AB}) = -\mathrm{Tr}(\hat{\rho}_{AB} \log \hat{\rho}_{AB}), \qquad (8.30)$$

where $\hat{\rho}_{AB}$ is the density operator for the two systems. We can also define von Neumann entropies for the A and B systems alone in terms of their reduced density operators:

$$S(A) = -\mathrm{Tr}_A(\hat{\rho}_A \log \hat{\rho}_A) = -\mathrm{Tr}_A[(\mathrm{Tr}_B \hat{\rho}_{AB}) \log (\mathrm{Tr}_B \hat{\rho}_{AB})],$$
$$S(B) = -\mathrm{Tr}_B(\hat{\rho}_B \log \hat{\rho}_B) = -\mathrm{Tr}_B[(\mathrm{Tr}_A \hat{\rho}_{AB}) \log (\mathrm{Tr}_A \hat{\rho}_{AB})].$$
$$(8.31)$$

These are clearly analogous to the expressions for $H(A)$ and $H(B)$ in eqn 1.41.

If the two systems are statistically independent, so that $\hat{\rho}_{AB} = \hat{\rho}_A \otimes \hat{\rho}_B$, then $S(A, B) = S(A) + S(B)$. This property is sometimes referred to as additivity. More generally, we find that the entropy is subadditive in that

$$S(A, B) \leq S(A) + S(B). \qquad (8.32)$$

This inequality follows directly from the positivity of the relative entropy (eqn 8.24):

$$\begin{aligned} S(\hat{\rho}_{AB} \| \hat{\rho}_A \otimes \hat{\rho}_B) &= \mathrm{Tr}_{AB}[\hat{\rho}_{AB}(\log \hat{\rho}_{AB} - \log \hat{\rho}_A \otimes \hat{\rho}_B)] \\ &= \mathrm{Tr}_{AB}(\hat{\rho}_{AB} \log \hat{\rho}_{AB}) - \mathrm{Tr}_{AB}(\hat{\rho}_{AB} \log \hat{\rho}_A \otimes \hat{I}_B) \\ &\quad - \mathrm{Tr}_{AB}(\hat{\rho}_{AB} \log \hat{I}_A \otimes \hat{\rho}_B) \\ &= -S(A, B) + S(A) + S(B). \end{aligned} \qquad (8.33)$$

Subadditivity is reminiscent of the inequality in eqn 1.42 for classical information.

We proved, in Section 1.3, that the classical information $H(A, B)$ is bounded from below. In particular, it must be greater than or equal to the larger of $H(A)$ and $H(B)$:

$$H(A, B) \geq \mathrm{Sup}(H(A), H(B)). \qquad (8.34)$$

This inequality does not hold, however, for the von Neumann entropy. Consider, in particular, a pure entangled state of the two systems,

$$|\psi\rangle_{AB} = \sum_n a_n |\lambda_n\rangle_A |\phi_n\rangle_B, \qquad (8.35)$$

here $\langle\lambda_n|\lambda_m\rangle = \delta_{nm} = \langle\phi_n|\phi_m\rangle$ so that eqn 8.35 is the Schmidt decomposition for the state. The von Neumann entropy for the two systems is zero, that is,

$$S(A, B) = 0 \tag{8.36}$$

but, because the state is entangled, each of the states $\hat{\rho}_A$ and $\hat{\rho}_B$ is mixed and the associated entropies will not be zero. Evaluating the partial traces over the B and A state spaces gives

$$\hat{\rho}_A = \sum_n |a_n|^2 |\lambda_n\rangle\langle\lambda_n|,$$
$$\hat{\rho}_B = \sum_n |a_n|^2 |\phi_n\rangle\langle\phi_n|. \tag{8.37}$$

The von Neumann entropy for each of these is the same:

$$S(A) = -\sum_n |a_n|^2 \log |a_n|^2 = S(B). \tag{8.38}$$

Clearly this must be true for all pure states of A and B. Here $S(A)$ and $S(B)$ are positive but $S(A, B)$ is zero and it is clear, therefore, that the classical inequality in eqn 8.34 does not apply to the von Neumann entropy. In its place we have the Araki–Lieb inequality

$$S(A, B) \geq |S(A) - S(B)|. \tag{8.39}$$

This is clearly satisfied for entangled pure states, for which both sides of the inequality are zero. A simple derivation of the Araki–Lieb inequality is presented in Appendix S. We can combine this inequality with the subadditivity condition to place both lower and upper bounds on $S(A, B)$:

$$|S(A) - S(B)| \leq S(A, B) \leq S(A) + S(B). \tag{8.40}$$

It is helpful to introduce von Neumann analogues of the mutual information $H(A : B)$ and of the conditional entropy $H(B|A)$. The von Neumann, or quantum, mutual information is defined as the difference between $S(A) + S(B)$ and $S(A, B)$:

$$S(A : B) = S(A) + S(B) - S(A, B). \tag{8.41}$$

This quantity, like its classical counterpart, is clearly symmetrical in A and B and it is also greater than or equal to zero, by virtue of subadditivity (eqn 8.32). The mutual information is restricted to be less than or equal to the lesser of $H(A)$ and $H(B)$. The quantum mutual information is restricted, by virtue of eqn 8.39, to the range

$$0 \leq S(A : B) \leq 2 \operatorname{Inf}\left(S(A), S(B)\right). \tag{8.42}$$

The mutual information is, as we have seen in Section 1.3, a measure of correlation and the same can be said of the von Neumann mutual information for quantum information. Indeed, this quantity is sometimes referred to as the index of correlation.

The von Neumann conditional entropy is defined by direct analogy with its classical counterpart, $H(B|A)$, to be

$$S(B|A) = S(A, B) - S(A). \tag{8.43}$$

This quantity, in contrast to its classical counterpart, is not restricted to be greater than or equal to zero. Indeed, it can take any value between $S(B)$ and $-S(A)$. The conditional entropy $H(B|A)$ has the simple and physically appealing interpretation as the information about A and B not already contained in A alone. To put it another way, if we write

$$H(A, B) = H(A) + H(B|A), \tag{8.44}$$

Properties of the von Neumann entropy   We summarize here the main properties of the von Neumann entropy:

(i) The von Neumann entropy associated with a density operator $\hat{\rho}$ is

$$S(\hat{\rho}) = -\mathrm{Tr}(\hat{\rho} \log \hat{\rho}).$$

It is zero only if the system is in a pure state; that is, $\hat{\rho} = |\psi\rangle\langle\psi|$.

(ii) We can place upper and lower bounds on $S\left(\sum_i p_i \hat{\rho}_i\right)$:

$$\sum_i p_i S(\hat{\rho}_i) \leq S\left(\sum_i p_i \hat{\rho}_i\right)$$

$$\leq \sum_i p_i S(\hat{\rho}_i)$$

$$- \sum_i p_i \log p_i.$$

(iii) The relative entropy is always greater than or equal to zero:

$$S(\hat{\sigma}\|\hat{\rho}) \geq 0.$$

It takes the value 0 only if $\hat{\sigma} = \hat{\rho}$.

(iv) The entropy for the state of two quantum systems is bounded by

$$|S(A) - S(B)| \leq S(AB)$$
$$\leq S(A) + S(B).$$

(v) Strong subadditivity:

$$S(ABC) + S(B) \leq$$
$$S(AB) + S(BC).$$

then the information associated with A and B is simply that associated with A plus that for B when A is known. It is tempting to apply the same interpretation to $S(B|A)$ and to write

$$S(A, B) = S(A) + S(B|A). \tag{8.45}$$

This suggests that the quantum information content of the state $\hat{\rho}_{AB}$ is that of $\hat{\rho}_A$ plus that for $\hat{\rho}_B$ when $\hat{\rho}_A$ is known. We shall accept, for now, this interpretation but acknowledge the need to explain how $S(B|A)$ can be negative and so *reduce* $S(A, B)$ compared with $S(A)$. We shall return to this problem in Section 8.5.

We conclude this section with one more inequality for the von Neumann entropy. This condition, referred to as strong subadditivity, relates to the state of three systems with density operator $\hat{\rho}_{ABC}$. Strong subadditivity states that

$$S(ABC) + S(B) \leq S(AB) + S(BC). \tag{8.46}$$

This is a stronger condition than subadditivity in that we can derive eqn 8.32 from it. All we need to do is to select a state with a density operator of the form

$$\hat{\rho}_{ABC} = \hat{\rho}_{AC} \otimes |\psi\rangle_B\,{}_B\langle\psi| \tag{8.47}$$

to obtain

$$S(AC) \leq S(A) + S(C). \tag{8.48}$$

It is also straightforward to show that the Araki–Lieb inequality (eqn 8.39) follows from strong subadditivity. The simplest mathematical derivation of eqn 8.46 is rather long and involved, so we shall not provide this proof. Instead we present, in the next section, an argument based on quantum state discrimination to indicate that it must be true.

8.3   Quantitative state comparison

We saw in Section 1.4 that the degree of similarity between bit strings determines the extent to which signal compression is possible and also the degree of resistance to noise of the encoded messages. It was helpful in designing coding schemes to have a quantitative measure, the Hamming distance, of the difference between strings. There are other measures in common use in classical information theory and these have been generalized to quantum information theory. The result is a variety of distinct measures of the difference between two possible quantum states. In this section, we introduce and describe the properties of three of these.

We start with the fidelity, F, defined for a pure state in eqn 5.48. The fidelity was introduced as the probability that the system provided will pass a test to determine whether it is in the desired state. To be specific, if the system has been prepared with density operator $\hat{\rho}$ and the desired pure state is $|\psi\rangle$, then a von Neumann measurement with the two projectors

$$\hat{P}_\psi = |\psi\rangle\langle\psi|,$$
$$\hat{P}_{\bar{\psi}} = \hat{I} - |\psi\rangle\langle\psi| \tag{8.49}$$

will give the result $|\psi\rangle$ with probability

$$\mathrm{Tr}\left(\hat{\rho}\hat{P}_\psi\right) = \langle\psi|\hat{\rho}|\psi\rangle = F. \tag{8.50}$$

We interpret this fidelity as the probability that the state we have prepared will behave as if it were in the pure state $|\psi\rangle$. The fidelity is used as a measure of the quality of the state or of the state preparation.

The extent to which we can distinguish between quantum states is a natural measure of the difference or distance between them, and so we can use F as a quantitative measure of the difference between $\hat{\rho}$ and $|\psi\rangle\langle\psi|$. If the fidelity is close to its maximum value of unity then $\hat{\rho}$ and $|\psi\rangle\langle\psi|$ are similar and can only be distinguished with difficulty (or low probability). If, however, F is small then it is easy to determine with confidence whether the system was prepared in the state $\hat{\rho}$ or the state $|\psi\rangle\langle\psi|$.

It is important to have quantitative measures for comparing mixed as well as pure states. In generalizing eqn 8.50 to mixed states it is natural to require the following four properties. (i) $0 \leq F \leq 1$, and $F(\hat{\rho}, \hat{\sigma}) = 1$ only if $\hat{\rho} = \hat{\sigma}$. (ii) The fidelity should be symmetrical in $\hat{\rho}$ and $\hat{\sigma}$. (iii) If $\hat{\sigma} = |\psi\rangle\langle\psi|$ then $F(\hat{\rho}, \hat{\sigma})$ should reduce to eqn 8.50. (iv) $F(\hat{\rho}, \hat{\sigma})$ should be invariant under unitary transformations, that is, $F(\hat{U}\hat{\rho}\hat{U}^\dagger, \hat{U}\hat{\sigma}\hat{U}^\dagger) = F(\hat{\rho}, \hat{\sigma})$, so that F is basis-independent. We show in Appendix T that the natural definition of the fidelity for mixed states

$$F(\hat{\rho}, \hat{\sigma}) = \left(\mathrm{Tr}\left|\hat{\rho}^{1/2}\hat{\sigma}^{1/2}\right|\right)^2$$
$$= \left(\mathrm{Tr}\sqrt{\hat{\rho}^{1/2}\hat{\sigma}\hat{\rho}^{1/2}}\right)^2. \tag{8.51}$$

Other forms of the fidelity   You will often also find the square root of eqn 8.51 referred to as the fidelity. The quantity $\mathrm{Tr}(\hat{\rho}\hat{\sigma})$ is also sometimes called the fidelity.

To calculate this quantity, we find the eigenvalues of the positive operator $\hat{\rho}^{1/2}\hat{\sigma}\hat{\rho}^{1/2}$, take the positive square root of each of these, sum them, and then square the result. The fidelity can be quite difficult to calculate, but it takes a simple form for the states of single qubits:

$$F = \text{Tr}\left(\hat{\rho}\hat{\sigma}\right) + 2\sqrt{\det\left(\hat{\rho}\right)\det\left(\hat{\sigma}\right)}, \tag{8.52}$$

where $\det\left(\hat{\rho}\right)$ denotes the determinant of the 2×2 matrix associated with $\hat{\rho}$. If $\hat{\sigma} = |\psi\rangle\langle\psi|$ then the corresponding determinant will be zero and

$$F = \text{Tr}\left(\hat{\rho}\hat{\sigma}\right) = \langle\psi|\hat{\rho}|\psi\rangle, \tag{8.53}$$

which agrees with eqn 8.50. At the other extreme, if $\hat{\sigma}$ is the maximally mixed state $\frac{1}{2}\hat{\text{I}}$ then

$$\begin{aligned}
F &= \frac{1}{2}\text{Tr}\left(\hat{\rho}\,\hat{\text{I}}\right) + 2\sqrt{\det\left(\hat{\rho}\right)\det\frac{1}{2}\left(\hat{\text{I}}\right)} \\
&= \frac{1}{2} + \sqrt{\det\left(\hat{\rho}\right)},
\end{aligned} \tag{8.54}$$

which is always greater than or equal to $\frac{1}{2}$. This is a consequence of the fact that $\frac{1}{2}\hat{\text{I}}$ has an overlap of $\frac{1}{2}$ with any qubit pure state.

Perhaps the most natural way to quantify the difference between quantum states is by our ability to discriminate between them. We saw in Section 4.4, in particular, that there exists an optimum strategy for discriminating between two (or more) candidate quantum states with minimum probability of error. A natural way to quantify the distance between the states $\hat{\rho}$ and $\hat{\sigma}$ is by the minimum-error probability for equal a priori probability,

$$P_e^{\min} = \frac{1}{2}\left(1 - \frac{1}{2}\text{Tr}\,|\hat{\rho} - \hat{\sigma}|\right), \tag{8.55}$$

where $\text{Tr}\,|\hat{\rho} - \hat{\sigma}|$ is the sum of the positive eigenvalues of $\hat{\rho} - \hat{\sigma}$ minus the sum of the negative eigenvalues. If this minimum-error probability is small then the states are rather distinct. If the error probability is large (close to $\frac{1}{2}$) then the difference will be correspondingly difficult to detect. These considerations lead us to define the trace distance between the states with density operators $\hat{\rho}$ and $\hat{\sigma}$ to be

$$D\left(\hat{\rho}, \hat{\sigma}\right) = \frac{1}{2}\text{Tr}\,|\hat{\rho} - \hat{\sigma}|. \tag{8.56}$$

Unlike the fidelity, the trace distance is a true *distance* in that it satisfies the natural conditions introduced in Section 1.4 for the Hamming distance,

Kolmogorov distance   The trace distance is the quantum analogue of a classical quantity, the Kolmogorov distance, between two probability distributions $P = \{p_i\}$ and $Q = \{q_i\}$:

$$D(P, Q) = \frac{1}{2}\sum_i |p_i - q_i|.$$

$$\begin{aligned}
&D\left(\hat{\rho}, \hat{\sigma}\right) \geq 0, \\
&D\left(\hat{\rho}, \hat{\sigma}\right) = 0 \quad \Leftrightarrow \quad \hat{\rho} = \hat{\sigma}, \\
&D\left(\hat{\rho}, \hat{\sigma}\right) = D\left(\hat{\sigma}, \hat{\rho}\right), \\
&D\left(\hat{\rho}, \hat{\sigma}\right) \leq D\left(\hat{\rho}, \hat{\rho}'\right) + D\left(\hat{\rho}', \hat{\sigma}\right).
\end{aligned} \tag{8.57}$$

he first three of these are apparent immediately from the definition
$D(\hat{\rho}, \hat{\sigma})$. The final one, however, needs a little more work. In order
to prove it, we first note that $\hat{\rho} - \hat{\sigma}$ has real eigenvalues which can be
positive, negative, or zero. Let $\hat{P}_+$ be the projector onto the eigenvectors with positive eigenvalues and $\hat{P}_-$ be the projector onto those with
negative eigenvalues. It necessarily follows, of course, that

$$\hat{P}_+\hat{P}_- = 0 = \hat{P}_-\hat{P}_+. \tag{8.58}$$

By using these projectors, we can rewrite $D(\hat{\rho}, \hat{\sigma})$ in the form

$$
\begin{aligned}
D(\hat{\rho}, \hat{\sigma}) &= \frac{1}{2}\mathrm{Tr}\left[\hat{P}_+(\hat{\rho} - \hat{\sigma})\right] - \frac{1}{2}\mathrm{Tr}\left[\hat{P}_-(\hat{\rho} - \hat{\sigma})\right] \\
&= \mathrm{Tr}\left[\hat{P}_+(\hat{\rho} - \hat{\sigma})\right],
\end{aligned} \tag{8.59}
$$

where we have used the fact that

$$\mathrm{Tr}\left[\hat{P}_+(\hat{\rho} - \hat{\sigma})\right] + \mathrm{Tr}\left[\hat{P}_-(\hat{\rho} - \hat{\sigma})\right] = \mathrm{Tr}(\hat{\rho} - \hat{\sigma}) = 0. \tag{8.60}$$

then follows that

$$
\begin{aligned}
D(\hat{\rho}, \hat{\sigma}) &= \mathrm{Tr}\left[\hat{P}_+(\hat{\rho} - \hat{\sigma})\right] \\
&= \mathrm{Tr}\left[\hat{P}_+(\hat{\rho} - \hat{\rho}')\right] + \mathrm{Tr}\left[\hat{P}_+(\hat{\rho}' - \hat{\sigma})\right] \\
&\le D(\hat{\rho}, \hat{\rho}') + D(\hat{\rho}', \hat{\sigma}),
\end{aligned} \tag{8.61}
$$

where the last line follows from the fact that $\hat{P}_+$ is the projector onto
the space of eigenstates with positive eigenvalues of $\hat{\rho} - \hat{\sigma}$, but not of
$\hat{\rho} - \hat{\rho}'$ or $\hat{\rho}' - \hat{\sigma}$.

The trace distance is usually easier to calculate than the fidelity and
takes a particularly simple form for a single qubit. We saw, in Section
4, that we can express any single-qubit state in terms of its (three-
dimensional) Bloch vector. Hence we can write our two single-qubit
density operators in the form

$$
\begin{aligned}
\hat{\rho} &= \frac{1}{2}\left(\hat{I} + \vec{r} \cdot \hat{\vec{\sigma}}\right), \\
\hat{\sigma} &= \frac{1}{2}\left(\hat{I} + \vec{s} \cdot \hat{\vec{\sigma}}\right).
\end{aligned} \tag{8.62}
$$

then follows that the trace distance is

$$
\begin{aligned}
D(\hat{\rho}, \hat{\sigma}) &= \frac{1}{4}\mathrm{Tr}\left|(\vec{r} - \vec{s}) \cdot \hat{\vec{\sigma}}\right| \\
&= \frac{1}{2}|\vec{r} - \vec{s}|,
\end{aligned} \tag{8.63}
$$

which is one-half of the distance between the two Bloch vectors. In
deriving this result, we have used the fact that the two eigenvalues of
$(\vec{r} - \vec{s}) \cdot \hat{\vec{\sigma}}$ are $\pm|\vec{r} - \vec{s}|$.

A number of simple and important inequalities are known for the trace distance. There is, in particular, a convexity condition in the form

$$D\left(\sum_i p_i \hat{\rho}_i, \sum_i q_i \hat{\sigma}_i\right) \leq \sum_i p_i D\left(\hat{\rho}_i, \hat{\sigma}_i\right) + D(P, Q), \qquad (8.64)$$

where $D(P, Q)$ is the Kolmogorov distance between the probability distributions $\{p_i\}$ and $\{q_i\}$:

$$D(P, Q) = \frac{1}{2} \sum_i |p_i - q_i|. \qquad (8.65)$$

From eqn 8.64 it follows that the trace distance is jointly convex in the two density operators:

$$D\left(\sum_i p_i \hat{\rho}_i, \sum_i p_i \hat{\sigma}_i\right) \leq \sum_i p_i D\left(\hat{\rho}_i, \hat{\sigma}_i\right). \qquad (8.66)$$

It can also be shown that no physically allowed operation can increase $D(\hat{\rho}, \hat{\sigma})$, so that

$$D\left(\sum_i \hat{A}_i \hat{\rho} \hat{A}_i^\dagger, \sum_i \hat{A}_i \hat{\sigma} \hat{A}_i^\dagger\right) \leq D(\hat{\rho}, \hat{\sigma}). \qquad (8.67)$$

There is a mathematical proof of this, but we can see that it must be true by referring back to the connection with minimum-error discrimination (see eqn 8.55). Clearly, no operation can *decrease* this minimum error probability or it would not be the minimum error probability! It then follows that no operation can *increase* $D(\hat{\rho}, \hat{\sigma})$. A simple example of this is the partial trace operation, for which we find

$$D(\hat{\rho}_A, \hat{\sigma}_A) \leq D(\hat{\rho}_{AB}, \hat{\sigma}_{AB}). \qquad (8.68)$$

This inequality has the simple meaning that throwing away information about the B system cannot make it easier to determine whether the A and B systems were prepared in the state $\hat{\rho}_{AB}$ or $\hat{\sigma}_{AB}$.

The fidelity and trace distance are appropriate measures of the difference between quantum states if we have only one copy, or perhaps a small number of copies, of the state. If we have a large number then, as with classical information theory, an entropic measure of distinguishability will probably be more appropriate. We find that it is the relative entropy, also known in this context as the Kullback–Leibler distance, that provides the natural quantity. To see how this arises, we consider the classical problem of discriminating between two distinct probabilities for two possible values associated with an event. Let these two pairs of probabilities be $P = \{p, 1-p\}$ and $Q = \{q, 1-q\}$. After observations of a large number N of events, we assign the first set of probabilities if the number of occurrences of the first outcome is $\approx Np$ and the second if the number is $\approx Nq$. In the unlikely event that the number of first outcomes is not close to either Np or Nq, we can consider the test to have

een inconclusive and make further observations. An error in assigning
ıe probability will occur if, for example, the true probabilities are q
ınd $1 - q$ but the number of occurrences of the first outcome is $\approx Np$.
Ve can estimate this probability by first noting that the probability for
ınding n occurrences of the first outcome (and $N - n$ of the second) if
ıe probabilities are q and $1 - q$ is

$$P(n|Q) = \frac{N!}{n!(N-n)!} q^n (1-q)^{N-n}. \qquad (8.69)$$

Ve can approximate the probability for then identifying the distribution
ş P given that it was in fact Q by the same approach as that used to
ɔtain Shannon's noisy coding theorem in Section 1.4. We use Stirling's
ɔproximation to write

$$\begin{aligned}
P(P|Q) &\approx \frac{N!}{(Np)![N(1-p)]!} q^{Np} (1-q)^{N(1-p)} \\
&\approx 2^{-Np \log p - N(1-p) \log(1-p)} 2^{Np \log q + N(1-p) \log(1-q)} \\
&= 2^{-NH(P\|Q)}. \qquad (8.70)
\end{aligned}$$

ımilarly, the probability for mistakenly identifying the set of probabil-
ıes P as Q is

$$P(Q|P) \approx 2^{-NH(Q\|P)} \neq 2^{-NH(P\|Q)}. \qquad (8.71)$$

he larger the value of $H(P\|Q)$, the smaller is the probability that Q
ıll be mistakenly identified as P. Equivalently, the larger the value
 $H(P\|Q)$, the smaller is the number N of observations required to
 entify the distribution as Q to any given level of confidence.

If we have a large number of copies of a quantum system, each of which
 prepared in the state $\hat{\rho}$ or $\hat{\sigma}$, the natural measure of distinguishability
 the quantum relative entropy. If we have N copies then the probability
ıat the optimal measurement will identify the state as $\hat{\sigma}$ given that it
as $\hat{\rho}$ is

$$P(\hat{\sigma}|\hat{\rho}) \approx 2^{-NS(\hat{\sigma}\|\hat{\rho})}. \qquad (8.72)$$

milarly, the probability that a state $\hat{\sigma}$ will be identified as $\hat{\rho}$ is given
 the different probability

$$P(\hat{\rho}|\hat{\sigma}) \approx 2^{-NS(\hat{\rho}\|\hat{\sigma})}. \qquad (8.73)$$

he quantum relative entropy is not a distance, in that it does not satisfy
ıe symmetry property of a distance:

$$S(\hat{\rho}\|\hat{\sigma}) \neq S(\hat{\sigma}\|\hat{\rho}). \qquad (8.74)$$

 does qualify, however, as a quantitative measure of the distinguisha-
lity of the states $\hat{\rho}$ and $\hat{\sigma}$. The larger the value of $S(\hat{\sigma}\|\hat{\rho})$, the easier it
 to identify the state correctly as $\hat{\rho}$. Equivalently, the larger the value
 $S(\hat{\sigma}\|\hat{\rho})$, the smaller is the number N of copies of the system required
 identify the state as $\hat{\rho}$ with any given degree of confidence.

Physical operations tend to make states less distinguishable and, as with the trace distance, this is reflected in the change of the relative entropy:

$$S\left(\sum_i \hat{A}_i \hat{\rho} \hat{A}_i^\dagger \| \sum_i \hat{A}_i \hat{\sigma} \hat{A}_i^\dagger\right) \leq S\left(\hat{\rho} \| \hat{\sigma}\right). \tag{8.75}$$

This means, in particular, that

$$S\left(\hat{\sigma}_A \| \hat{\rho}_A\right) \leq S\left(\hat{\sigma}_{AB} \| \hat{\rho}_{AB}\right), \tag{8.76}$$

which reflects the fact that discarding information about the B system cannot make it easier to discriminate between the states $\hat{\rho}_{AB}$ and $\hat{\sigma}_{AB}$. A closely related idea is the joint convexity of the quantum relative entropy,

$$S\left(p_1 \hat{\sigma}_1 + p_2 \hat{\sigma}_2 \| p_1 \hat{\rho}_1 + p_2 \hat{\rho}_2\right) \leq p_1 S\left(\hat{\sigma}_1 \| \hat{\rho}_1\right) + p_2 S\left(\hat{\sigma}_2 \| \hat{\rho}_2\right), \tag{8.77}$$

so that mixing states makes states less distinct.

We conclude our discussion of the quantum relative entropy by showing how the property of strong subadditivity (eqn 8.46) follows as a consequence of eqn 8.76. Let $\hat{\rho}_{ABC}$ and $(d_A)^{-1}\hat{I} \otimes \hat{\rho}_{BC}$, where $\hat{\rho}_{BC} = \text{Tr}_A \hat{\rho}_{ABC}$ and d_A is the dimension of the A state space, be two possible states of three quantum systems, which we label A, B, and C. The quantum relative entropy for these states has the form

$$S\left(\hat{\rho}_{ABC} \| \frac{1}{d_A}\hat{I} \otimes \hat{\rho}_{BC}\right) = \text{Tr}_{ABC}\left(\hat{\rho}_{ABC} \log \hat{\rho}_{ABC}\right) + \log d_A$$
$$- \text{Tr}_{BC}\left(\hat{\rho}_{BC} \log \hat{\rho}_{BC}\right)$$
$$= -S(ABC) + S(BC) + \log d_A. \tag{8.78}$$

It follows from eqn 8.76 that

$$S\left(\hat{\rho}_{ABC} \| \frac{1}{d_A}\hat{I} \otimes \hat{\rho}_{BC}\right) \geq S\left(\hat{\rho}_{AB} \| \frac{1}{d_A}\hat{I} \otimes \hat{\rho}_B\right)$$
$$= -S(AB) + S(B) + \log d_A. \tag{8.79}$$

Combining eqns 8.78 and 8.79 then gives

$$-S(ABC) + S(BC) \geq -S(AB) + S(B), \tag{8.80}$$

which we recognize as the condition for strong subadditivity in eqn 8.46.

It is useful to be able to quantify the quality of a device, such as a quantum gate or circuit, designed to implement a desired unitary transformation

$$\hat{\rho} \rightarrow \hat{\rho}' = \hat{U}\hat{\rho}\hat{U}^\dagger. \tag{8.81}$$

The device, inevitably, will be imperfect and the actual transformation will be an operation

$$\hat{\rho} \rightarrow \hat{\rho}'' = \sum_i \hat{A}_i \hat{\rho} \hat{A}_i^\dagger. \tag{8.82}$$

We can apply any of our measures for state comparison to $\hat{\rho}'$ and $\hat{\rho}''$, and these will reflect the difference between the desired unitary transformation in eqn 8.81 and the operation in eqn 8.82. In order to do this we need to choose an input state, and it is reasonable to choose the pure state $\hat{\rho} = |\psi\rangle\langle\psi|$ for which the output states $\hat{\rho}'$ and $\hat{\rho}''$ are as different as possible. The fidelity, for example, of the device is then

$$F = \text{Inf}_{|\psi\rangle} \langle\psi|\hat{U}^\dagger \hat{\rho}'' \hat{U}|\psi\rangle$$
$$= \text{Inf}_{|\psi\rangle} \sum_i \left|\langle\psi|\hat{U}^\dagger \hat{A}_i|\psi\rangle\right|^2, \qquad (8.83)$$

where the state $|\psi\rangle$ is chosen to give the minimum value. This quantity will take its maximum value of unity only if the device is perfect.

As an example, let us suppose that we require a device to perform as a Pauli-X gate and perform the unitary transformation $\hat{\sigma}_x$. In reality, however, the gate produces the unitary transformation

$$\hat{V} = \cos\theta\hat{\sigma}_x + i\sin\theta\hat{I}, \qquad (8.84)$$

so that the gate fidelity is

$$F = \text{Inf}_{|\psi\rangle} \left|\langle\psi|\hat{\sigma}_x \left(\cos\theta\hat{\sigma}_x + i\sin\theta\hat{I}\right)|\psi\rangle\right|^2$$
$$= \text{Inf}_{|\psi\rangle} \left|\langle\psi|\cos\theta\hat{I} + i\sin\theta\hat{\sigma}_x|\psi\rangle\right|^2. \qquad (8.85)$$

The minimum value occurs if we choose for $|\psi\rangle$ any equally weighted superposition of the eigenstates of $\hat{\sigma}_x$, for example the state $|0\rangle$:

$$F = \left|\langle 0|\cos\theta\hat{I} + i\sin\theta\hat{\sigma}_x|0\rangle\right|^2 = \cos^2\theta. \qquad (8.86)$$

The fidelity for a desired operation is sometimes also defined in terms of the average performance of the gate by considering its action on all possible input states. This average fidelity is useful in determining the threshold for some quantum operations such as cloning, as described in Appendix F.

8.4   Measures of entanglement

Entanglement is, as we have seen, a quintessentially quantum phenomenon and plays a central role in our subject. It makes possible teleportation and quantum dense coding and underlies the increased speed of quantum algorithms. Entangled systems, especially when shared, are very much a resource for quantum communications and information processing. As such, it is important to be able to quantify the amount of entanglement associated with any given state. We shall find that this task is reasonably straightforward for pure states. It is rather more challenging for mixed entangled states.

We can use any of the quantities introduced in the preceding section to compare the state of interest with the unentangled states. Let us denote by $\hat{\rho}^{\text{unent}}$ the unentangled states, which, for bipartite systems, have the general form

$$\hat{\rho}^{\text{unent}} = \sum_i p_i \hat{\rho}_A^i \otimes \hat{\rho}_B^i, \tag{8.87}$$

where the p_i are positive and $\sum_i p_i = 1$. Comparing the given state with these unentangled states then provides a quantitative measure of the entanglement. The simplest measure of entanglement is the distance between our state and the *nearest* unentangled state,

$$D_{\text{ent}}(\hat{\rho}) = \text{Inf}_{\hat{\rho}^{\text{unent}}} \frac{1}{2} \left| \hat{\rho} - \hat{\rho}^{\text{unent}} \right|, \tag{8.88}$$

where the state $\hat{\rho}^{\text{unent}}$ is chosen to give the minimum distance. In this way we identify the unentangled state most similar to our entangled state and quantify the entanglement by our ability to discriminate, using a minimum-error measurement, between the given state and the most similar unentangled state. It is straightforward to calculate this distance if our entangled state is pure. To do this we first write the Schmidt decomposition of the state,

$$|\psi\rangle_{AB} = \sum_n a_n |\lambda_n\rangle_A |\phi_n\rangle_B. \tag{8.89}$$

The unentangled state that is nearest to $|\psi\rangle$ will clearly be a mixture of product states of the form $|\lambda_n\rangle_A |\phi_n\rangle_B$. It is easy to see that the nearest unentangled state has the form

$$\hat{\rho}^{\text{unent}} = \sum_n |a_n|^2 |\lambda_n\rangle_A {}_A\langle\lambda_n| \otimes |\phi_n\rangle_B {}_B\langle\phi_n|, \tag{8.90}$$

which has the same diagonal elements as $|\psi\rangle_{AB}$ in the Schmidt basis. Consider, for example, the two-qubit entangled state

$$|\psi\rangle = \cos\theta |0\rangle \otimes |0\rangle + \sin\theta |1\rangle \otimes |1\rangle. \tag{8.91}$$

The nearest unentangled state is

$$\hat{\rho}^{\text{unent}} = \cos^2\theta |0\rangle\langle 0| \otimes |0\rangle\langle 0| + \sin^2\theta |1\rangle\langle 1| \otimes |1\rangle\langle 1| \tag{8.92}$$

and the associated distance is

$$D_{\text{ent}} = |\cos\theta \sin\theta|. \tag{8.93}$$

This is clearly zero only if $\cos\theta = 0$ or $\sin\theta = 0$, in which case the state $|\psi\rangle$ is a product state. It takes its maximum value of $\frac{1}{2}$ when the state coincides with one of the maximally entangled Bell states.

If we adopt the fidelity as our measure of entanglement then we find a very different form for the most similar unentangled state. In this case our measure of entanglement takes the form

$$F_{\text{ent}}(\hat{\rho}) = \text{Sup}_{\hat{\rho}^{\text{unent}}} \left(\text{Tr}\sqrt{\hat{\rho}^{1/2} \hat{\rho}^{\text{unent}} \hat{\rho}^{1/2}} \right)^2, \tag{8.94}$$

here the state $\hat{\rho}^{\text{unent}}$ is chosen to give the maximum fidelity. In adopting this fidelity-based measure, we are seeking the unentangled state most likely to pass as the given entangled state. For the pure state in eqn 8.89, we find that

$$F_{\text{ent}} = \text{Sup}_{\hat{\rho}^{\text{unent}}} \langle \psi | \hat{\rho}^{\text{unent}} | \psi \rangle. \tag{8.95}$$

The maximization is straightforward; we need only choose for our unentangled state the pure product state $|\lambda_n\rangle_A |\phi_n\rangle_B$ for which $|a_n|$ is the greatest:

$$F_{\text{ent}} = \text{Sup}_n |\langle \psi | \lambda_n, \phi_n \rangle|^2 = \text{Sup}_n |a_n|^2. \tag{8.96}$$

This means, in particular, that the *distance* between the Bell state $|\Psi^-\rangle$ and an unentangled state is minimized by choosing

$$\hat{\rho}^{\text{unent}} = \frac{1}{2} |01\rangle\langle 01| + \frac{1}{2} |10\rangle\langle 10|, \tag{8.97}$$

but that the *fidelity* with an unentangled state is maximized by choosing the product state

$$\hat{\rho}^{\text{unent}} = |01\rangle\langle 01|. \tag{8.98}$$

These give $D_{\text{ent}} = \frac{1}{2}$ and $F_{\text{ent}} = \frac{1}{2}$.

Perhaps the most important and widely used measures of entanglement are based on the von Neumann entropy. We can associate these with tasks that we might like to perform given a large number of copies of our entangled systems. In seeking appropriate entropic measures, we can be guided by the following natural properties. (i) If $\hat{\rho}$ is unentangled then $E(\hat{\rho}) = 0$. (ii) Any local unitary transformations should leave $E(\hat{\rho})$ unchanged:

$$E\left(\hat{U}_A \otimes \hat{U}_B \, \hat{\rho} \, \hat{U}_A^\dagger \otimes \hat{U}_B^\dagger \right) = E(\hat{\rho}). \tag{8.99}$$

(iii) Any local operations, including measurements, should not, on average, increase the entanglement:

$$\sum_i p_i E(\hat{\rho}_i) \leq E(\hat{\rho}), \tag{8.100}$$

where the $\hat{\rho}_i$ are the states produced by the operations,

$$\hat{\rho}_i = \frac{1}{p_i} \hat{A}_i \otimes \hat{B}_i \, \hat{\rho} \, \hat{A}_i^\dagger \otimes \hat{B}_i^\dagger, \tag{8.101}$$

where

$$p_i = \text{Tr}_{AB} \left(\hat{\rho} \, \hat{A}_i^\dagger \hat{A}_i \otimes \hat{B}_i^\dagger \hat{B}_i \right). \tag{8.102}$$

The relative entropy of entanglement is defined to be the smallest value of the quantum relative entropy for our state $\hat{\rho}$ and an unentangled state $\hat{\rho}^{\text{unent}}$:

$$E_{\text{RE}}(\hat{\rho}) = \text{Inf}_{\hat{\rho}^{\text{unent}}} S(\hat{\rho} \| \hat{\rho}^{\text{unent}}). \tag{8.103}$$

This quantity will clearly be zero if $\hat{\rho}$ is an unentangled state and will be positive otherwise. It is also clear that $E_{\text{RE}}(\hat{\rho})$ is unchanged by local unitary transformations. The final property is also true but it requires a bit more work to show this. For the pure state in eqn 8.89, the relative entropy of entanglement has the simple form

$$E_{\text{RE}}\left(|\psi\rangle\langle\psi|\right) = -\text{Inf}_{\hat{\rho}^{\text{unent}}}\langle\psi| \log\left(\hat{\rho}^{\text{unent}}\right)|\psi\rangle. \qquad (8.104)$$

It is straightforward to show that the required form of $\hat{\rho}^{\text{unent}}$ is the same as that found in our discussion of the distance measure of entanglement, given in eqn 8.90. This leads to the appealingly simple result

$$E_{\text{RE}}\left(|\psi\rangle\langle\psi|\right) = S\left(\hat{\rho}_A\right) = S\left(\hat{\rho}_B\right). \qquad (8.105)$$

For a pair of qubits, this ranges from unity for a maximally entangled state to zero for a product state.

There are other entropic measures of entanglement, including, notably, the entanglement of distillation and the entanglement of formation. These arise in the theory of quantum communications and will be described in the following section.

8.5   Quantum communications theory

The coding theorems of Shannon provide the fundamental limits for the transmission of information using a classical communication channel. In quantum communication, however, we encode quantum information in the state of the physical system to be transmitted. The existence of non-orthogonal states, incompatible observables, and entangled states leads, as we have seen, to intrinsically quantum phenomena such as quantum key distribution and teleportation. A quantitative description of these requires a quantum theory of communication.

A quantum communication device consists of a quantum channel, as introduced in Section 3.2, through which qubits or other quantum systems can be sent from Alice to Bob, and this may be accompanied, as in quantum key distribution, by a classical communication channel. Figure 8.1 is a schematic representation of such a communication system. In the preparation event A Alice selects, with probability $P(a_i)$, one of the messages a_i and encodes this as the quantum state $\hat{\rho}_i$ of a suitable quantum system such as the polarization of a photon. This quantum state may be affected by noise, including absorption losses, before arriving at Bob. Bob may store the system for future use or extract some information by performing a measurement. In the former situation, the channel should be characterized by comparing the state received with that sent using, for example, one of the quantities described in Section 8.3. In the latter, Bob needs to make a choice of measurement and this will, in general, correspond to a generalized measurement described by a set of probability operators $\{\hat{\pi}_j\}$. The properties of this quantum communication channel are then determined by the conditional probabilities $P(b_j|a_i)$, which, in the absence of noise on the channel, are

$$P(b_j|a_i) = \text{Tr}\left(\hat{\pi}_j\hat{\rho}_i\right). \qquad (8.106)$$

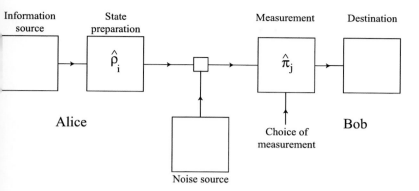

Fig. 8.1 Schematic representation of a quantum communication channel.

We shall assume that the classical channel, if one is required, has been optimized to transmit classical information without error, in accordance with Shannon's theorems.

It is perhaps most natural to start by asking how well our quantum channel communicates classical information. For a classical channel, each physical bit has only two distinct states, corresponding to the logical values 0 and 1, and this limits the capacity of the channel (naturally enough) to one bit for each physical bit. A qubit, however, can be prepared in either of the orthogonal states $|0\rangle$ and $|1\rangle$, or in any superposition of these. Given this wide variety of possible states, it is at least conceivable that a quantum channel might allow us to achieve more than one bit per qubit. That this is not the case is a consequence of the fundamental limitations on our ability to discriminate between non-orthogonal states, as described in Section 4.4. It is not possible, therefore, to predict the outcome of Bob's measurement, even for a *noiseless* quantum channel, and in this way the quantum channel transmits classical information in a similar manner to a classical *noisy* channel in that

$$P(b_j|a_i) \neq \delta_{ij}. \tag{8.107}$$

We found in Section 1.4 that Shannon's noisy-channel coding theorem provides the limit to the channel capacity in this situation through the mutual information:

$$C = \operatorname{Sup} H(A:B). \tag{8.108}$$

Here the mutual information is

$$H(A:B) = \sum_{ij} P(a_i)\operatorname{Tr}(\hat{\pi}_j\hat{\rho}_i) \log\left[\frac{\operatorname{Tr}(\hat{\pi}_j\hat{\rho}_i)}{\operatorname{Tr}(\hat{\pi}_j\hat{\rho})}\right], \tag{8.109}$$

where, as usual, $\hat{\rho}$ is the prior density operator

$$\hat{\rho} = \sum_i P(a_i)\hat{\rho}_i. \tag{8.110}$$

The maximization in eqn 8.108 is more complicated than its classical counterpart in that, in addition to varying the preparation probabilities

$P(a_i)$, we also need to vary the probability operators associated with Bob's choice of measurement. It is perhaps for this reason that there are very few sets of signal states $\{\hat{\rho}_i\}$ for which C is known.

One simple example is a qubit channel in which we use the two non-orthogonal pure states discussed in Section 4.4:

$$|\psi_1\rangle = \cos\theta|0\rangle + \sin\theta|1\rangle,$$
$$|\psi_2\rangle = \cos\theta|0\rangle - \sin\theta|1\rangle, \tag{8.111}$$

where $0 \leq \theta \leq \pi/4$. In this case the mutual information is maximized if Alice selects these with equal probability, so that $P(a_1) = P(a_2) = \frac{1}{2}$, and Bob performs a measurement optimized to give the minimum probability of error. The channel capacity in this case is

$$C = \frac{1}{2}\left[1 + \sin(2\theta)\right]\log\left[1 + \sin(2\theta)\right] + \frac{1}{2}\left[1 - \sin(2\theta)\right]\log\left[1 - \sin(2\theta)\right]. \tag{8.112}$$

This takes its minimum value of zero when $\theta = 0$ and the two states are identical. It has a maximum value of one bit when $\theta = \pi/4$ and the states are orthogonal.

It is possible to place an upper bound on the mutual information and, through this, on the channel capacity. If the signal states are $\hat{\rho}_i$ and these are selected with probabilities $P(a_i)$ then

$$H(A:B) \leq S(\hat{\rho}) - \sum_i P(a_i)S(\hat{\rho}_i) = \chi. \tag{8.113}$$

This inequality is due to Holevo, and the quantity χ is referred to as the Holevo bound. We can obtain this by comparing $H(A:B)$ with the von Neumann mutual information $S(A:B)$ introduced in Section 8.2. In order to proceed, let us suppose that Alice and Bob share a pair of systems prepared in the correlated (but not entangled) state

$$\hat{\rho}_{AB} = \sum_i P(a_i)|a_i\rangle_A {}_A\langle a_i| \otimes \hat{\rho}_{Bi}, \tag{8.114}$$

where the states $\{|a_i\rangle\}$ are mutually orthogonal: $\langle a_i|a_j\rangle = \delta_{ij}$. It is clear that this system will operate in the same way as our communication channel if Alice makes a measurement on her system in the $\{|a_i\rangle\}$ basis. She will get the result corresponding to the state $|a_i\rangle$ with probability $P(a_i)$ and the state of Bob's system will then be $\hat{\rho}_i$, so that Bob receives one of the states $\hat{\rho}_i$ with the same probability $P(a_i)$. The observed degree of correlation, as expressed in the value of $H(A:B)$, cannot exceed that which is already present between the A and B systems in the state $\hat{\rho}_{AB}$ and it follows, therefore, that

$$
\begin{aligned}
H(A:B) &\leq S(A:B) \\
&= S(A) + S(B) - S(A,B) \\
&= S\left(\sum_i P(a_i)\hat{\rho}_i\right) - \sum_i P(a_i)S(\hat{\rho}_i) \\
&= \chi, \tag{8.115}
\end{aligned}
$$

hich is the Holevo bound.

We began our discussion of the capacity of a quantum channel by
uggesting that it might be possible to convey more than one bit on
ach qubit. That this is not possible is, perhaps, the most fundamental
onsequence of the Holevo bound. Clearly, χ will take its maximum
alue if the signal states $\hat{\rho}_i$ are all pure, so that $S(\hat{\rho}_i) = 0$, and they are
elected so that $\hat{\rho}$ is the maximally mixed state. This leads to the limit

$$\chi \leq \log d, \tag{8.116}$$

here d is the dimension of the state space of the system sent from
lice to Bob. It follows, in particular, that $\chi \leq 1$ bit for each qubit
id this then is the upper limit for the single-qubit capacity. We can
adily reach this value by selecting, with equal probability, one of two
thogonal qubit states and instructing Bob to make a von Neumann
easurement in this basis.

The Holevo bound can be thought of as a consequence of quantum
omplementarity or, equivalently, the limits imposed by quantum the-
y on our ability to discriminate between non-orthogonal states. To
ustrate this idea let us suppose that, as in the BB84 protocol for
uantum key distribution, Alice prepares each qubit in one of the four
ates $|0\rangle$, $|1\rangle$, $2^{-1/2}(|0\rangle + |1\rangle)$ and $2^{-1/2}(|0\rangle - |1\rangle)$. If Bob could discrim-
ate between these with certainty then the channel capacity would be
o bits per qubit, but this exceeds the Holevo bound of one bit. The
olevo bound, therefore, is a manifestation of the complementarity of
compatible observables, in this case $\hat{\sigma}_z$ and $\hat{\sigma}_x$.

The fact that a single qubit can carry at most one bit of information
es not mean, of course, that qubits are simply equivalent to bits.
e have seen this already in quantum key distribution but it might be
lpful to have a further, more direct, example. Let us suppose that
ice has two bits, A and B, which she wishes to convey to Bob. Bob
ll use one of these but Alice does not know which. The problem is
at Alice has at her disposal only a single bit to send to Bob. All she
n do is to select at random one of her bit values and send this to Bob.
she guesses correctly then Bob will have the desired bit value, but if
r guess was incorrect then the transmitted bit value will be correct
th probability 1/2. It follows that this strategy will provide Bob with
e correct desired bit value with probability 3/4. If Alice has a single
bit, however, then she can do better that this. She encodes each of
e four possible bit pairs onto the state of the qubit according to the
heme

$$
\begin{aligned}
00 &\rightarrow |0\rangle, \\
11 &\rightarrow |1\rangle, \\
01 &\rightarrow \frac{1}{\sqrt{2}}(|0\rangle + |1\rangle), \\
10 &\rightarrow \frac{1}{\sqrt{2}}(|0\rangle - |1\rangle).
\end{aligned}
\tag{8.117}
$$

Each of these pure signal states will be sent with probability $1/4$, and it follows that the Holevo bound in this case will be $\chi = 1$ bit. Bob assigns the desired bit value on the basis of the result of a measurement of one of two observables: $2^{-1/2}(\hat{\sigma}_z + \hat{\sigma}_z)$ if he needs the first qubit and $2^{-1/2}(\hat{\sigma}_z - \hat{\sigma}_z)$ if he needs the second. The probability that this gives the correct bit value is then

$$
\begin{aligned}
P_c = {} & \frac{1}{4}\langle 0|\frac{1}{2}\left[\hat{I} + \frac{1}{\sqrt{2}}(\hat{\sigma}_z + \hat{\sigma}_x)\right]|0\rangle + \frac{1}{4}\langle 1|\frac{1}{2}\left[\hat{I} - \frac{1}{\sqrt{2}}(\hat{\sigma}_z + \hat{\sigma}_x)\right]|1\rangle \\
& + \frac{1}{8}\left(\langle 0| + \langle 1|\right)\frac{1}{2}\left[\hat{I} - \frac{1}{\sqrt{2}}(\hat{\sigma}_z - \hat{\sigma}_x)\right]\left(|0\rangle + |1\rangle\right) \\
& + \frac{1}{8}\left(\langle 0| - \langle 1|\right)\frac{1}{2}\left[\hat{I} + \frac{1}{\sqrt{2}}(\hat{\sigma}_z - \hat{\sigma}_x)\right]\left(|0\rangle - |1\rangle\right) \\
= {} & \frac{1}{2}\left(1 + \frac{1}{\sqrt{2}}\right) \approx 0.854,
\end{aligned}
\tag{8.118}
$$

which clearly exceeds the single-bit value of 0.75.

The fact that non-orthogonal quantum states have a non-zero overlap suggests the possibility that a quantum channel based on such states might exhibit a level of redundancy and that removing this should allow us to send the quantum information using a smaller number of qubits. As an illustration, let us consider a set of three qubits, each of which is prepared in one of the two non-orthogonal states in eqn 8.111. There are clearly $2^3 = 8$ possible states of the three qubits, and if each of these is selected with equal probability then we can send the required states to Bob using three classical bits; all we need to do is to identify the sequence of states and leave Bob to use this information to construct the states for himself. The eight states are equiprobable, and it follows from Shannon's noiseless coding theorem that a minimum of three bits is required. Invoking the Holevo bound might lead us to infer that a minimum of three qubits is then required if a quantum channel is used. Suprisingly, however, it is possible to send the same information using fewer than three qubits. Our three-qubit state is

$$
\begin{aligned}
|\psi_j\rangle \otimes |\psi_k\rangle \otimes |\psi_l\rangle = {} & \cos^3\theta|000\rangle \\
& + \cos^2\theta\sin\theta\left((-1)^{j-1}|100\rangle\right. \\
& \left. + (-1)^{k-1}|010\rangle + (-1)^{l-1}|001\rangle\right) \\
& + \cos\theta\sin^2\theta\left((-1)^{j-k}|110\rangle\right. \\
& \left. + (-1)^{j-l}|101\rangle + (-1)^{k-l}|011\rangle\right) \\
& + \sin^3\theta(-1)^{j+k+l-1}|111\rangle.
\end{aligned}
\tag{8.119}
$$

Let us suppose that θ is small so that

$$
|\langle\psi_{1,2}|0\rangle| \gg |\langle\psi_{1,2}|1\rangle|.
\tag{8.120}
$$

It then follows that the probability amplitudes for the states $|000\rangle$, $|100\rangle$, $|010\rangle$, and $|001\rangle$ are very much larger than those for the other four states. We can compress the state to just two qubits by performing a unitary transformation that exchanges the states $|001\rangle$ and $|110\rangle$ whilst leaving the other states unchanged:

$$\hat{U} = |001\rangle\langle110| + |110\rangle\langle001| + \hat{I} \otimes \hat{I} \otimes \hat{I}$$
$$-|001\rangle\langle001| - |110\rangle\langle110|. \qquad (8.121)$$

If we then measure the third qubit in the computational basis we shall, with high probability, find the result corresponding to the state $|0\rangle$, and the state of the remaining two qubits will then be

$$|\psi_{jkl}\rangle = \frac{1}{\sqrt{1 + 2\sin^2\theta}} \left[\cos\theta|00\rangle + \sin\theta\left((-1)^{j-1}|10\rangle\right.\right.$$
$$\left.\left.+(-1)^{k-1}|01\rangle + (-1)^{l-1}|11\rangle\right)\right]. \qquad (8.122)$$

This state carries the three integers j, k, and l in the phases of the amplitudes for the states $|10\rangle$, $|01\rangle$, and $|11\rangle$. If this compressed two-qubit state is sent to Bob then he can supply a third qubit prepared in the state $|0\rangle$, and then invert the unitary transformation given in eqn 8.121 and so prepare a state which has a high fidelity with the initial uncompressed state in eqn 8.119.

The compression of quantum states becomes possible with a probability approaching unity in the limit of large numbers of qubits, in much the same way that Shannon's noiseless coding theorem applies, strictly, in the limit of long messages. It was shown by Schumacher that it is the von Neumann entropy that limits the extent to which such quantum compression is possible. Consider a sequence of N qubits, each of which has been prepared in one of the two states $|\psi_1\rangle$ and $|\psi_2\rangle$ defined in eqn 8.111. If each qubit is equally likely to be prepared in either state, then there are clearly 2^N equiprobable states:

$$|\psi_{j_1}\rangle \otimes |\psi_{j_2}\rangle \otimes \cdots |\psi_{j_N}\rangle\otimes = \left(\cos\theta|0\rangle - (-1)^{j_1}\sin\theta|1\rangle\right)$$
$$\otimes \left(\cos\theta|0\rangle - (-1)^{j_2}\sin\theta|1\rangle\right) \otimes \cdots$$
$$\otimes \left(\cos\theta|0\rangle - (-1)^{j_N}\sin\theta|1\rangle\right). \qquad (8.123)$$

It is instructive to begin our analysis of Schumacher compression by thinking about the effect of performing a measurement on this state. If we were to measure each qubit in the computational basis then we would find that the number of qubits giving the value 0 would be close to $N\cos^2\theta$ and that the number giving the value 1 would be close to $N\sin^2\theta$. To do so, of course, would destroy the superpositions and so erase any record of the information-bearing phases labelled by the integers j_n. If, however, we perform a collective von Neumann measurement on all of the qubits to determine the number of qubits in the state $|0\rangle$ and the number in the state $|1\rangle$ *without* determining the state of each individual qubit then we shall find that n qubits are in the state $|1\rangle$, where $n \approx N\sin^2\theta$. The post-measurement state will then be an equally weighted superposition of all of the states with n ones and $N - n$ zeros:

$$|\Psi_n\rangle = (-1)^n \sqrt{\frac{n!(N-n)!}{N!}} \sum_{k \in n \text{ ones}} (-1)^{\sum'_\ell j_\ell}|k\rangle, \qquad (8.124)$$

where the integers k are bit strings associated with the qubit states as described in Section 7.2. Each of the states $|k\rangle$ has a positive or negative amplitude depending on which of the 2^N possible states in eqn 8.123 was prepared; the primed sum runs over all of the indices j_n for which the state of the corresponding qubit contributes $|1\rangle$ to the qubit string $|k\rangle$. As with the compressed state in eqn 8.122, these positive and negative amplitudes retain information about the form of the original state.

The N-qubit state in eqn 8.124 is a superposition of only

$$W = \frac{N!}{n!(N-n)!} \approx 2^{NH(\cos^2 \theta)} \tag{8.125}$$

states $|k\rangle$, where

$$H(x) = -x \log x - (1-x) \log(1-x). \tag{8.126}$$

It follows that we can transform this state into one of about $NH(\cos^2 \theta)$ qubits. All we require is an N-qubit unitary transformation that transforms the state $|k\rangle$ corresponding to the smallest value of k into the approximately $NH(\cos^2 \theta)$-qubit state $|0\cdots00\rangle$, the state corresponding to the next smallest into the state $|0\cdots01\rangle$, and so on. At the end of this process about $N\left[1 - H(\cos^2 \theta)\right]$ qubits will be in the state $|0\rangle$ whatever the initial state and these can be discarded.

The final step in deriving Schumacher's noiseless coding theorem is to note that we do not need to perform a measurement of the number of zeros and ones in the string of qubits. Indeed, to do so would be to destroy the information in the initial state contained in the superposition of states corresponding to different numbers of zeros. It suffices that we know that if we were to perform such a measurement then the value obtained for the number of zeros would be, with very high probability, close to $N \cos^2 \theta$. This means that we can devise a suitable unitary transformation to encode the large-amplitude parts of our initial state (those for which the number of zeros n is sufficiently close to $N \cos^2 \theta$) with only slightly more than $H(N \cos^2 \theta)$ qubits. This is because, as in our derivation of Shannon's corresponding theorem, the total number of states $|k\rangle$ with appreciable amplitudes, corresponding to all likely values of n, is

$$W = 2^{N\left[H(\cos^2 \theta)+\delta\right]}, \tag{8.127}$$

where δ tends to zero as $N \to \infty$. An efficient unitary transformation will leave close to, but not more than, $N\left[1 - H(\cos^2 \theta)\right]$ qubits in the state $|0\rangle$ and these can therefore be discarded. The remaining set of approximately $NH(\cos^2 \theta)$ qubits forms the compressed quantum state. In this way, Alice can compress any of the initial 2^N states of her N qubits onto just $NH(\cos^2 \theta)$ qubits and send these to Bob. Bob can perform the decompression with near unit fidelity by supplying $N\left[1 - H(\cos^2 \theta)\right]$ qubits, each prepared in the state $|0\rangle$, and then inverting Alice's compressing unitary transformation.

A more precise statement of Schumacher's noiseless coding theorem is that a quantum state of many qubits cannot be compressed to fewer than

$H(\cos^2\theta)$ qubits and then reconstructed but that it can be compressed
nearly this length and then reconstructed with near unit fidelity. The
priori state of each qubit is

$$\hat{\rho} = \frac{1}{2}\left(|\psi_1\rangle\langle\psi_1| + |\psi_2\rangle\langle\psi_2|\right)$$
$$= \cos^2\theta|0\rangle\langle0| + \sin^2\theta|1\rangle\langle1|, \qquad (8.128)$$

e von Neumann entropy for which is

$$S\left(\hat{\rho}\right) = -\cos^2\theta\log\cos^2\theta - \sin^2\theta\log\sin^2\theta$$
$$= H\left(\cos^2\theta\right). \qquad (8.129)$$

his leads to the more general and more elegant statement that the state
N qubits can be compressed to one of only slightly more than

$$W = 2^{NS(\hat{\rho})} \qquad (8.130)$$

ates of about $NS(\hat{\rho})$ qubits and then later decompressed with near
it fidelity. Thus Schumacher's noiseless coding theorem has the same
rm as Shannon's but with the classical information $H(A)$ replaced by
e von Neumann entropy $S(\hat{\rho})$.

Comparing Schumacher's noiseless coding theorem with Shannon's
ves a simple interpretation for the inequality in eqn 8.20. If, however,
e messages $\{a_i\}$ are selected with probabilities $\{P(a_i)\}$ then Shannon's
eorem tells us that optimal coding allows us to compress an N-bit
essage into one of $NH(A)$ bits. If the messages are encoded onto
antum states $\{\hat{\rho}_i\}$ then we can compress the resulting state of N qubits
to one of just $NS(\hat{\rho})$ qubits. The fact that $NS(\hat{\rho})$ is typically less than
$H(A)$ is a manifestation of the additional quantum redundancy, which
a consequence of the non-orthogonality of the signal states.

If Schumacher's noiseless coding theorem is the quantum analogue of
annon's then it is natural to ask whether there is also a quantum ver-
on of his noisy-channel coding theorem. We saw in Section 6.4 how it
possible to protect qubits against at least some errors by construct-
g quantum codewords from multiple qubits, and would like a bound
the number of ancillary qubits required to provide error-free quan-
m information processing. We would also like to know the channel
pacity for a noisy quantum channel and thereby determine the max-
um rate at which classical information can be transmitted through
quantum channel. It is not yet possible to provide answers to either
these questions with the degree of generality provided by Shannon's
isy-channel coding theorem. We can say something about the channel
pacity, however, if we restrict Alice to preparing only a product state
a large number of quantum systems by selecting from a given set of
nsity operators $\{\hat{\rho}_i\}$ for each system. In this case the channel capac-
cannot exceed that given by the Holevo–Schumacher–Westmoreland
SW) bound:

$$C \le \mathrm{Sup}\left[S\left(\sum_i P(a_i)\hat{\rho}_i'\right) - \sum_i P(a_i)S\left(\hat{\rho}_i'\right)\right], \qquad (8.131)$$

where $\hat{\rho}_i'$ is the state produced by the action of the noisy channel on the state $\hat{\rho}_i$ prepared by Alice. If the action of the channel is associated with the effect operators $\hat{A}_j$ then

$$\hat{\rho}_i \rightarrow \hat{\rho}_i' = \sum_j \hat{A}_j \hat{\rho}_i \hat{A}_j^\dagger. \tag{8.132}$$

If we are interested in the most efficient way of communicating along the noisy channel then the maximization implicit in eqn 8.131 needs to be carried out over the preparation probabilities and also the signal states $\hat{\rho}_i$. We can arrive at the HSW bound by the same line of reasoning that led us to the Holevo bound. Let us suppose that Alice prepares two quantum systems in the mixed state given in eqn 8.114 and then sends the B system through the quantum channel to Bob. This will leave the two systems in the state

$$\begin{aligned}
\hat{\rho}_{AB}' &= \sum_i P(a_i)|a_i\rangle_{A\ A}\langle a_i| \otimes \hat{\rho}_{Bi}' \\
&= \sum_{ij} P(a_i)|a_i\rangle_{A\ A}\langle a_i| \otimes \hat{A}_j \hat{\rho}_{Bi} \hat{A}_j^\dagger.
\end{aligned} \tag{8.133}$$

The mutual information extracted by Alice and Bob from this state can hardly exceed that already present in this state, and this leads us to the bound

$$\begin{aligned}
H(A:B) &\leq S(A:B) \\
&= S\left(\sum_i P(a_i)\hat{\rho}_i'\right) - \sum_i P(a_i)S(\hat{\rho}_i').
\end{aligned} \tag{8.134}$$

The channel capacity is the greatest possible value of $H(A:B)$, and maximizing this gives the HSW bound in eqn 8.131.

In quantum information science we think of entanglement, and especially distributed entanglement, as a resource that is useful for tasks such as quantum key distribution, teleportation, and distributed quantum computation. We introduced, in Section 5.3, the ebit as a unit of shared entanglement; if Alice and Bob each have a qubit and these have been prepared in a maximally entangled state then they share one ebit. It is important to be able also to quantify the entanglement for other, non-maximally entangled states. The natural way to do this, in the context of quantum communications, is by our ability to manipulate entangled states using only local operations and classical communications. The entanglement of formation, $E_F(\hat{\rho})$, is an asymptotic measure of the number of ebits required to prepare a number of shared copies of any given entangled state. The entanglement of distillation, $E_D(\hat{\rho})$, is associated with the reverse process; it represents the number of maximally entangled qubit pairs that can be prepared (or distilled) from a number of entangled qubits prepared in a non-maximally entangled state.

It may not be obvious that we can concentrate entanglement so as to prepare maximally entangled qubits from non-maximally entangled ones.

o see that this is possible, let us suppose that Alice and Bob share two
airs of qubits, each prepared in the state $\cos\theta|0\rangle_A|0\rangle_B + \sin\theta|1\rangle_A|1\rangle_B$,
o that the state of their two pairs of qubits is

$$|\psi_2\rangle = (\cos\theta|0\rangle_A|0\rangle_B + \sin\theta|1\rangle_A|1\rangle_B) \otimes (\cos\theta|0\rangle_A|0\rangle_B + \sin\theta|1\rangle_A|1\rangle_B). \tag{8.135}$$

lice can perform a collective von Neumann measurement on her two
ıbits to determine if they have the same or different values in the
omputational basis without determining the value of $\hat{\sigma}_z$ for each qubit.
her qubits are different then the resulting state of the four qubits will
e

$$|\tilde{\psi}_2\rangle = \frac{1}{\sqrt{2}}(|01\rangle_A|01\rangle_B + |10\rangle_A|10\rangle_B). \tag{8.136}$$

lice and Bob can readily transform this into a maximally entangled
ate of two qubits by each performing a unitary transformation on their
vo qubits so that

$$|01\rangle \to |00\rangle,$$
$$|10\rangle \to |10\rangle, \tag{8.137}$$

ıd then each discarding the second qubit. This leaves the remaining
vo qubits in the maximally entangled state $|\Phi^+\rangle_{AB}$.
The entanglement of distillation applies when Alice and Bob have a
rge number, N, of similarly prepared non-maximally entangled pairs
qubits:

$$|\psi_N\rangle = (\cos\theta|0\rangle_A|0\rangle_B + \sin\theta|1\rangle_A|1\rangle_B)^{\otimes N}$$
$$= \sum_{k=0}^{2^N-1} \cos^{N-\kappa(k)}\theta \sin^{\kappa(k)}\theta|k\rangle_A|k\rangle_B, \tag{8.138}$$

here the state $|k\rangle_A$ represents the state of Alice's N qubits associated
.th the integer k, as described in Section 7.2. The function $\kappa(k)$ is
e number of ones in the binary representation of k. Each of Alice's
ıbits will be in the state $|1\rangle$ with probability $\sin^2\theta$ and the state $|0\rangle$
.th probability $\cos^2\theta$. If N is very large then the number of ones in
e bit string will be very close to the average value:

$$\kappa \approx N\sin^2\theta. \tag{8.139}$$

Alice performs a collective projective measurement on all of her qubits
determine the number of qubits in the state $|1\rangle$ then she will find a
lue of κ close to $N\sin^2\theta$. The resulting state will then be

$$|\tilde{\psi}_N\rangle = \sqrt{\frac{\kappa!(N-\kappa!)}{N!}} \sum_{k\in\kappa} |k\rangle_A|k\rangle_B$$
$$\approx 2^{-NH(\cos^2\theta)} \sum_{k\in\kappa} |k\rangle_A|k\rangle_B, \tag{8.140}$$

ıere the sum is over all values for which k has the measured num-
r of ones (κ). This N-qubit state can be mapped onto one of about

$NH(\cos^2\theta)$ qubits by means of local unitary transformations to map the $2^{NH(\cos^2\theta)}$ states $|k\rangle$ appearing in eqn 8.140 onto states corresponding to the first $2^{NH(\cos^2\theta)}$ integers, so that all but $NH(\cos^2\theta)$ qubits are left in the state $|0\rangle$. This procedure leaves Alice and Bob with

$$NH(\cos^2\theta) = NS(\hat\rho_A) = NE_{\mathrm{D}}(\hat\rho) \qquad (8.141)$$

maximally entangled pairs of qubits, each in the state $|\Phi^+\rangle$. Here $S(\hat\rho_A)$ is the von Neumann entropy for any one of Alice's qubits (or Bob's, of course) in the initial state. For a pure state, the entanglement of distillation E_{D} is equal to the relative entropy of entanglement and is given simply by the von Neumann entropy for either Alice's or Bob's system.

The entanglement of formation quantifies the number of maximally entangled qubit pairs required by Alice and Bob in order for them to be able to create pairs in a desired non-maximally entangled state. It is clear that one ebit suffices to generate each required pair, as Alice can prepare, locally, an entangled state $\cos\theta|00\rangle + \sin\theta|11\rangle$, and then consume the available ebit, shared with Bob, to teleport the state of the first qubit to Bob. If Alice needs to prepare a large number N of such non-maximally entangled pairs, so that she and Bob share the state in eqn 8.138, she can first prepare, locally, the state $(\cos\theta|00\rangle + \sin\theta|11\rangle)^{\otimes N}$ and then use Schmacher compression to imprint this state on approximately $NH(\cos^2\theta)$ maximally entangled qubit pairs. This $2NH(\cos^2\theta)$-qubit state can be shared with Bob by teleportation if they share $NH(\cos^2\theta)$ ebits and a classical channel. Finally, the Schumacher compression can be undone by means of a local unitary transformation performed by both Alice and Bob. The minimum number of ebits required to form N shared copies of the non-maximally entangled state $\cos\theta|0\rangle_A|0\rangle_B + \sin\theta|1\rangle_A|1\rangle_B$ is, therefore,

$$NH(\cos^2\theta) = NS(\hat\rho_A) = NE_{\mathrm{F}}(\hat\rho). \qquad (8.142)$$

We see that for pure states the entanglement of formation E_{F} is equal to the entanglement of distillation. It follows that the processes of forming non-maximally entangled states and of distilling maximally entangled states are, essentially, reversible. This is not strictly true, of course, but becomes an ever better approximation as N increases.

It would be useful to know the entanglement of distillation, the entanglement of formation, and indeed the relative entropy of entanglement for mixed states. Very little is known at present about these quantities. It is clear, however, that working with mixed states should not allow us to create or increase entanglement using only local operations and classical communications. If we start with N ebits, use these to form non-maximally, perhaps mixed-state entangled pairs, and then distill these into ebits, then we should not have (at least on average) more entangled pairs than we had to start with. We can use E_{F} and E_{D} to follow this process:

$$N \text{ ebits} \rightarrow \frac{N}{E_{\mathrm{F}}} \text{ non-max. ent. pairs} \rightarrow \frac{NE_{\mathrm{D}}}{E_{\mathrm{F}}} \text{ ebits.} \qquad (8.143)$$

follows, therefore, that

$$E_D\left(\hat{\rho}\right) \le E_F\left(\hat{\rho}\right).\tag{8.144}$$

We also have an implicit expression for the entanglement of formation. If AB is the state of Alice and Bob's system then we consider all possible pure-state decompositions of this state, that is, all ensembles of states $|\psi_i\rangle_{AB}$ such that

$$\hat{\rho}_{AB} = \sum_i p_i|\psi_i\rangle_{AB\ AB}\langle\psi_i|.\tag{8.145}$$

The entanglement of formation is then

$$E_F\left(\hat{\rho}_{AB}\right) = \mathrm{Inf}\sum_i p_i S\left(\hat{\rho}_i^A\right),\tag{8.146}$$

where $\hat{\rho}_i^A = \mathrm{Tr}_B(|\psi_i\rangle_{AB\ AB}\langle\psi_i|)$ and the minimization is carried out over all possible ensembles given by eqn 8.145. Wootters has derived from this an explicit expression for the entanglement of formation for an arbitrary two-qubit state. His expression is given in Appendix U

Quantum information theory, as we have stated, is not yet complete, and fundamental results continue to appear. It is appropriate, therefore, to conclude with a recent discovery by Horodecki, Oppenheim, and Winter (2005) which has revealed the significance of the von Neumann conditional entropy

$$S(B|A) = S(A,B) - S(A).\tag{8.147}$$

The analogous classical entropy, $H(B|A)$, is always greater than or equal to zero but $S(B|A)$ can take negative values:

$$\begin{aligned}0 &\le H(B|A) \le H(B),\\ -S(A) &\le S(B|A) \le S(B).\end{aligned}\tag{8.148}$$

The von Neumann conditional entropy is the number of qubits that Bob needs to send to Alice so that Alice can construct their joint state $\hat{\rho}_{AB}$. We can illustrate this idea using three simple examples. In the first, Bob has a qubit in the maximally mixed state and Alice has a qubit in the state $|0\rangle$:

$$\hat{\rho}_{AB} = |0\rangle_{A\ A}\langle 0| \otimes \frac{1}{2}\hat{I}_B,\tag{8.149}$$

so that $S(B|A) = 1$. Bob must send his single qubit to Alice in order for her to be able to construct the state. If this seems strange then consider that Bob's qubit might be maximally mixed by virtue of being entangled with a further qubit held by a third party, Claire, so that the combined pure state is

$$|\psi\rangle_{ABC} = |0\rangle_A \frac{1}{\sqrt{2}}\left(|0\rangle_B|0\rangle_C + |1\rangle_B|1\rangle_C\right).\tag{8.150}$$

Clearly, the entanglement between Bob's qubit and Claire's is only transferred to Alice if Bob's qubit is sent (or teleported) to Alice.

In our second example, Alice and Bob share the correlated, but unentangled, state

$$\hat{\rho}_{AB} = \frac{1}{2}\left(|00\rangle_{AB}\,_{AB}\langle 00| + |11\rangle_{AB}\,_{AB}\langle 11|\right), \qquad (8.151)$$

for which $S(B|A) = 0$. This suggests that only classical communication is required to transfer the state to Alice. To see that this is indeed the case let us suppose, once again, that the mixed state is a consequence of an entangled pure state shared with Claire:

$$|\psi\rangle_{ABC} = \frac{1}{\sqrt{2}}\left(|0\rangle_A|0\rangle_B|0\rangle_C + |1\rangle_A|1\rangle_B|1\rangle_C\right). \qquad (8.152)$$

Bob can transfer his part of the state to Alice by first performing a measurement of $\hat{\sigma}_x$ on his qubit so that the state of Alice's and Claire's qubits is

$$|\psi\rangle_{AC} = \frac{1}{\sqrt{2}}\left(|0\rangle_A|0\rangle_C \pm |1\rangle_A|1\rangle_C\right). \qquad (8.153)$$

If Bob sends the result of his measurement to Alice then she can use this to provide a phase shift to her qubit, if necessary, to obtain the state

$$|\psi'\rangle_{AC} = \frac{1}{\sqrt{2}}\left(|0\rangle_A|0\rangle_C + |1\rangle_A|1\rangle_C\right). \qquad (8.154)$$

Adding an extra qubit in the state $|0\rangle_{A'}$ and performing a CNOT operation, with qubit A as the control and qubit A' as the target, gives the desired state

$$\hat{U}_{\text{CNOT}}|\psi'\rangle_{AC}|0\rangle_{A'} = \frac{1}{\sqrt{2}}\left(|0\rangle_A|0\rangle_C|0\rangle_{A'} + |1\rangle_A|1\rangle_C|1\rangle_{A'}\right). \qquad (8.155)$$

Only one (classical) bit of information needs to be communicated from Bob to Alice and no transfer of quantum information is required.

In our third and final example, Alice and Bob share a maximally entangled state

$$|\psi\rangle_{AB} = \frac{1}{\sqrt{2}}\left(|0\rangle_A|0\rangle_B + |1\rangle_A|1\rangle_B\right) \qquad (8.156)$$

so that $S(B|A) = -1$. No communication is necessary, as Alice can prepare copies of the state locally without assistance from Bob. The fact that they share an ebit, however, means that they can use this state to teleport a further qubit. If they share one maximally entangled qubit and one pair in the state given in eqn 8.149, for example, then $S(B|A) = 0$ for the combined state of the four qubits. The meaning is clear; they can use the maximally entangled qubit, for which $S(B|A) = -1$, to teleport the state of Bob's other qubit (in the maximally mixed state). No additional qubits are required, and this is reflected in the fact that $S(B|A) = 0$ for the four-qubit state.

uggestions for further reading

ruß, D. and Leuchs, G. (eds) (2007). *Lectures on quantum information theory*. Wiley-VCH, Weinheim.

iósi, L. (2007). *A short course in quantum information theory*. Springer-Verlag, Berlin.

ayashi, H. (2006). *Quantum information: an introduction*. Springer-Verlag, Berlin.

olevo, A. S. (2001). *Statistical structure of quantum theory*. Springer-Verlag, Berlin.

orodecki, M., Oppenheim, J., and Winter, A. (2005). Partial quantum information. *Nature* **436**, 673.

eger, G. (2007). *Quantum information: an overview*. Springer, New York.

ielsen, M. A. and Chuang, I. L. (2000). *Quantum computation and quantum information*. Cambridge University Press, Cambridge.

lenio, M. B. and Virmani, S. (2007). An introduction to entanglement measurements. *Quantum Information and Computation* **7**, 1.

uelle, D. (1999). *Statistical mechanics: rigorous results*. Imperial College Press, London.

enholm, S. and Suominen, K.-A. (2005). *Quantum approach to informatics*. Wiley, Hoboken, NJ.

edral, V. (2006). *Introduction to quantum information science*. Oxford University Press, Oxford.

ehrl, A. (1978). General properties of entropy. *Reviews of Modern Physics* **50**, 221.

xercises

(8.1) Calculate the von Neumann entropy for a qubit prepared in an equally weighted mixture of the states $|0\rangle$ and $2^{-1/2}(|0\rangle + |1\rangle)$.

(8.2) Calculate the von Neumann entropy for the two-qubit mixed Werner state defined in eqn 2.120.

(8.3) A single-qubit state has $\langle \hat{\sigma}_x \rangle = s$. Find the most general forms for the corresponding density operator

   (a) with the maximum von Neumann entropy;
   (b) with the minimum von Neumann entropy.

(8.4) Prove the relations in eqns 8.5 and 8.6 using the diagonal representation of $\hat{\rho}$ given in eqn 8.4.

(8.5) Illustrate the concavity property stated in eqn 8.12 by calculating both sides of the inequality for the density operators

$$\hat{\rho}_1 = \frac{1}{2}\left(\hat{I} + \frac{1}{2}\hat{\sigma}_x\right), \quad \hat{\rho}_2 = \frac{1}{2}\left(\hat{I} + \frac{1}{2}\hat{\sigma}_y\right).$$

(8.6) Prove the general concavity condition given in eqn 8.16 from eqn 8.12.

(8.7) The Fano entropy is defined to be minus the logarithm of the purity:

$$\mathcal{F}(\hat{\rho}) = -\log\left[\text{Tr}\left(\hat{\rho}^2\right)\right].$$

   (a) Find the maximum and minimum values of $\mathcal{F}$ and compare these with the corresponding values for the von Neumann entropy.
   (b) Show that

$$S(\hat{\rho}) \geq \mathcal{F}(\hat{\rho}).$$

(8.8) Check the inequality in eqn 8.22 for the two equiprobable single-qubit density operators

$$\hat{\rho}_1 = \frac{1}{2} \begin{pmatrix} 1 & 0 \\ 0 & 1 \end{pmatrix}, \quad \hat{\rho}_2 = \frac{1}{3} \begin{pmatrix} 2 & 1 \\ 1 & 1 \end{pmatrix}.$$

(8.9) Use the positivity of the quantum relative entropy stated in eqn 8.24 to derive the inequality in eqn 8.8.

(8.10) (a) Under what conditions will the quantum relative entropy $S(\hat{\sigma}\|\hat{\rho})$ take the value infinity?
   (b) If $S(\hat{\sigma}\|\hat{\rho}) = \infty$, does it necessarily follow that $S(\hat{\rho}\|\hat{\sigma}) = \infty$?

(8.11) Use Klein's inequality (eqn R.10) to derive the positivity of the quantum relative entropy (eqn 8.24).

(8.12) Use Lagrange's method of undetermined multipliers to determine the minimum value of the quantum relative entropy expressed in nats, $S_e(\hat{\sigma}\|\hat{\rho})$, and the conditions for this minimum to occur:

   (a) by varying $\hat{\sigma}$;
   (b) by varying $\hat{\rho}$.

(8.13) Consider the combination of von Neumann entropies

$$S_p = S\left[p\hat{\sigma} + (1-p)\hat{\rho}\right] - pS(\hat{\sigma}) - (1-p)S(\hat{\rho}),$$

where $0 \le p \le 1$.

   (a) Show that $S_p \ge 0$
   (b) Evaluate the derivative of S_p with respect to p at $p = 0$ and $p = 1$ and compare these expressions with $S(\hat{\sigma}\|\hat{\rho})$ and $S(\hat{\rho}\|\hat{\sigma})$.

(8.14) Calculate the entropies $S(A,B)$, $S(A)$, and $S(B)$ for the states

   (a) $|\Psi^+\rangle = 2^{-1/2}\left(|01\rangle_{AB} + |10\rangle_{AB}\right)$;
   (b) $|GHZ\rangle = 2^{-1/2}\left(|000\rangle_{ABC} + |111\rangle_{ABC}\right)$;
   (c) $|W\rangle = \alpha|001\rangle_{ABC} + \beta|010\rangle_{ABC} + \gamma|100\rangle_{ABC}$.

(8.15) Evaluate the von Neumann entropies $S(A,B)$, $S(A)$, and $S(B)$ for the Werner mixed state in eqn 2.120. For which values of p is

   (a) $S(A,B) = S(A) + S(B)$;
   (b) $S(A,B) = |S(A) - S(B)|$?

(8.16) Derive, for the Werner state in eqn 2.120, a condition on p such that

$$S(A,B) \ge \text{Sup}\left(S(A), S(B)\right).$$

How does this condition compare with the value $p = \frac{2}{3}$, above which the state is not entangled?

(8.17) Under what conditions is

   (a) $S(A:B) = 0$;
   (b) $S(A:B) = 2\,\text{Inf}\left(S(A), S(B)\right)$?

(8.18) Find a state of two harmonic oscillators, or single field modes, for which $S(A:B)$ takes its maximum value given that the oscillators share *precisely* n quanta.

(8.19) (a) Find a state of two harmonic oscillators, or single field modes, for which $S(A:B)$ takes its maximum value given that each of the oscillators has an *average* of $\bar{n}/2$ quanta.
   (b) Will the state be any different if we simply require that the sum of the number of quanta in the two oscillators is, on average, $\bar{n}$?

(8.20) Use the strong subadditivity condition stated in eqn 8.46 to derive the Araki–Lieb inequality (eqn 8.39).

(8.21) By considering a pure state of *four* quantum systems or otherwise, show that the strong subadditivity condition stated in eqn 8.46 is equivalent to

$$S(C) + S(B) \ge S(AB) + S(AC).$$

(8.22) Calculate the fidelity for the single-qubit states

   (a) $\hat{\rho} = \frac{1}{2}(\hat{I} + \vec{r}\cdot\hat{\vec{\sigma}})$ and $|0\rangle$;
   (b) $\hat{\rho} = \frac{1}{2}\hat{I}$ and any pure state $|\psi\rangle$.

(8.23) Calculate the fidelity for the Werner state in eqn 2.120 and the states

   (a) $|\Psi^-\rangle$;
   (b) $|0\rangle \otimes |1\rangle$;
   (c) $|0\rangle \otimes |0\rangle$.

(8.24) For the mixed-state fidelity given in eqn 8.51, show that

   (a) it is invariant under unitary transformations;
   (b) that if $\hat{\sigma} = |\psi\rangle\langle\psi|$ then it reduces to eqn 8.50.

(8.25) Show that the mixed-state fidelity given in eqn 8.51 is symmetric in $\hat{\rho}$ and $\hat{\sigma}$.
   [*Hint:* you might start by showing that the two positive operators $\hat{\rho}^{1/2}\hat{\sigma}\hat{\rho}^{1/2}$ and $\hat{\sigma}^{1/2}\hat{\rho}\hat{\sigma}^{1/2}$ have the same eigenvalues, even though they may have different eigenstates.]

(8.26) Show, by considering commuting density operators, that the *classical* fidelity has the form

$$F(P, Q) = \left(\sum_i \sqrt{p_i q_i}\right)^2.$$

Without considering the quantum formula, find the maximum and minimum values of $F(P, Q)$.

27) Show that the fidelity for a single qubit takes the form of eqn 8.52.

28) Calculate the fidelity for the single-qubit density operators

$$\hat{\rho} = \begin{pmatrix} \frac{1}{2} & \alpha \\ \alpha^* & \frac{1}{2} \end{pmatrix},$$

$$\hat{\sigma} = \begin{pmatrix} p & \beta \\ \beta^* & 1-p \end{pmatrix}.$$

(a) For which values of p is the fidelity (i) a minimum; (ii) a maximum?

(b) For which values of β is the fidelity (i) a minimum; (ii) a maximum?

29) Calculate the distance between the states given in the previous question.

(a) For which values of p is the distance (i) a minimum; (ii) a maximum?

(b) For which values of β is the distance (i) a minimum; (ii) a maximum?

30) Derive the inequality in eqn 8.64.

31) Calculate the two distances appearing in the inequality in eqn 8.67 for the density matrices in eqn 8.62 and for the transformation in eqn 4.91.

32) Show that for two pure states,

$$D(\hat{\rho}, \hat{\sigma}) = \sqrt{1 - F(\hat{\rho}, \hat{\sigma})}.$$

33) We wish to determine whether a die is fair, with the probabilities for each of the numbers being $\frac{1}{6}$, or loaded, with $P(6) = \frac{1}{4}$ and $P(1) = P(2) = \cdots = P(5) = \frac{3}{20}$. Estimate the number of rolls of the die required so that:

(a) An indication that the die is fair is correct with a probability of at least 0.999999.

(b) An indication that the die is loaded is correct with a probability of at least 0.999999.

34) Derive the following properties of the quantum relative entropy:

(a)

$$S(\hat{\rho}_A \otimes \hat{\rho}_B \| \hat{\sigma}_A \otimes \hat{\sigma}_B) = S(\hat{\rho}_A \| \hat{\sigma}_A) + S(\hat{\rho}_B \| \hat{\sigma}_B),$$

(b)

$$S(p\hat{\sigma}_1 + (1-p)\hat{\sigma}_2 \| \hat{\rho}) \le pS(\hat{\sigma}_1 \| \hat{\rho}) + (1-p)S(\hat{\sigma}_2 \| \hat{\rho}),$$

(c)

$$\sum_i p_i S(\hat{\rho}_i \| \hat{\sigma}) = \sum_i p_i S(\hat{\rho}_i \| \hat{\rho}) + S(\hat{\rho} \| \hat{\sigma}),$$

where $\hat{\rho} = \sum_i p_i \hat{\rho}_i$.

(8.35) A device has been designed to implement the single-qubit unitary transformation $\hat{U} = \hat{\sigma}_z$. It performs an operation characterized by the three effect operators

$$\hat{A}_0 = (1 - p - q)^{1/2}\hat{\sigma}_z,$$
$$\hat{A}_1 = p^{1/2}\hat{\sigma}_x,$$
$$\hat{A}_2 = q^{1/2}\hat{\sigma}_y.$$

Calculate the gate fidelity.

(8.36) A device has been designed to implement a single-qubit unitary transformation $\hat{U}$ but produces a different single-qubit unitary transformation $\hat{U}'$. Show that using a two-qubit Bell state will always provide the necessary minimum in the definition of the gate fidelity in eqn 8.83.

(8.37) Evaluate the distance between the entangled pure state in eqn 8.91 and the mixed state

$$\hat{\rho} = p|00\rangle\langle 00| + (1-p)|11\rangle\langle 11|.$$

Show that the minimum value occurs for $p = \cos^2 \theta$.

(8.38) Evaluate the fidelity for the entangled pure state in eqn 8.91 and the mixed state

$$\hat{\rho} = p|00\rangle\langle 00| + (1-p)|11\rangle\langle 11|.$$

Find the value of p for which the fidelity is maximized.

(8.39) Given the maximally entangled Bell state $|\Psi^-\rangle$, find the most general forms for $\hat{\rho}^{\text{unent}}$ for which

(a) the distance is minimized;

(b) the fidelity is maximized.

(8.40) Suggest why we do not use $S(\hat{\rho}^{\text{unent}} \| \hat{\rho})$ as a measure of entanglement.

(8.41) Prove that the relative entropy of entanglement for any pure state is equal to the von Neumann entropy of either of the entangled subsystems.

(8.42) The trine states given in eqn 4.60 are to be used in a noiseless quantum communication channel. Calculate the mutual information for each of the following:

(a) Equal probabilities for each of the three states and the generalized measurement described by the probability operators in eqn 4.61.

(b) Equal probabilities for each of the three states and the generalized measurement described by the probability operators in eqn 4.66.

(c) Zero probability for one of the states and equal probability for the others, and the optimum measurement.

(8.43) Complete the derivation of the Holevo bound given in eqn 8.115 by confirming that $S(A) + S(B) - S(A, B) = \chi$.

(8.44) Calculate the Holevo bound for the two mixed signal states

$$\hat{\rho}_1 = q|0\rangle\langle 0| + (1 - q)|0\rangle\langle 0|,$$
$$\hat{\rho}_2 = (1 - q)|0\rangle\langle 0| + q|0\rangle\langle 0|,$$

where $0 \le q \le \frac{1}{2}$. Hence find the channel capacity and compare this with that found for the binary symmetric channel in Section 1.4.

(8.45) Reconcile, if you can, the idea of quantum dense coding, in which two bits are encoded on a single qubit, with the Holevo bound for a single qubit.

(8.46) Alice has three bits, A, B, and C, to convey to Bob, who will use one of them.

(a) If she has only one bit at her disposal, what is the greatest probability that Bob will get the correct bit value?

(b) Show that she can do better than this if she has at her disposal a single qubit to send to Bob.

(8.47) Design a quantum circuit to induce the unitary transformation given in eqn 8.121. Is there an alternative transformation for performing the desired compression that is simpler to implement?

(8.48) The compression and restoration of the state in eqn 8.119 will be effective if θ is sufficiently small.

(a) What is the probability that the compression step will be successful?

(b) We can quantify the decompression step by the fidelity of the decompressed state as compared with the initial state. Calculate this fidelity for the state prepared by Bob from the state given in eqn 8.122.

(8.49) Determine a bound on the extent to which it is possible to compress a long sequence of qubit states in which

(a) each qubit is prepared with equal probability in one of the three trine states in eqn 4.60;

(b) each qubit is prepared in the state $|\psi_1\rangle$ with probability p and in the state $|\psi_2\rangle$ with probability $1 - p$, where the states are defined in eqn 8.111.

(8.50) A large number of *pairs* of qubits is prepared so that each pair is, with equal probability, in one of the two Bell states $|\Psi^-\rangle$ and $|\Psi^+\rangle$. To what extent can the state be compressed? If it can be compressed, devise an efficient method for doing this.

(8.51) Calculate the channel capacity for a qubit channel in which the input states are transformed as

$$\hat{\rho}_i \rightarrow (1 - p)\hat{\rho}_i + p\hat{\sigma}_z\hat{\rho}_i\hat{\sigma}_z.$$

What coding scheme reaches this capacity?

(8.52) Evaluate the HSW bound for the qubit channel

$$\hat{\rho}_i \rightarrow (1 - p)\hat{\rho}_i + \frac{p}{3}\left(\hat{\sigma}_x\hat{\rho}_i\hat{\sigma}_x + \hat{\sigma}_y\hat{\rho}_i\hat{\sigma}_y + \hat{\sigma}_z\hat{\rho}_i\hat{\sigma}_z\right).$$

Can you find a set of states and a measurement strategy for Bob that gives this value for the mutual information?

(8.53) Show that when Alice's collective measurement on the state in eqn 8.135 reveals that the qubits are the same in the computational basis, the resulting state is less strongly entangled than initially. On average, that is, given both possible measurement outcomes, will the entanglement increase, decrease, or stay the same?

(8.54) Calculate the concurrence for the two-qubit state

$$\hat{\rho} = \frac{1}{2}\left(\hat{I} + \vec{r} \cdot \hat{\vec{\sigma}}\right) \otimes \frac{1}{2}\left(\hat{I} + \vec{s} \cdot \hat{\vec{\sigma}}\right).$$

(8.55) Calculate the concurrence for Werner's mixed state in eqn 2.120 and hence evaluate the entanglement of formation for this state.

The equivalence of information and entropy

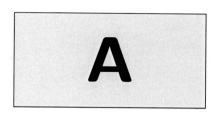

In this appendix we present, essentially verbatim, Shannon's proof that information has the same mathematical form as entropy.

We seek to define the information associated with an as yet unknown outcome of an event. We let the event have n possible outcomes and let these be associated with the probabilities $p_1, p_2, \cdots, p_n$. We require a quantity $H(p_1, p_2, \cdots, p_n)$ that reflects the amount of choice involved in the selection of the event or, equivalently, our degree of uncertainty as to the outcome. Clearly, the greater the uncertainty, the greater the amount of information to be acquired on determining the outcome. It is reasonable to require H to have the following properties:

1. H should be a continuous function of the probabilities p_i.

2. If all of the p_i are equal, so that $p_i = \frac{1}{n}$, then H should be a monotonically increasing function of n. This is reasonable, as with equally likely events there is more choice when there are more possible outcomes.

3. If a choice is broken down into successive choices, then the original H should be the probability-weighted sum of the individual values of H. The meaning of this is illustrated in Fig. A.1. In the first probability tree we have one event with three possible outcomes, the probabilities for which are $p_1 = \frac{1}{2}$, $p_2 = \frac{1}{3}$, and $p_3 = \frac{1}{6}$. In the second we have two possible outcomes, each with probability $\frac{1}{2}$, and if the second of these occurs then we have a second event, with two outcomes, having probabilities $\frac{2}{3}$ and $\frac{1}{3}$. The final results have the same probabilities in the two cases and we require, therefore, that

$$H\left(\frac{1}{2}, \frac{1}{3}, \frac{1}{6}\right) = H\left(\frac{1}{2}, \frac{1}{2}\right) + \frac{1}{2}H\left(\frac{2}{3}, \frac{1}{3}\right). \qquad (A.1)$$

The coefficient $\frac{1}{2}$ is the weighting factor introduced because the second event only occurs half of the time.

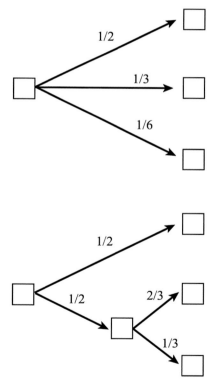

Fig. A.1 Two probability trees with the same final probabilities.

Let $H(\frac{1}{n}, \frac{1}{n}, \cdots, \frac{1}{n}) = A(n)$. From property 3, it follows that we can decompose a single event with s^m equally likely probabilities (of s^{-m}) into a series of m events, each with s equally likely outcomes, and that this gives

$$A(s^m) = mA(s). \qquad (A.2)$$

We can treat an event with t^n equally likely outcomes in the same way:

$$A(t^n) = nA(t). \tag{A.3}$$

We can choose n to be arbitrarily large and also find a value of m such that

$$s^m \leq t^n \leq s^{m+1}. \tag{A.4}$$

Taking the logarithms and dividing by $n \log s$ then gives

$$\frac{m}{n} \leq \frac{\log t}{\log s} \leq \frac{m}{n} + \frac{1}{n}$$

$$\Rightarrow \left| \frac{m}{n} - \frac{\log t}{\log s} \right| < \epsilon, \tag{A.5}$$

where $\epsilon = \frac{1}{n}$ is arbitrarily small. From the monotonic property 2 it follows that

$$A(s^m) \leq A(t^n) \leq A(s^{m+1})$$

$$\Rightarrow \quad mA(s) \leq nA(t) \leq (m+1)A(s). \tag{A.6}$$

Dividing this by $nA(s)$ then gives

$$\frac{m}{n} \leq \frac{A(t)}{A(s)} \leq \frac{m}{n} + \frac{1}{n}$$

$$\Rightarrow \left| \frac{m}{n} - \frac{A(t)}{A(s)} \right| < \epsilon. \tag{A.7}$$

Combining eqns A.6 and A.7 then leads us to conclude that

$$\left| \frac{A(t)}{A(s)} - \frac{\log t}{\log s} \right| < 2\epsilon$$

$$\Rightarrow A(t) = K \log t, \tag{A.8}$$

where the constant K needs to be positive in order to satisfy property 2.

Suppose now that the n possible outcomes do not have equal probabilities but that the probabilities are commensurable, in that we can write $p_i = n_i / \sum_j n_j$. We can consider an event with $\sum_j n_j$ equiprobable outcomes as first an event with n possible outcomes with probabilities p_i, followed by a second event such that if the first event has the ith outcome then the second event has one of n_i equally probable outcomes. We can use property 3 once again to calculate $A(\sum_j n_j)$ in two different ways:

$$K \log \sum_j n_j = H(p_1, p_2, \cdots, p_n) + K \sum_i p_i \log n_i. \tag{A.9}$$

It follows that

$$H(p_1, p_2, \cdots, p_n) = K \left(\sum_i p_i \log \sum_j n_j - \sum_i p_i \log n_i \right)$$

$$= -K \log \frac{n_i}{\sum_j n_j}$$

$$= -K \sum_i p_i \log p_i. \tag{A.10}$$

the p_i are incommensurable, then they may be approximated by rationals and the same expression must hold by our continuity assumption (property 1). Thus eqn A.10 for the information H holds in general.

Lagrange multipliers

We are often faced with needing to extremize a function of a number of variables by varying these variables but only under the restrictions imposed by one or more constraints. A powerful approach to this is provided by Lagrange's method of undetermined multipliers. We can apply the method to any number of variables, but it is perhaps simplest to explain it by reference to just two. Suppose that we need to find the stationary points of a function $f(x, y)$ of the two variables x and y, subject to the constraint that $g(x, y) = 0$. In principle, we could use the constraint to write y in terms of x and then find the points at which $df/dx = 0$. Equivalently, for f to be stationary, the total differential df must be zero:

$$df = \frac{\partial f}{\partial x} dx + \frac{\partial f}{\partial y} dy = 0. \tag{B.1}$$

Without the constraint, this would lead to the familiar conditions for stationarity, that

$$\frac{\partial f}{\partial x} = 0, \quad \frac{\partial f}{\partial y} = 0. \tag{B.2}$$

The constraint, however, means that the differentials dx and dy are not independent, but rather they are related by the total differential of g:

$$dg = \frac{\partial g}{\partial x} dx + \frac{\partial g}{\partial y} dy = 0. \tag{B.3}$$

We can multiply eqn B.3 by a parameter λ and add it to eqn B.1 to give

$$d(f + \lambda g) = \left(\frac{\partial f}{\partial x} + \lambda \frac{\partial g}{\partial x} \right) dx + \left(\frac{\partial f}{\partial y} + \lambda \frac{\partial g}{\partial y} \right) dy = 0. \tag{B.4}$$

We choose λ such that

$$\frac{\partial f}{\partial x} + \lambda \frac{\partial g}{\partial x} = 0, \tag{B.5}$$

which then implies that

$$\frac{\partial f}{\partial y} + \lambda \frac{\partial g}{\partial y} = 0. \tag{B.6}$$

Solving eqns B.5 and B.6, together with the constraint $g(x, y) = 0$, gives the required stationary point or points. We note that these equations are the same as would have been arrived at had we extremized the function

$$F(x, y) = f(x, y) + \lambda g(x, y) \tag{B.7}$$

with respect to the independent variables x and y. In doing so we have, in effect, allowed these variables to be treated as independent by introducing the additional variable λ, the value of which will be fixed by imposing the constraint $g(x, y) = 0$ on our solution.

We can summarize the method of Lagrange multipliers as follows. To find a maximum or minimum of a function $f(x_1, x_2, \cdots, x_n)$ subject to the constraints $g_1 = 0, g_2 = 0, \cdots, g_m = 0$, we first form the function

$$F(x_1, x_2, \cdots, x_n) = f(x_1, x_2, \cdots, x_n) + \sum_{j=1}^{m} \lambda_j g_j (x_1, x_2, \cdots, x_n). \quad \text{(B.8)}$$

Next we vary F, treating the variables $x_1, x_2, \cdots, x_n$ as independent. Solving the resulting equations together with the constraint equations determines the values of the Lagrange multipliers and the extrema of f.

We can illustrate the power of this method by deriving a useful property of the relative entropy. Consider an event A that can have the possible outcomes $\{a_i\}$ and suppose that there are two probability distributions for these, $P(a_i)$ and $Q(a_i)$. The relative entropy is then defined to be

$$H_e\left(P\|Q\right) = -\sum_i P(a_i) \ln\left(\frac{Q(a_i)}{P(a_i)}\right)$$
$$= \sum_i P(a_i) \left(\ln P(a_i) - \ln Q(a_i)\right). \quad \text{(B.9)}$$

Note that this quantity is not symmetric in the two probability distributions. We have chosen here to express this quantity in nats, as we shall be performing an analytic extremization. The relative entropy has the useful property that

$$H_e(P\|Q) \geq 0, \quad \text{(B.10)}$$

with the equality holding if and only if the two probability distributions are identical: $P(a_i) = Q(a_i)$, $\forall i$. We can prove this by first noting that the relative entropy is not bounded from above; if for any outcome a_j, say, we have $Q(a_i) = 0$ but $P(a_i) \neq 0$ then $H_e(P\|Q)$ will be positive and infinitely big. The second step is to use the method of Lagrange multipliers to find the single extremum, which will be the minimum value. We find this minimum by varying $H_e(P\|Q)$ subject to the constraint that $\sum_i Q(a_i) = 1$, corresponding to the fact that the $Q(a_i)$ are probabilities. The variation then gives

$$d\left(H_e(P\|Q) + \lambda\left[\sum_i Q(a_i) - 1\right]\right) = \sum_i dQ(a_i)\left(-\frac{P(a_i)}{Q(a_i)} + \lambda\right)$$
$$= 0. \quad \text{(B.11)}$$

This is required to be zero for arbitrary small variations $dQ(a_i)$, and this tells us that the minimum value occurs for $P(a_i) = \lambda Q(a_i)$. The fact that the $P(a_i)$ and $Q(a_i)$ are probabilities, constrained to sum to unity, tells us that the Lagrange multiplier λ takes the value unity. The single extremum of $H_e(P\|Q)$ is its minimum and occurs when the two probability distributions are identical. This minimum value is clearly zero: $H_e(P\|P) = 0$ and hence the inequality in eqn B.10 is proven.

Lagrange's method can also be applied, with care, to some operator optimization problems. A simple but important example is the derivation of the thermal state as that with the maximum von Neumann entropy for a given mean energy $\bar{E}$. The density operator $\hat{\rho}$ is required to satisfy two constraints:

$$\mathrm{Tr}\left(\hat{\rho}\right) = 1,$$
$$\mathrm{Tr}\left(\hat{\rho}\hat{H}\right) = \bar{E}. \tag{B.12}$$

It is simplest to work in natural units, in which the von Neumann entropy takes the form

$$S_e\left(\hat{\rho}\right) = -\mathrm{Tr}\left(\hat{\rho}\ln\hat{\rho}\right), \tag{B.13}$$

and then subject the quantity

$$\tilde{S} = S_e + \lambda\left[1 - \mathrm{Tr}\left(\hat{\rho}\right)\right] + \beta\left[\bar{E} - \mathrm{Tr}\left(\hat{\rho}\hat{H}\right)\right] \tag{B.14}$$

to arbitrary variations of $\hat{\rho}$. Varying $\hat{\rho}$ and setting the variation of $\tilde{S}$ to zero gives

$$d\tilde{S} = \mathrm{Tr}\left[\left(-\ln\hat{\rho} - \hat{I}(1+\lambda) - \beta\hat{H}\right)d\hat{\rho}\right], \tag{B.15}$$

the solution of which is

$$\hat{\rho} = e^{(1+\lambda)}e^{-\beta\hat{H}}. \tag{B.16}$$

We can determine the value of λ by requiring the trace to be unity:

$$\hat{\rho} = \frac{e^{-\beta\hat{H}}}{\mathrm{Tr}\left(e^{-\beta\hat{H}}\right)}. \tag{B.17}$$

The value of β is determined by the mean-energy condition and is usually expressed in terms of the temperature as $\beta = (k_B T)^{-1}$.

Stirling's approximation

Stirling's approximation for large factorials can be derived by appealing to the integral representation of the gamma function,

$$N! = \Gamma(N+1)$$
$$= \int_0^\infty t^N e^{-t} dt. \tag{C.1}$$

If we change the integration variable to $\tau = N^{-1/2}t - N^{1/2}$ then this becomes

$$N! = \int_{-\sqrt{N}}^\infty \left(N + \tau\sqrt{N}\right)^N \exp\left[-\left(N + \tau\sqrt{N}\right)\right] \sqrt{N} d\tau$$
$$= e^{-N} N^{N+1/2} \int_{-\sqrt{N}}^\infty e^{-\tau\sqrt{N}} \left(1 + \frac{\tau}{\sqrt{N}}\right)^N d\tau$$
$$= e^{-N} N^{N+1/2} \int_{-\sqrt{N}}^\infty \exp\left[-\tau\sqrt{N} + N\ln\left(1 + \frac{\tau}{\sqrt{N}}\right)\right] d\tau. \tag{C.2}$$

We can split the range of integration into two, one part from $-\sqrt{N}$ to $\sqrt{N}$ and another part from $\sqrt{N}$ to ∞, and replace the logarithm by its Maclaurin expansion in the first part to give

$$\frac{N!}{e^{-N} N^{N+1/2}} = \int_{-\sqrt{N}}^{\sqrt{N}} \exp\left[-\tau\sqrt{N} + N\left(\frac{\tau}{\sqrt{N}} - \frac{\tau^2}{2N} + \cdots\right)\right] d\tau$$
$$+ \int_{\sqrt{N}}^\infty \exp\left[-\tau\sqrt{N} + N\ln\left(1 + \frac{\tau}{\sqrt{N}}\right)\right] d\tau. \tag{C.3}$$

For large N, the second integral is very small and so we can obtain our approximation by neglecting it:

$$\frac{N!}{e^{-N} N^{N+1/2}} \approx \int_{-\infty}^\infty \exp\left(-\frac{\tau^2}{2}\right) d\tau = \sqrt{2\pi}. \tag{C.4}$$

It follows that

$$N! \approx \sqrt{2\pi} N^{N+1/2} e^{-N}. \tag{C.5}$$

This approximation is, in fact, very much better than its derivation might suggest: for $N = 10$, the exact value is $10! = 3\,628\,800$, while the approximation gives about $3\,598\,700$; for $N = 3$, we find $3! = 6$ with the approximate value being 5.84; and even for $N = 1$, the approximate value is the surprisingly accurate 0.92.

Stirling's approximation is often quoted for the logarithm of the factorial:

$$\ln(N!) \approx \left(N + \frac{1}{2}\right) \ln N - N + \frac{1}{2} \ln(2\pi), \qquad \text{(C.6)}$$

or, to order N,

$$\ln(N!) \approx N \ln N - N. \qquad \text{(C.7)}$$

For logarithms in base 2, this becomes

$$\log(N!) \approx N \log N - \frac{N}{\ln 2}. \qquad \text{(C.8)}$$

The Schmidt decomposition

The connection between the Schmidt decomposition and the reduced density operators provides a method for obtaining the Schmidt decomposition of an entangled pure state of two systems. It also constitutes a proof that an entangled state can be written in the form eqn 2.105. Consider a general pure state of our two quantum systems expanded in terms of the orthonormal states $\{|a_i\rangle\}$ and $\{|b_j\rangle\}$,

$$|\psi\rangle = \sum_{ij} c_{ij}|a_i\rangle \otimes |b_j\rangle. \tag{D.1}$$

From this we can form the reduced density operator $\hat{\rho}_a$ and diagonalize it to obtain its eigenstates $|\lambda_n\rangle$. If we write the states $|a_i\rangle$ as a superposition of these eigenstates

$$|a_i\rangle = \sum_n u_{in}|\lambda_n\rangle, \tag{D.2}$$

then our state becomes

$$|\psi\rangle = \sum_{ijn} u_{in}c_{ij}|\lambda_n\rangle \otimes |b_j\rangle$$
$$= \sum_{nj} d_{nj}|\lambda_n\rangle \otimes |b_j\rangle. \tag{D.3}$$

The reduced density operator for the first system can be obtained by evaluating the trace of $|\psi\rangle\langle\psi|$ in the $\{|b_j\rangle\}$ basis, which gives

$$\hat{\rho}_a = \sum_{njk} d_{nj}d_{kj}^*|\lambda_n\rangle\langle\lambda_k|. \tag{D.4}$$

The requirement that the $|\lambda_n\rangle$ are the eigenstates of $\hat{\rho}_a$ means that

$$\sum_j d_{nj}d_{kj}^* = |a_n|^2\delta_{nk}. \tag{D.5}$$

This condition also ensures that the states $|\phi_n\rangle = \sum_j (d_{nj}/a_n)|b_j\rangle$ are orthonormal so that

$$|\psi\rangle = \sum_n a_n|\lambda_n\rangle \otimes |\phi_n\rangle. \tag{D.6}$$

Number theory for cryptography

Modern cryptographic protocols rely for their security on the properties of large numbers. Mathematics recognizes different types of numbers, including the positive integers $(1, 2, 3, \cdots)$, the non-negative integers (the positive integers plus zero), the integers $(0, \pm 1, \pm 2, \cdots)$, the rationals (ratios of integers), and the irrationals. We need to consider only the non-negative integers, and present here some of their properties which have greatest relevance for cryptography. Our discussion draws heavily on material from the introductory texts by Hunter and by Buchmann (see suggestions for further reading in Chapter 3).

E.1 Division properties

If a and b are non-negative integers with $b \neq 0$ then there are unique integers q and r such that

$$a = qb + r \qquad \text{(E.1)}$$

and

$$0 \leq r < b. \qquad \text{(E.2)}$$

These conditions are called the principal division identity for the integers. They represent the division of a by b. The integer r is called the principal remainder, or simply the remainder, of a with respect to b, and q is called the quotient in the division. If $r = 0$ then $a = qb$, and we say that a is divisible by b, or that a is a multiple of b. We also say that b is a divisor or factor of a, or that b divides a. If b divides a then we write a.

E.2 Least common multiple and greatest common divisor

If $a_1, a_2, \cdots, a_n$ are positive integers then any integer which is divisible by each of them is a common multiple of $a_1, a_2, \cdots, a_n$. The smallest of these is the least common multiple (lcm), which we denote $\mathrm{lcm}(a_1, a_2, \cdots, a_n)$. For example,

$$\mathrm{lcm}(4, 6, 20) = 60. \qquad \text{(E.3)}$$

Any integer that divides each of the integers $a_1, a_2, \cdots, a_n$ is called a common divisor of these integers. The largest of these is the greatest common divisor (gcd). For example,

$$\gcd(4, 6, 20) = 2. \tag{E.4}$$

An efficient method for finding the greatest common divisor of a pair of integers is the Euclidean algorithm. Suppose that we are seeking the greatest common divisor of the integers a_1 and a_2 and let $a_1 > a_2$. It follows from eqns E.1 and E.2 that

$$a_1 = a_2 q_1 + a_3, \qquad 0 \le a_3 < a_2. \tag{E.5}$$

If $a_3 = 0$ then $a_2 | a_1$ and $\gcd(a_1, a_2) = a_2$. If $a_3 > 0$ then

$$a_2 = a_3 q_2 + a_4, \qquad 0 \le a_4 < a_3. \tag{E.6}$$

If $a_4 = 0$ then $\gcd(a_1, a_2) = a_3$. If $a_4 > 0$ then

$$a_3 = a_4 q_3 + a_5, \qquad 0 \le a_5 < a_4. \tag{E.7}$$

Continuing in this way, we must get to $a_{k+1} = 0$ for some k, as the remainders become ever smaller with each iteration. The final step gives

$$a_{k-1} = a_k q_{k-1} \Rightarrow a_k = \gcd(a_1, a_2). \tag{E.8}$$

As simple example, we can use the Euclidean algorithm to find the greatest common divisor of 35 and 98:

$$98 = 35 \times 2 + 28,$$
$$35 = 28 \times 1 + 7,$$
$$28 = 7 \times 4,$$
$$\Rightarrow \gcd(35, 98) = 7. \tag{E.9}$$

E.3   Prime numbers

A positive integer p which is greater than 1 is called a prime number, or simply a prime, if 1 and p are its only (positive) divisors. The first few primes are $2, 3, 5, 7, 11, 13, 17, 19, 23, \cdots$. The primes have been subject to special study, and many of their properties are known. Among the most important are:

That there is an infinity of primes was proven by Euclid, who reasoned as follows. If there is not then, there is a largest prime p_n and then we can arrange the primes in ascending order as the sequence $p_1, p_2, \cdots, p_n$. The integer $N = p_1 p_2 p_3 \cdots p_n + 1$, however, is not divisible by any of the primes up to p_n and so must either itself be a prime or be divisible by at least one prime bigger than p_n. Hence we have a contradiction and there cannot be a largest prime.

(i) There is an infinity of primes.
(ii) We do not have a general formula for predicting which numbers are prime, but we do have some information about their distribution. In particular, the prime number theorem states that if $\pi(x)$ is the number of primes less than the positive integer x then

$$\lim_{x \to \infty} \left(\pi(x) \frac{\ln x}{x} \right) = 1. \tag{E.10}$$

It follows that if x is large then $\pi(x)$ is approximately $x / \ln x$.

(iii) Any integer greater than 1 can be expressed as a product of primes. This product is unique apart from the order of the primes. This important result is often called the fundamental theorem of arithmetic.

E.4   Relatively prime integers and Euler's φ-function

the integers $a_1, a_2, \cdots, a_n$ have greatest common divisor 1 then they
re said to be relatively prime, or coprime. If we have just two integers
a and a_2 and $\gcd(a_1, a_2) = 1$ then we also say that a_1 is prime to a_2
and a_2 is prime to a_1).

If N is a positive integer then we denote by $\varphi(N)$ the number of
integers less than or equal to N which are prime to N. If p is a prime
then $\varphi(p) = p - 1$. If M and N are relatively prime integers then
$(MN) = \varphi(M)\varphi(N)$.

E.5   Congruences

We say that a is congruent to b modulo M and write

$$a \equiv b \bmod M \qquad (E.11)$$

M divides $b - a$. Another way to state this is that a is congruent to b
modulo M if a and b have the same principal remainder on division by
M:

$$a = cM + r,$$
$$b = dM + r. \qquad (E.12)$$

We state, without proof, an important theorem commonly referred to as
Euler's theorem or as Fermat's little theorem. If $\gcd(a, M) = 1$ then

$$a^{\varphi(M)} \equiv 1 \bmod M. \qquad (E.13)$$

In particular, if p is a prime number then

$$a^{p-1} \equiv 1 \bmod p. \qquad (E.14)$$

E.6   Primitive root modulo p

An integer g is a primitive root mod p if the p numbers $g^a \bmod p$ for
$a = 1, 2, \cdots, p$ are all different. This means that each of the values
$1, \cdots, p-1$ corresponds to a unique value of a. The number of primitive
roots mod p is $\varphi(p-1)$. For example, if $p = 13$ then there are $\varphi(12) = 4$
primitive roots mod 13; these are $2, 6, 7$, and 11. We list here, by way
of illustration, the values of $7^a \bmod 13$:

$$7^1 \bmod 13 = 7, \qquad 7^2 \bmod 13 = 10,$$
$$7^3 \bmod 13 = 5, \qquad 7^4 \bmod 13 = 9,$$
$$7^5 \bmod 13 = 11, \qquad 7^6 \bmod 13 = 12,$$
$$7^7 \bmod 13 = 6, \qquad 7^8 \bmod 13 = 3,$$
$$7^9 \bmod 13 = 8, \qquad 7^{10} \bmod 13 = 4,$$
$$7^{11} \bmod 13 = 2, \qquad 7^{12} \bmod 13 = 1. \qquad (E.15)$$

Note that the first of these is trivial and that the last is a consequence of eqn E.14. The order of the remaining values, however, is far from obvious.

E.7 Diffie–Hellman cryptosystem

Alice and Bob first agree on a large prime number, p, and an associated primitive root mod p, g. Alice chooses an integer a $(2 \leq p \leq p-2)$ and computes the value

$$A = g^a \bmod p. \tag{E.16}$$

Alice would be unwise, of course, to use $a = 1$ or $a = p-1$, as the corresponding value of A (g or 1) is trivial in that an eavesdropper would have no difficulty in determining a.

This she sends to Bob. Similarly, Bob selects an integer b $(2 \leq b \leq p-2)$ and sends to Alice the value

$$B = g^b \bmod p. \tag{E.17}$$

Bob generates the key by raising A to the power $b \bmod p$:

$$\mathcal{K} = A^b \bmod p = [g^a \bmod p]^b \bmod p. \tag{E.18}$$

We can simplify this by noting that we can write

$$g^a = kp + g^a \bmod p, \tag{E.19}$$

for some positive integer k, so that

$$g^{ab} = [g^a \bmod p]^b + \sum_{\ell=1}^{b} \frac{b!}{\ell!(b-\ell)!} (g^a \bmod p)^{b-\ell} k^\ell p^\ell$$

$$\Rightarrow g^{ab} \bmod p = [g^a \bmod p]^b \bmod p. \tag{E.20}$$

It follows that

$$\mathcal{K} = A^b \bmod p = g^{ab} \bmod p, \tag{E.21}$$

and Alice can generate the same key as

$$\mathcal{K} = B^a \bmod p = g^{ab} \bmod p. \tag{E.22}$$

E.8 RSA cryptosystem

The RSA cryptosystem requires each receiver (Bob) to generate a public key, which he publishes, and a private key, which he keeps secret. He starts by generating (randomly) two independent and distinct primes p and q, and then computes their product

$$N = pq. \tag{E.23}$$

The next step is to choose an integer e in the range $1 < e < \varphi(N) = (p-1)(q-1)$ with $\gcd(e, \varphi(N)) = 1$. This integer is the encryption exponent and, together with N, forms the public key. Bob's private key an integer, d, the decryption exponent; this is required to lie in the range $1 < d < \varphi(N)$ and to satisfy the condition

$$de \equiv 1 \bmod \varphi(N). \tag{E.24}$$

hat a suitable decryption exponent exists is a consequence of the re-uirement that $\gcd(e, \varphi(N)) = 1$.

At its simplest level, the RSA cryptosystem can be used to encode a aintext in the form of an integer M in the range $0 \leq M < N$. If Alice ishes to send this message to Bob she looks up his public key (e, N) ad uses this to generate the ciphertext

$$C = M^e \bmod N. \tag{E.25}$$

n efficient algorithm called 'fast exponentiation' makes this encryption rocess easy to achieve.

The decryption step uses the private key, or decryption exponent, d, hich should be known only to the intended recipient (Bob). It relies the important theorem (proven in the margin) that

$$(M^e \bmod N)^d \equiv M^{ed} \bmod N = M \tag{E.26}$$

r any integer $0 \leq M < N$. This is the required original plaintext.

A simple example (taken from Buchmann) may help to illustrate the ocesses involved. Suppose that Bob were to choose the (unrealistically nall) primes $p = 11$ and $q = 23$. The product of these is $N = 253$ and e associated Euler function is $\varphi(N) = 10 \times 22 = 220$. If we choose $= 3$ (the smallest possible value) then we can generate $d = 147$ by an tension of the Euclidean algorithm. If we wish to encipher the number $= 165$ then the encryption step gives the ciphertext

$$C = 165^3 \bmod 253 = 110. \tag{E.27}$$

he decryption step gives

$$C^d \bmod N = 110^{147} \bmod 253 = 165, \tag{E.28}$$

hich is the original plaintext.

Proof It follows from eqn E.24 that there is an integer ℓ for which

$$de = 1 + \ell\varphi(N) = 1 + \ell(p-1)(q-1).$$

This implies that

$$\begin{aligned} M^{ed} &= M^{1+\ell(p-1)(q-1)} \\ &= M\left(M^{p-1}\right)^{\ell(q-1)} \\ &\equiv M \bmod p, \end{aligned}$$

where the final step follows from eqn E.14. Naturally, it is also true that $M^{ed} \equiv M \bmod q$. The primes p and q are distinct and therefore

$$M^{ed} \equiv M \bmod N = M,$$

where the final step follows from the fact that $0 \leq M < N$.

Quantum copying

In Section 3.2 we showed that it is impossible to copy perfectly the unknown state of a qubit. This is the content of the no-cloning theorem. An obvious question, of course, is to ask what is the best that can be done. A range of copying strategies for quantum states have been devised, optimized for a variety of figures of merit. Here we present four of these: cloning based on a measurement, the cloning transformation discussed by Wootters and Zurek in proving the no-cloning theorem, the optimal symmetric cloning of Bužek and Hillery, and, finally, the perfect but probabilistic cloning of Duan and Guo.

Our task is to create a copy of a qubit prepared in an unknown pure state. We write this general (pure) state in the form

$$|\psi\rangle = \cos\left(\frac{\theta}{2}\right)|0\rangle + e^{i\varphi}\sin\left(\frac{\theta}{2}\right)|1\rangle, \tag{F.1}$$

where the angles θ and φ are the polar coordinates of the corresponding point on the Bloch sphere. We would like to get as close as we can to the transformation

$$|\psi\rangle \otimes |B\rangle \rightarrow |\psi\rangle \otimes |\psi\rangle. \tag{F.2}$$

We shall use, as a figure of merit, the fidelity both of the copy and the original qubit, as compared with the original state. The fidelity, as defined in Section 8.3, is simply the probability that the post-cloning states will pass as true copies of the original. In order not to produce a biased result, we shall average over all possible initial pure states by integrating over the surface of the Bloch sphere.

The first idea that comes to mind is simply to identify the state as well as we can and then to make a copy of the state corresponding to the measurement result. A measurement of any of the three spin components represented by the Pauli operators will give the result $+1$ or -1 and we can associate the measurement result with the corresponding eigenvector. As we are averaging over all of the possible states, we can simply chose to measure $\hat{\sigma}_z$. Performing this measurement on the state in eqn F.1 will give the result $+1$ with probability $\cos^2(\theta/2)$ and the result -1 with probability $\sin^2(\theta/2)$. If we get the result $+1$ (or -1) then we prepare a new qubit in the state $|0\rangle$ (or $|1\rangle$ respectively). The probability that this 'copy' will pass as the original is given by the squared modulus of the overlap between this eigenstate and the states in eqn F.1. It follows that the fidelity is

$$F(\theta, \varphi) = \cos^4\left(\frac{\theta}{2}\right) + \sin^4\left(\frac{\theta}{2}\right). \tag{F.3}$$

This fidelity takes the value unity if $\theta = 0$ or π, corresponding to the initial state being an eigenstate of $\hat{\sigma}_z$, but is only $\frac{1}{2}$ if $\theta = \pi/2$. As the state is unknown, our figure of merit is obtained by averaging over all initial pure states:

$$F = \frac{1}{4\pi} \int_0^{2\pi} d\varphi \int_0^\pi \sin\theta \, d\theta \, F(\theta, \varphi) = \frac{2}{3}. \tag{F.4}$$

This figure is better than the value of $\frac{1}{2}$ which would result from simply guessing the state, but it is not the largest possible value.

In proving the no-cloning theorem, we considered a copying transformation of the form

$$\begin{aligned}
|0\rangle \otimes |B\rangle &\rightarrow |0\rangle \otimes |0\rangle, \\
|1\rangle \otimes |B\rangle &\rightarrow |1\rangle \otimes |1\rangle,
\end{aligned} \tag{F.5}$$

so that perfect copies result if the initial state was either $|0\rangle$ or $|1\rangle$. Applying this to copy the state in eqn F.1 leads to

$$|\psi\rangle \otimes |B\rangle \rightarrow \cos\left(\frac{\theta}{2}\right)|0\rangle \otimes |0\rangle + e^{i\varphi}\sin\left(\frac{\theta}{2}\right)|1\rangle \otimes |1\rangle. \tag{F.6}$$

The fidelity for each of the copies, as compared with $|\psi\rangle$, is the same as given in eqn F.3. It follows that the average fidelity achieved by copying in this way is the same as that achieved by measuring the qubit and then preparing a copy in the eigenstate associated with the measurement outcome.

The above schemes copy some states better than others, but we can construct transformations that correspond to copying all possible pure states equally well. The optimal symmetric cloning operation, derived by Bužek and Hillery, incorporates both an ancillary qubit in the blank state $|B\rangle$ and a state for the copying machine, which we denote $|Q\rangle$. The optimal transformation has the form

$$\begin{aligned}
|0\rangle \otimes |B\rangle \otimes |Q\rangle &\rightarrow \sqrt{\frac{2}{3}}|0\rangle \otimes |0\rangle \otimes |q\rangle + \sqrt{\frac{1}{3}}|\Psi^+\rangle \otimes |q^\perp\rangle, \\
|1\rangle \otimes |B\rangle \otimes |Q\rangle &\rightarrow \sqrt{\frac{2}{3}}|1\rangle \otimes |1\rangle \otimes |q^\perp\rangle + \sqrt{\frac{1}{3}}|\Psi^+\rangle \otimes |q\rangle,
\end{aligned} \tag{F.7}$$

where $|\Psi^+\rangle$ is the Bell state given in eqn 2.108 and the states $|q\rangle$ and $|q^\perp\rangle$ are orthogonal states of the copying machine: $\langle q|q^\perp\rangle = 0$. If this symmetric optimal cloning operation is used to copy a qubit in our general pure state given in eqn F.1, then both the qubit and the ancilla are left in the mixed state

$$\hat{\rho} = \frac{5}{6}|\psi\rangle\langle\psi| + \frac{1}{6}|\psi^\perp\rangle\langle\psi^\perp|, \tag{F.8}$$

where $|\psi^\perp\rangle$ is the qubit state orthogonal to $|\psi\rangle$: $\langle\psi|\psi^\perp\rangle = 0$. This means that the fidelity for both the original qubit and the ancilla, compared with the desired original state, is $\frac{5}{6}$. This is the largest allowed value

r a symmetric, that is, state-independent, cloning operation. It is
raightforward to show, moreover, that the state produced is orthogonal
the two-qubit state $|\psi^\perp\rangle \otimes |\psi^\perp\rangle$ and hence that at least one of the
ıbits will be left in the correct state.

Only two orthogonal states of the cloning machine are relevant for
e optimal symmetric cloning transformation, and so we can replace
e machine by a third qubit. It is interesting to map the states of the
achine onto the qubit states

$$|q\rangle = |1\rangle,$$
$$|q^\perp\rangle = -|0\rangle. \tag{F.9}$$

we do this then the cloning operation in eqn F.7 becomes

$$|0\rangle \otimes |B\rangle \otimes |Q\rangle \rightarrow \sqrt{\frac{2}{3}}|0\rangle \otimes |0\rangle \otimes |1\rangle - \sqrt{\frac{1}{3}}|\Psi^+\rangle \otimes |0\rangle,$$
$$|1\rangle \otimes |B\rangle \otimes |Q\rangle \rightarrow -\sqrt{\frac{2}{3}}|1\rangle \otimes |1\rangle \otimes |0\rangle + \sqrt{\frac{1}{3}}|\Psi^+\rangle \otimes |1\rangle. \tag{F.10}$$

follows that the cloning operation produces the state

$$\rangle \otimes |B\rangle \otimes |Q\rangle \rightarrow \sqrt{\frac{2}{3}}|\psi\rangle \otimes |\psi\rangle \otimes |\psi^\perp\rangle - \sqrt{\frac{1}{6}}(|\psi\rangle \otimes |\psi^\perp\rangle + |\psi^\perp\rangle \otimes |\psi\rangle) \otimes |\psi\rangle. \tag{F.11}$$

follows, in turn, that the reduced density operator for the third qubit
also a mixture of the states $|\psi\rangle$ and $|\psi^\perp\rangle$:

$$\hat{\rho} = \frac{2}{3}|\psi^\perp\rangle\langle\psi^\perp| + \frac{1}{3}|\psi\rangle\langle\psi|. \tag{F.12}$$

e have seen in Chapter 6 that the quantum NOT operation, which
oduces the state $|\psi^\perp\rangle$ from any unknown state $|\psi\rangle$, cannot be imple-
ɛnted perfectly. Here we see, however, that the optimal symmetric
ɔning transformation realizes this operation on the third qubit with
fidelity of $\frac{2}{3}$. This matches the performance of the optimal universal
ɔT gate described in Section 6.2.

A range of cloning schemes have been devised which are optimized for
pying a particular set of states. Among these is the perfect cloning
ɪeme of Duan and Guo, which is designed to create perfect clones of
her of a pair of non-orthogonal states. There is no conflict with the
-cloning theorem, as a device of this type will only create clones with
n-unit probability. Let the qubit to be copied be in one of the two
n-orthogonal states $|\psi_1\rangle$ and $|\psi_2\rangle$. Without loss of generality, we shall
ssume that the overlap of these states, $\langle\psi_1|\psi_2\rangle$, is real and positive.
ɛ also introduce an ancillary qubit in the blank state $|B\rangle$, and a third
bit, prepared in the state $|Q\rangle$. A unitary transformation preserves the
erlap between states and we can use this property to obtain an upper
ɔund on the probability of successful cloning. Let the unitary operator
ɛ on the initial states to produce

$$\hat{U}|\psi_1\rangle \otimes |B\rangle \otimes |Q\rangle = a|\psi_1\rangle \otimes |\psi_1\rangle \otimes |q\rangle + \sqrt{1 - a^2}|\Phi_1\rangle \otimes |q^\perp\rangle,$$
$$\hat{U}|\psi_2\rangle \otimes |B\rangle \otimes |Q\rangle = a|\psi_2\rangle \otimes |\psi_2\rangle \otimes |q\rangle + \sqrt{1 - a^2}|\Phi_2\rangle \otimes |q^\perp\rangle,$$

$$\text{(F.13)}$$

where a is the square root of the probability that the cloning operation is successful. We can determine whether or not we have been successful by measuring the state of the third qubit. Unitarity requires that the overlap of the states in eqn F.13 is $\langle\psi_1|\psi_2\rangle$:

$$a^2\langle\psi_1|\psi_2\rangle^2 + (1 - a^2)\langle\Phi_1|\Phi_2\rangle = \langle\psi_1|\psi_2\rangle. \qquad \text{(F.14)}$$

The maximum probability for success is the maximum value of a^2 that is consistent with this condition, and this occurs for $\langle\Phi_1|\Phi_2\rangle = 1$:

$$P_{\text{Clone}} \leq a^2_{\text{Max}} = \frac{1}{1 + \langle\psi_1|\psi_2\rangle}. \qquad \text{(F.15)}$$

If the cloning is unsuccessful, then the state of the original qubit and the blank is independent of the initial state.

Quantized field modes

the quantum theory of light, the electric and magnetic fields, like
her observables, are operators. It is convenient and, for our purposes,
fficient to consider only a small number of modes of the field. A single
ode is characterized by its frequency, its spatial distribution, and its
olarization. The complex electric field operator for our single mode is

$$\hat{\vec{E}} = \mathcal{E}(\vec{r})\vec{\epsilon}\,\hat{a}e^{-i\omega t}, \tag{G.1}$$

 here $\mathcal{E}(\vec{r})$ contains the spatial mode profile, $\vec{\epsilon}$ is the polarization vector,
d ω is the angular frequency of the mode. The operator $\hat{a}$ embodies
e quantum nature of the field operator: if we replaced it by a complex
mplitude then we would recover the classical description of the field.
The electromagnetic energy for the mode is simply that for a harmonic
cillator of frequency ω, and this leads to the Hamiltonian

$$\hat{H} = \hbar\omega\left(\hat{a}^\dagger\hat{a} + \frac{1}{2}\hat{I}\right). \tag{G.2}$$

is similarity with the harmonic oscillator suggests that we should
uantize' the field by imposing the commutation relation

$$[\hat{a}, \hat{a}^\dagger] = \hat{I}. \tag{G.3}$$

e energy eigenstates for our field mode or, equivalently, for a harmonic
cillator may be obtained directly from this commutation relation. We
rt by assuming that there exists an energy eigenstate, which we shall
note by $|n\rangle$, with eigenenergy E_n:

$$\hat{H}|n\rangle = E_n|n\rangle. \tag{G.4}$$

follows that the state $\hat{a}^\dagger|n\rangle$ is also an energy eigenstate, as

$$\begin{aligned}
\hat{H}\hat{a}^\dagger|n\rangle &= \hbar\omega\left(\hat{a}^\dagger\hat{a}\hat{a}^\dagger + \frac{1}{2}\hat{a}^\dagger\right)|n\rangle \\
&= \hbar\omega\left[\hat{a}^\dagger\left(\hat{a}^\dagger\hat{a} + \hat{I}\right) + \frac{1}{2}\hat{a}^\dagger\right]|n\rangle \\
&= (E_n + \hbar\omega)\,\hat{a}^\dagger|n\rangle. \tag{G.5}
\end{aligned}$$

e state $\hat{a}^\dagger|n\rangle$ is, therefore, an energy eigenstate with eigenenergy
$+ \hbar\omega$. It follows, by induction, that there exists a ladder of energy

eigenstates $\hat{a}^{\dagger m}|n\rangle$ with eigenenergies $E_n + m\hbar\omega$, where $m = 0, 1, 2, \cdots$. The state $\hat{a}|n\rangle$ is also an energy eigenstate:

$$\hat{H}\hat{a}|n\rangle = \hbar\omega \left(\hat{a}^{\dagger}\hat{a}\hat{a} + \frac{1}{2}\hat{a} \right) |n\rangle$$

$$= \hbar\omega \left[\left(\hat{a}\hat{a}^{\dagger} - \hat{1} \right) \hat{a} + \frac{1}{2}\hat{a} \right] |n\rangle$$

$$= (E_n - \hbar\omega) \hat{a}^{\dagger}|n\rangle. \qquad (G.6)$$

It then follows that the ladder of energy eigenstates also extends downwards in energy, with eigenstates $\hat{a}^{m}|n\rangle$ with eigenenergies $E_n - m\hbar\omega$.

The energy of our harmonic oscillator is bounded from below by $\frac{1}{2}\hbar\omega$. We can see this by noting that the expectation value of $\hat{H}$ in an arbitrary normalized pure state $|\psi\rangle$ is

$$\langle\psi|\hat{H}|\psi\rangle = \left(\langle\psi|\hat{a}^{\dagger}\rangle \left(\hat{a}|\psi\rangle \right) + \frac{1}{2}\hbar\omega$$

$$\geq \frac{1}{2}\hbar\omega. \qquad (G.7)$$

Here we have used the positivity condition in eqn 2.3 applied to the state $\hat{a}|\psi\rangle$. This lower bound is only compatible with our ladder of eigenenergies if there exists a ground state, $|0\rangle$, which satisfies the equation

$$\hat{a}|0\rangle = 0. \qquad (G.8)$$

It follows that the eigenenergy of the ground state is $\frac{1}{2}\hbar\omega$:

$$\hat{H}|0\rangle = \left(\hat{a}^{\dagger}\hat{a} + \frac{1}{2}\hat{1} \right) |0\rangle$$

$$= \frac{1}{2}\hbar\omega|0\rangle. \qquad (G.9)$$

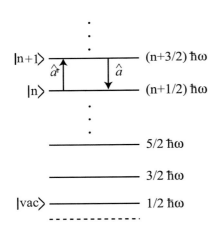

$|n+1\rangle$ — (n+3/2) $\hbar\omega$

$|n\rangle$ — (n+1/2) $\hbar\omega$

— 5/2 $\hbar\omega$

— 3/2 $\hbar\omega$

$|\text{vac}\rangle$ — 1/2 $\hbar\omega$

Fig. G.1 The ladder of energy levels.

Strictly, the action of the number operator on the number states only determines the action of $\hat{a}$ and $\hat{a}^{\dagger}$ on the number states up to an arbitrary phase, and the choices of these phases, embodied in eqn G.10, constitute a conventional choice.

The allowed eigenenergies are $\left(n + \frac{1}{2} \right) \hbar\omega$ and we associate the number n with the number of photons occupying the mode; the state $|0\rangle$ is the zero-photon, or *vacuum*, state. The integer n is also the eigenvalue of the number operator $\hat{n} = \hat{a}^{\dagger}\hat{a}$, and it follows that the actions of $\hat{a}$ and $\hat{a}^{\dagger}$ on the number state $|n\rangle$ are

$$\hat{a}^{\dagger}|n\rangle = \sqrt{n+1}|n+1\rangle,$$

$$\hat{a}|n\rangle = \sqrt{n}|n-1\rangle. \qquad (G.10)$$

We commonly refer to the operators $\hat{a}$ and $\hat{a}^{\dagger}$ as the annihilation and creation operators, respectively, because their effect is to remove or annihilate a photon or to add or create a photon in the mode. The use of $|0\rangle$ to represent the vacuum state might cause confusion in quantum information problems because we have already used this symbol to denote a qubit state. For this reason, we shall use $|\text{vac}\rangle$ to denote the ground state of the electromagnetic field. We summarize the quantum properties of our single-mode field in Fig. G.1.

If we have two modes of the same frequency that overlap at a beam splitter then the fields are superposed in the same way as for the classical fields described in Section 3.3. We can describe this in terms of the annihilation operators for the input and output modes as depicted in Fig. G.2. The output annihilation operators are related to those for the input modes by the same relationships as for the classical field amplitudes, given in eqn 3.57:

$$\hat{a}_{1\,\text{out}}^{H,V} = t_1\hat{a}_{1\,\text{in}}^{H,V} + r_1\hat{a}_{2\,\text{in}}^{H,V},$$
$$\hat{a}_{2\,\text{out}}^{H,V} = t_2\hat{a}_{2\,\text{in}}^{H,V} + r_2\hat{a}_{1\,\text{in}}^{H,V}. \qquad (\text{G.11})$$

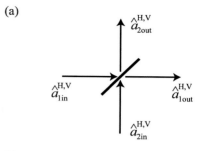

In this quantum treatment, the properties of the reflection and transmission coefficients can be obtained by imposing the commutation relations

$$\left[\hat{a}_{k\,\text{in}}^{i}, \hat{a}_{\ell\,\text{in}}^{j\dagger}\right] = \delta_{ij}\delta_{k\ell}\hat{I} = \left[\hat{a}_{k\,\text{out}}^{i}, \hat{a}_{\ell\,\text{out}}^{j\dagger}\right], \qquad (\text{G.12})$$

where $i,j = H,V$ and $k,\ell = 1,2$. For a polarizing beam splitter, the annihilation operators are related by

$$\hat{a}_{1\,\text{out}}^{H} = \hat{a}_{1\,\text{in}}^{H}, \qquad \hat{a}_{1\,\text{out}}^{V} = \hat{a}_{2\,\text{in}}^{V},$$
$$\hat{a}_{2\,\text{out}}^{H} = \hat{a}_{2\,\text{in}}^{H}, \qquad \hat{a}_{2\,\text{out}}^{V} = \hat{a}_{1\,\text{in}}^{V}. \qquad (\text{G.13})$$

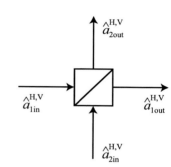

As an example of the use of the quantum theory of light, we consider the interference at a beam splitter between a pair of single photons with the same polarization. If two horizontally polarized photons are allowed to overlap in this way then our initial state is $\hat{a}_{1\,\text{in}}^{H\dagger}\hat{a}_{2\,\text{in}}^{H\dagger}|\text{vac}\rangle$. We can obtain the probability amplitudes for the photons to appear in the output modes by rewriting this state in terms of the output-mode creation operators. To do this, we first invert the relationships in eqn G.11 and then take the Hermitian conjugate to obtain the input creation operators in terms of those for the outputs:

$$\hat{a}_{1\,\text{in}}^{H,V\dagger} = t_1\hat{a}_{1\,\text{out}}^{H,V\dagger} + r_2\hat{a}_{2\,\text{out}}^{H,V\dagger},$$
$$\hat{a}_{2\,\text{in}}^{H,V\dagger} = t_2\hat{a}_{2\,\text{out}}^{H,V\dagger} + r_1\hat{a}_{1\,\text{out}}^{H,V\dagger}. \qquad (\text{G.14})$$

It follows that our output state is

$$\hat{a}_{1\,\text{in}}^{H\dagger}\hat{a}_{2\,\text{in}}^{H\dagger}|\text{vac}\rangle = \left(t_1\hat{a}_{1\,\text{out}}^{H,V\dagger} + r_2\hat{a}_{2\,\text{out}}^{H,V\dagger}\right)\left(t_2\hat{a}_{2\,\text{out}}^{H,V\dagger} + r_1\hat{a}_{1\,\text{out}}^{H,V\dagger}\right)|\text{vac}\rangle$$
$$= t_1r_1\sqrt{2}|2_1, 0_2\rangle + (t_1t_2 + r_2r_1)|1_1, 1_2\rangle$$
$$+ r_2t_2\sqrt{2}|0_1, 2_2\rangle, \qquad (\text{G.15})$$

where $|m_1, n_2\rangle$ denotes the state with m photons in output mode 1 and n in output mode 2. The transmission and reflection coefficients satisfy the condition $t_1^*r_1 + r_2^*t_2 = 0$, and this means that

$$\arg(r_1) - \arg(t_1) + \arg(r_2) - \arg(t_2) = (2\ell + 1)\pi, \qquad (\text{G.16})$$

so that the amplitude for a single photon emerging in each output mode displays destructive interference:

$$t_1t_2 + r_2r_1 = t_1t_2\left(1 - \frac{|r_1r_2|}{|t_1t_2|}\right) = t_1t_2\left(1 - \frac{|r|^2}{|t|^2}\right). \qquad (\text{G.17})$$

Fig. G.2 Beam splitters with input and output annihilation operators. (a) The polarization-insensitive beam splitter. (b) The polarizing beam splitter.

If the probabilities of transmission and reflection are both $\frac{1}{2}$ then this amplitude is zero, and both photons leave in the same output mode. This is an intrinsically quantum interference effect, which was first demonstrated experimentally by Hong, Ou, and Mandel. We can view it as a consequence of the bosonic nature of photons; if the same experiment were to be performed with a pair of fermions then, because of the Pauli exclusion principle, precisely one particle would leave in each output mode.

Position and momentum eigenstates

The position and momentum operators satisfy the commutation relation

$$[\hat{x}, \hat{p}] = i\hbar. \tag{H.1}$$

The operators and their eigenstates satisfy the eigenvalue equations

$$\hat{x}|x\rangle = x|x\rangle,$$
$$\hat{p}|p\rangle = p|p\rangle, \tag{H.2}$$

and the familiar wavefunction description of the state is $\psi(x) = \langle x|\psi\rangle$, with the momentum wavefunction given by $\phi(p) = \langle p|\psi\rangle$. The eigenstates for each observable are not orthogonal in the same sense as for observables with discrete eigenvalues, but rather they satisfy *delta function orthogonality*:

$$\langle x|x'\rangle = \delta(x - x'),$$
$$\langle p|p'\rangle = \delta(p - p'). \tag{H.3}$$

The continuity of the eigenvalues also leads us to modify the completeness condition, which we express in terms of an integral:

$$\int dx|x\rangle\langle x| = \hat{\mathrm{I}} = \int dp|p\rangle\langle p|. \tag{H.4}$$

Taking the matrix elements of eqn H.1 in the basis of the position eigenstates $|x\rangle$, we find

$$\langle x|\,[\hat{x}, \hat{p}]\,|x'\rangle = (x - x')\langle x|\hat{p}|x'\rangle$$
$$= i\hbar\delta(x - x'). \tag{H.5}$$

follows that

$$\langle x|\hat{p}|x'\rangle = i\hbar\frac{\delta(x - x')}{x - x'} = -i\hbar\frac{d}{dx}\delta(x - x'). \tag{H.6}$$

This result, together with the resolution of the identity in eqn H.4, leads to the familiar differential representation of the momentum operator,

$$\langle x|\hat{p}|\psi\rangle = \int_{-\infty}^{\infty} dx'\langle x|\hat{p}|x'\rangle\langle x'|\psi\rangle$$
$$= -i\hbar\int_{-\infty}^{\infty} dx'\frac{d}{dx}\delta(x - x')\langle x'|\psi\rangle$$
$$= -i\hbar\frac{d}{dx}\psi(x). \tag{H.7}$$

Derivatives of a delta function
The derivative of a delta function is, like the delta function itself, defined via its value on integration:

$$\int_{-\infty}^{\infty} f(x)\frac{d}{dx}\delta(x)dx = -\left.\frac{df(x)}{dx}\right|_{x=0},$$

where we have used integration by parts. An alternative representation of the derivative of the delta function can be found by considering

$$\int_{-\infty}^{\infty} xg(x)\,\frac{d}{dx}\delta(x)\,dx$$
$$= -\left.\frac{d}{dx}[xg(x)]\right|_{x=0} = -g(0).$$

Comparing these two forms leads us to the identity

$$\frac{d}{dx}\delta(x) = -\frac{\delta(x)}{x}.$$

The representation of $\hat{x}|\psi\rangle$ is simply $x\psi(x)$, as

$$\langle x|\hat{x}|\psi\rangle = x\langle x|\psi\rangle = x\psi(x). \tag{H.8}$$

The Fourier relationship between the position and momentum representations follows on applying eqn H.7 to the momentum eigenstates:

$$\langle x|\hat{p}|p\rangle = p\langle x|p\rangle$$
$$= -i\hbar\frac{d}{dx}\langle x|p\rangle. \tag{H.9}$$

Solving this for $\langle x|p\rangle$ gives

$$\langle x|p\rangle = (2\pi\hbar)^{-1/2}\exp\left(i\frac{px}{\hbar}\right), \tag{H.10}$$

where the normalization has been determined by imposing the condition

$$\langle p'|\int dx\,|x\rangle\langle x|p\rangle = \delta(p-p'). \tag{H.11}$$

It follows that the momentum and position wavefunctions are related by

$$\phi(p) = \langle p|\int dx\,|x\rangle\langle x|\psi\rangle$$
$$= \int \frac{dx}{\sqrt{2\pi\hbar}}\exp\left(i\frac{px}{\hbar}\right)\psi(x). \tag{H.12}$$

The differential representation in eqn H.7 tells us that we can shift the average position of our wavefunction by means of a unitary transformation generated by the momentum operator. We start by noting that we can write a Taylor series in the concise form

$$\psi(x+x_0) = \exp\left(x_0\frac{d}{dx}\right)\psi(x)$$
$$= \exp\left(i\frac{x_0\hat{p}}{\hbar}\right)\psi(x). \tag{H.13}$$

It follows in the same way that the position operator acts to generate shifts in the momentum:

$$\phi(p+p_0) = \exp\left(p_0\frac{d}{dp}\right)\phi(p)$$
$$= \exp\left(-i\frac{p_0\hat{x}}{\hbar}\right)\phi(p). \tag{H.14}$$

We can shift both the position and the momentum by means of the unitary displacement operator

$$\hat{D}(x_0,p_0) = \exp\left[i\frac{(x_0\hat{p}-p_0\hat{x})}{\hbar}\right]. \tag{H.15}$$

The position and momentum operators do not commute, so we cannot simply factorize this operator into a product of a position-shifting operator and a momentum-shifting operator. The commutator of these

perators is quite simple, however, and this means that we can rewrite
he displacement operator as a product of a position shift, a momentum
hift, and a phase factor:

$$\hat{D}(x_0, p_0) = \exp\left(i\frac{x_0\hat{p}}{\hbar}\right) \exp\left(-i\frac{p_0\hat{x}}{\hbar}\right) \exp\left(\frac{ix_0p_0}{2\hbar}\right)$$

$$= \exp\left(-i\frac{p_0\hat{x}}{\hbar}\right) \exp\left(i\frac{x_0\hat{p}}{\hbar}\right) \exp\left(-\frac{ix_0p_0}{2\hbar}\right). \quad \text{(H.16)}$$

Necessary conditions for a minimum-error POM

Section 4.4, we gave the conditions for a POM $\{\hat{\pi}_j\}$ to maximize the probability for correctly identifying a state. These conditions were

$$\hat{\pi}_j \left(p_j \hat{\rho}_j - p_k \hat{\rho}_k \right) \hat{\pi}_k = 0, \qquad \forall j, k \tag{I.1}$$

d that

$$\sum_i p_i \hat{\rho}_i \hat{\pi}_i - p_j \hat{\rho}_j \geq 0, \qquad \forall j. \tag{I.2}$$

ie latter condition requires the operator

$$\hat{\Gamma} = \sum_i p_i \hat{\rho}_i \hat{\pi}_i \tag{I.3}$$

be Hermitian so that it can also be positive. We saw in Section 4.4 at if the inequality in eqn I.2 holds then the POM gives the minimum ror. It follows that this inequality is a sufficient condition for the nimum-error POM. Our tasks in this appendix are to show that this equality is also a *necessary* condition and to show that the condition eqn I.1 is also necessary. The analysis presented here is based on that S. M. Barnett and S. Croke, *Journal of Physics A: Mathematical and eoretical* **42**, 062001 (2009).

We start by introducing the Hermitian operators

$$\hat{G}_j = \sum_i p_i \frac{1}{2} \{\hat{\rho}_i, \hat{\pi}_i\} - p_j \hat{\rho}_j, \tag{I.4}$$

iere the operators $\hat{\pi}_i$ comprise a minimum-error POM. Note the intro-ction of the anticommutator so as to ensure rather than presuppose rmiticity. We shall show that it is necessary for all of these operators be positive if the measurement is a minimum-error strategy. To see s, let us suppose that for one state $\hat{\rho}_0$, the corresponding operator has iegative eigenvalue $-\lambda$ and a corresponding eigenstate $|\lambda\rangle$, so that

$$\hat{G}_0|\lambda\rangle = -\lambda|\lambda\rangle. \tag{I.5}$$

the existence of this negative eigenvalue means that there exists a)M with a greater probability for correctly identifying the state then it lows that the positivity of $\hat{G}_0$ is a *necessary* condition for a minimum-or POM.

Consider the primed POM with probability operators

$$\hat{\pi}_i' = \left(\hat{I} - \varepsilon\hat{P}_\lambda\right)\hat{\pi}_i\left(\hat{I} - \varepsilon\hat{P}_\lambda\right) + \varepsilon(2 - \varepsilon)\hat{P}_\lambda\delta_{i0}, \tag{I.6}$$

where $\hat{P}_\lambda = |\lambda\rangle\langle\lambda|$ and $\varepsilon \ll 1$. It is clear that these elements form a POM, as the operators $\left(\hat{I} - \varepsilon\hat{P}_\lambda\right)\hat{\pi}_i\left(\hat{I} - \varepsilon\hat{P}_\lambda\right)$ and $\hat{P}_\lambda$ are clearly positive and $\sum_i \hat{\pi}_i' = \hat{I}$. The probability that this primed measurement will correctly identify the state is

$$\begin{aligned}
P'_{\text{corr}} &= \sum_i p_i \text{Tr}\left(\hat{\rho}_i\hat{\pi}_i'\right) \\
&= \sum_i p_i \text{Tr}\left[\hat{\rho}_i\left(\hat{I} - \varepsilon\hat{P}_\lambda\right)\hat{\pi}_i\left(\hat{I} - \varepsilon\hat{P}_\lambda\right)\right] + \varepsilon(2 - \varepsilon)p_0\langle\lambda|\hat{\rho}_0|\lambda\rangle \\
&= P_{\text{corr}} - 2\varepsilon\sum_i p_i\langle\lambda|\frac{1}{2}\{\hat{\rho}_i, \hat{\pi}_i\}|\lambda\rangle + 2\varepsilon p_0\langle\lambda|\hat{\rho}_0|\lambda\rangle + O(\varepsilon^2) \\
&= P_{\text{corr}} + 2\varepsilon\lambda + O(\varepsilon^2).
\end{aligned} \tag{I.7}$$

This is greater than P_{corr} and so contradicts the assumption that $\{\hat{\pi}_i\}$ is a minimum-error POM. There is nothing special about the state $\hat{\rho}_0$, of course, and therefore if any of the $\hat{G}_j$ has a negative eigenvalue then the corresponding POM is not a minimum-error measurement. It follows that the positivity of all of the $\hat{G}_j$ is a *necessary* condition for the POM $\{\hat{\pi}_i\}$ to represent a minimum-error measurement.

We also need to show that $\hat{\Gamma}$ is Hermitian so that

$$\hat{G}_j = \hat{\Gamma} - p_j\hat{\rho}_j. \tag{I.8}$$

To see that this is the case, we need only note that

$$\sum_i \text{Tr}\left(\hat{G}_i\hat{\pi}_i\right) = 0 \tag{I.9}$$

If any two operators $\hat{B}$ and $\hat{C}$ are positive (and therefore Hermitian) then

$$\text{Tr}\left(\hat{B}\hat{C}\right) = 0 \Rightarrow \hat{B}\hat{C} = 0 = \hat{C}\hat{B}.$$

To see this is true, we can write $\hat{C} = \sum_n C_n|C_n\rangle\langle C_n|$, where C_n and $|C_n\rangle$ are the (positive) eigenvalues and the eigenvectors of $\hat{C}$. It then follows that

$$\text{Tr}\left(\hat{B}\hat{C}\right) = \sum_n C_n\langle C_n|\hat{B}|C_n\rangle.$$

Each term in the sum is positive or zero and hence the trace can only be zero if $\hat{B}|C_n\rangle = 0 \; \forall n$, which means that $\hat{B}\hat{C} = 0$. The condition $\hat{C}\hat{B} = 0$ follows on commuting the operators under the trace.

and, because both $\hat{G}_i$ and $\hat{\pi}_i$ are positive operators, this means that $\text{Tr}\left(\hat{G}_i\hat{\pi}_i\right) = 0$, which implies that $\hat{G}_i\hat{\pi}_i = 0$. Summing this over i then gives

$$\sum_i \frac{1}{2}\left(p_i\hat{\rho}_i\hat{\pi}_i - p_i\hat{\pi}_i\rho_i\right) = \frac{1}{2}\left(\hat{\Gamma} - \hat{\Gamma}^\dagger\right) = 0, \tag{I.10}$$

so that $\hat{\Gamma}$ is necessarily Hermitian.

It remains only to demonstrate the necessity of the condition in eqn I.1. We can show this by noting that the positivity condition in eqn I.2 together with

$$\sum_i \text{Tr}\left[\left(\hat{\Gamma} - p_i\hat{\rho}_i\right)\hat{\pi}_i\right] = 0 \tag{I.11}$$

means that

$$\begin{aligned}
\left(\hat{\Gamma} - p_k\hat{\rho}_k\right)\hat{\pi}_k &= 0, \\
\hat{\pi}_j\left(\hat{\Gamma} - p_j\hat{\rho}_j\right) &= 0.
\end{aligned} \tag{I.12}$$

If we premultiply the first of these by $\hat{\pi}_j$, postmultiply the second by $\hat{\pi}_k$, and take the difference then we are led to eqn I.1.

Complete positivity

J

We stated in Section 4.5 that the most general allowed transformation of a density operator has the form

$$\hat{\rho} \rightarrow \hat{\rho}' = \sum_i \hat{A}_i \hat{\rho} \hat{A}_i^\dagger. \tag{J.1}$$

There are a variety of proofs of this important result in the literature, all of which rely on the property of complete positivity. Complete positivity means that if the system of interest is entangled with another then the transformation must map the density operator for any such state onto another (positive) density operator. We follow here the analysis of S. Croke *et al.*, *Annals of Physics* **323**, 893 (2008).

It is convenient to work with a matrix representation of our density operator and to write a general linear transformation as a relationship between matrix elements in the form

$$\rho'_{\alpha\beta} = \sum_{kl} \mathcal{L}_{kl}^{\alpha\beta} \rho_{kl}, \tag{J.2}$$

where $\rho'_{\alpha\beta} = \langle\alpha|\hat{\rho}'|\beta\rangle$, $\rho_{kl} = \langle k|\hat{\rho}|l\rangle$, and the states $|\alpha\rangle, |\beta\rangle, \cdots$ form a complete orthonormal basis, as do the states $|k\rangle, |l\rangle, \cdots$. For definiteness, we consider a system with an N-dimensional state space so that all the indices run from 1 to N and the density matrices are $N \times N$. The transformation $\mathcal{L}$ in eqn J.2 may be considered as a mapping on indices of the form

$$\mathcal{L} : \{kl\} \Rightarrow \{\alpha\beta\}. \tag{J.3}$$

It is also useful to think of the four-index object $\mathcal{L}_{kl}^{\alpha\beta}$ as an associated mapping of the form

$$\mathcal{L}_{\text{ass}} : \{\beta l\} \Rightarrow \{\alpha k\}. \tag{J.4}$$

This associated mapping plays an important role in establishing complete positivity.

The properties of the density operator constrain the form of $\mathcal{L}$. Firstly we require that the transformed density operator is Hermitian, so that $\rho'_{\alpha\beta} = \rho'^*_{\beta\alpha}$:

Operator form We can relate $\mathcal{L}$ to our original general linear operator transformation in eqn 4.73,

$$\hat{\rho} \rightarrow \sum_i \hat{A}_i \hat{\rho} \hat{B}_i,$$

by taking matrix elements:

$$\mathcal{L}_{kl}^{\alpha\beta} = \sum_i \langle\alpha|\hat{A}_i|k\rangle \langle l|\hat{B}_i|\beta\rangle.$$

$$\sum_{kl} \mathcal{L}_{kl}^{\alpha\beta} \rho_{kl} = \sum_{kl} \mathcal{L}_{kl}^{\beta\alpha*} \rho_{kl}^* = \sum_{kl} \mathcal{L}_{kl}^{\beta\alpha*} \rho_{lk}$$

$$\Rightarrow \mathcal{L}_{kl}^{\alpha\beta} = \mathcal{L}_{lk}^{\beta\alpha*}. \tag{J.5}$$

Secondly, we require that the trace of the transformed density matrix is unity,

$$\mathrm{Tr}\rho' = \sum_\alpha \sum_{kl} \mathcal{L}_{kl}^{\alpha\alpha} \rho_{kl} = 1$$

$$\Rightarrow \sum_\alpha \mathcal{L}_{kl}^{\alpha\alpha} = \delta_{kl}, \tag{J.6}$$

so that $\sum_\alpha \mathcal{L}_{kl}^{\alpha\alpha}$ is the identity matrix. The third important property is that we require the transformation to be completely positive; if the system is entangled with an ancillary system then the transformation must map the density operator for any such composite state onto another (positive) density operator.

In order to establish complete positivity, it is useful to consider a pure entangled state of the original system S, on which our transformation acts, and an ancillary system A. We write the state in its Schmidt decomposition

$$|\psi\rangle_{SA} = \sum_k c_k |k\rangle_S |\chi_k\rangle_A, \tag{J.7}$$

so that the states $\{|k\rangle\}$ and $\{|\chi_k\rangle\}$ are orthonormal sets. It then follows that the diagonal forms of the reduced density operators for the system and ancilla are

$$\hat{\rho}_S = \sum_k |c_k|^2 |k\rangle\langle k|, \qquad \hat{\rho}_A = \sum_k |c_k|^2 |\chi_k\rangle\langle\chi_k|, \tag{J.8}$$

so that $\rho_{kl} = |c_k|^2 \delta_{kl}$. The matrix elements in the S state space for the state in eqn J.7 are

$$\rho_{kl} = \langle k||\psi\rangle\langle\psi||l\rangle = c_k c_l^* |\chi_k\rangle\langle\chi_l|, \tag{J.9}$$

which is an operator in the A state space. It follows that the S-space matrix elements of the transformed density operator are also operators on the A space and have the form

$$\rho'_{\alpha\beta} = \sum_{kl} \mathcal{L}_{kl}^{\alpha\beta} c_k c_l^* |\chi_k\rangle\langle\chi_l|. \tag{J.10}$$

Consider the special case of the maximally entangled state for which $c_k = N^{-1/2}$. For this state, the transformed density operator has the form

$$\hat{\rho}'_{SA} = \frac{1}{N} \sum_{\alpha\beta kl} |\alpha\rangle_{SS}\langle\beta| \otimes |\chi_k\rangle_{AA}\langle\chi_l|. \tag{J.11}$$

This must, of course, be a positive operator, and it follows that a necessary condition for the transformation $\mathcal{L}$ to be positive is that the associated operator

$$\hat{\Lambda} = \sum_{\alpha\beta kl} \mathcal{L}_{kl}^{\alpha\beta} |\alpha\rangle_{SS}\langle\beta| \otimes |\chi_k\rangle_{AA}\langle\chi_l| \tag{J.12}$$

must be positive. The positivity of this operator is also a sufficient condition. To see this we recall that we require that the transformed nsity operator is positive for any initial state. For a general (pure) tangled state as given in eqn J.7, this means

$$_A\langle\phi|\left(\sum_{\alpha\beta kl}\mathcal{L}_{kl}^{\alpha\beta}c_kc_l^*|\alpha\rangle_{SS}\langle\beta|\otimes|\chi_k\rangle_{AA}\langle\chi_l|\right)|\phi\rangle_{SA}\geq 0,\forall|\phi\rangle_{SA}. \quad (J.13)$$

we define the state

$$|\phi'\rangle_{SA}=\left(\sum_{\alpha k}c_k|\alpha\rangle_{SS}\langle\alpha|\otimes|\chi_k\rangle_{AA}\langle\chi_k|\right)|\phi\rangle_{SA}, \quad (J.14)$$

en we can rewrite our positivity condition (eqn J.13) in the form

$$_{SA}\langle\phi'|\hat{\Lambda}|\phi'\rangle_{SA}\geq 0, \quad (J.15)$$

ich clearly holds as $\hat{\Lambda}$ is a positive operator.

We have established that a necessary and sufficient condition for our ear mapping $\mathcal{L}$ to be completely positive (and therefore physically ceptable) is that the associated mapping given in eqn J.4 should be sitive. This condition is equivalent to the positivity of the operator

It remains to establish the relationship between the positivity of $\hat{\Lambda}$ d the required operator form of our general transformation given in n J.1. To do so requires us to examine the associated transformation a little more detail. We can consider the associated transformation ven in eqn J.4 as acting on column vectors

$$\mathbf{x}=\begin{pmatrix}\vdots\\x_{\alpha k}\\\vdots\end{pmatrix}, \quad (J.16)$$

th adjoint row vectors of the form

$$\mathbf{x}^\dagger=(\cdots x_{\alpha k}^*\cdots). \quad (J.17)$$

e inner product is defined in the natural way as

$$(\mathbf{y},\mathbf{x})=\mathbf{y}^\dagger\mathbf{x}=\sum_{\alpha k}y_{\alpha k}^*x_{\alpha k}, \quad (J.18)$$

that for the transformed vector $\mathcal{L}\mathbf{x}$ we have the inner product

$$(\mathbf{y},\mathcal{L}\mathbf{x})=\sum_{\alpha\beta kl}y_{\alpha k}^*\left(\mathcal{L}_{kl}^{\alpha\beta}x_{\beta l}\right). \quad (J.19)$$

follows from the properties of $\mathcal{L}_{kl}^{\alpha\beta}$ that

$$(\mathbf{y},\mathcal{L}\mathbf{x})=(\mathbf{x},\mathcal{L}\mathbf{y})^*=(\mathcal{L}\mathbf{y},\mathbf{x}), \quad (J.20)$$

that the associated transformation is Hermitian with respect to this ner product.

It follows from the Hermiticity of the associated transformation that we can find a complete set of eigenvectors and eigenvalues:

$$\sum_{\beta l} \mathcal{L}_{kl}^{\alpha\beta} B_{\beta l}^{(\nu)} = \lambda_\nu B_{\alpha k}^{(\nu)}, \tag{J.21}$$

where the eigenvalues λ_ν are real, and the eigenvectors are orthogonal and can be chosen to be normalized so that

$$\sum_{\alpha k} B_{\alpha k}^{(\nu)*} B_{\alpha k}^{(\mu)} = \delta_{\nu\mu}. \tag{J.22}$$

It follows from the positivity of the associated transformation, moreover, that the eigenvalues λ_ν are positive (strictly, ≥ 0). We can expand operators using these eigenvectors as a basis and we find, in particular, that

$$\mathcal{L}_{kl}^{\alpha\beta} = \sum_\nu \lambda_\nu B_{\alpha k}^{(\nu)} B_{\beta l}^{(\nu)*}. \tag{J.23}$$

Similarly, we can expand the identity and thereby establish completeness:

$$\sum_\nu B_{\alpha k}^{(\nu)} B_{\beta l}^{(\nu)*} = \delta_{\alpha\beta} \delta_{kl}. \tag{J.24}$$

Finally, we can use the expansion in eqn J.23 to write our matrix elements for the transformed density operator in the form

$$\rho'_{\alpha\beta} = \sum_{\nu k l} \lambda_\nu B_{\alpha k}^{(\nu)} B_{\beta l}^{(\nu)*} \rho_{kl} = \sum_{\nu k l} \lambda_\nu B_{\alpha k}^{(\nu)} \rho_{kl} B_{l\beta}^{(\nu)\dagger}. \tag{J.25}$$

The positivity of the eigenvalues suggests the alternative forms $A_{\alpha k}^{(\nu)} = \sqrt{\lambda_\nu} B_{\alpha k}^{(\nu)}$, so that

$$\rho'_{\alpha\beta} = \sum_{\nu k l} A_{\alpha k}^{(\nu)} \rho_{kl} A_{l\beta}^{(\nu)\dagger} \tag{J.26}$$

or, in a basis-independent form,

$$\hat{\rho}' = \sum_\nu \hat{A}^{(\nu)} \hat{\rho} \hat{A}^{(\nu)\dagger}, \tag{J.27}$$

which is the required form for our general operation.

We can use the required positivity of the associated transformation to check if a desired transformation is allowed: if the associated transformation is positive then it is, but if it is negative then it is not. A simple example is the transpose of a qubit density operator. If we write our initial density operator as the column vector

$$\rho = \begin{pmatrix} \rho_{00} \\ \rho_{01} \\ \rho_{10} \\ \rho_{11} \end{pmatrix}, \tag{J.28}$$

then the transpose operation is enacted by the matrix

$$\mathcal{L} = \begin{pmatrix} 1 & 0 & 0 & 0 \\ 0 & 0 & 1 & 0 \\ 0 & 1 & 0 & 0 \\ 0 & 0 & 0 & 1 \end{pmatrix}. \tag{J.29}$$

or this simple case, the matrix for the associated transformation has he same form, $\mathcal{L}_{\text{ass}} = \mathcal{L}$. The matrix has a single negative eigenvalue nd so is not positive. It follows that the transposition operation is not hysical.

Hardy's theorem

Hardy's theorem, like that of Greenberger, Horne, and Zeilinger, is a demonstration of the conflict between quantum theory and local realism which does not rely on an inequality. It has the advantage that, as with the violation of Bell's inequality, it requires only an entangled state of two qubits, albeit a non-maximally entangled state.

Consider a pair of qubits, one held by Alice and the other by Bob, prepared in the pure state

$$|\text{Hardy}\rangle = \frac{1}{\sqrt{3}} \left(|0\rangle_A |0\rangle_B + |1\rangle_A |0\rangle_B + |0\rangle_A |1\rangle_B \right). \qquad (K.1)$$

We proceed by noting that this state can also be written in the form

$$|\text{Hardy}\rangle = \sqrt{\frac{2}{3}} |0'\rangle_A |0\rangle_B + \frac{1}{\sqrt{3}} |0\rangle_A |1\rangle_B$$
$$= \frac{1}{\sqrt{3}} |1\rangle_A |0\rangle_B + \sqrt{\frac{2}{3}} |0\rangle_A |0'\rangle_B, \qquad (K.2)$$

where $|0'\rangle = 2^{-1/2}(|0\rangle + |1\rangle)$ is the eigenstate of $\hat{\sigma}_x$ with eigenvalue $+1$. The following statements follow directly from the form of $|\text{Hardy}\rangle$:

(i) If both Alice and Bob measure the observable corresponding to $\hat{\sigma}_z$ then at least one of them will get the result $+1$, corresponding to the state $|0\rangle$.

(ii) If Alice measures $\hat{\sigma}_z$ and gets the value $+1$ then a measurement by Bob of $\hat{\sigma}_x$ will, with certainty, get the value $+1$, corresponding to the state $|0'\rangle$.

(iii) If Bob measures $\hat{\sigma}_z$ and gets the value $+1$ then a measurement by Alice of $\hat{\sigma}_x$ will, with certainty, get the value $+1$.

Local realistic ideas lead us to treat as simultaneously real the values ± 1 of the observables corresponding to the operators $\hat{\sigma}_z$ and $\hat{\sigma}_x$. The values of these, which we denote σ_z and σ_x, respectively, should be independent of any choice of an observation carried out on the other qubit. This leads us to express the above three experimentally testable properties as the following probabilities:

$$P\left(\sigma_z^A = -1, \sigma_z^B = -1\right) = 0,$$
$$P\left(\sigma_x^B = +1 | \sigma_z^A = +1\right) = 1,$$
$$P\left(\sigma_x^A = +1 | \sigma_z^B = +1\right) = 1. \qquad (K.3)$$

The first of these tells us that at least one of the properties σ_z^A and σ_z^B must take the value $+1$, and the following two then tell us that at least one of the properties σ_x^A and σ_x^B must take the value $+1$. It is a prediction of local realism, therefore, that σ_x^A and σ_x^B cannot *both* take the value -1:

$$P\left(\sigma_x^A = -1, \sigma_x^B = -1\right) = 0. \tag{K.4}$$

A quantum mechanical treatment, however, shows that measurements by both Alice and Bob of $\hat{\sigma}_x$ can *both* give the result 1 with probability

$$P\left(\sigma_x^A = -1, \sigma_x^B = -1\right) = |_A\langle 1'|_B\langle 1'|\text{Hardy}\rangle|^2$$
$$= \frac{1}{12}, \tag{K.5}$$

where the state $|1'\rangle = (|0\rangle - |1\rangle)/\sqrt{2}$ is the eigenstate of $\hat{\sigma}_x$ with eigenvalue -1. This non-zero value constitutes a conflict between quantum theory and local realism.

Hardy's demonstration of non-locality is more general than that presented here. In particular, a conflict with local realism of this form can be demonstrated for any non-maximally entangled pure state of two qubits.

Universal gates

e seek a set of simple gates that is complete in that a quantum circuit
rmed from such gates allows us to construct any desired multiqubit
itary transformation. One example of such a complete set of gates is
e combination of single-qubit gates and CNOT gates. We can prove
s in two stages, by establishing first that any unitary matrix can be
composed into a product of two-level unitary matrices and then that
y such two-level unitary matrix can be realized using only single-qubit
d CNOT gates. Our analysis follows closely that given by Nielsen and
uang.

Our first task is to show that a unitary matrix can be decomposed into
roduct of two-level unitary matrices, that is, matrices that couple only
o states and leave the remainder unchanged. Consider, as a starting
int, the 3×3 unitary matrix

We are not emphasising here the role of
unitary matrices as quantum operators,
and so do not use a hat on U.

$$U = \begin{pmatrix} a & b & c \\ d & e & f \\ g & h & j \end{pmatrix}.$$

(L.1)

require, first, a unitary matrix U_1 such that $U_1 U$ has zero for one of
off-diagonal elements. A suitable matrix is

$$U_1 = \left(|a|^2 + |d|^2\right)^{-1/2} \begin{pmatrix} a^* & d^* & 0 \\ -d & a & 0 \\ 0 & 0 & \left(|a|^2 + |d|^2\right)^{1/2} \end{pmatrix},$$

(L.2)

that

$$U_1 U = \begin{pmatrix} a' & b' & c' \\ 0 & e' & f' \\ g' & h' & j' \end{pmatrix}.$$

(L.3)

can insert a second off-diagonal zero in the same column as the first
the action of a second unitary matrix,

$$U_2 = \left(|a'|^2 + |g'|^2\right)^{-1/2} \begin{pmatrix} a'^* & 0 & g'^* \\ 0 & \left(|a'|^2 + |g'|^2\right)^{1/2} & 0 \\ -g' & 0 & a' \end{pmatrix},$$

(L.4)

that

$$U_2 U_1 U = \begin{pmatrix} a'' & b'' & c'' \\ 0 & e'' & f'' \\ 0 & h'' & j'' \end{pmatrix}.$$

(L.5)

e product of a sequence of unitary matrices is itself unitary, and it
n follows from the unitarity of $U_2 U_1 U$ that the second and third

These conclusions follow directly from the unitarity conditions

$$(U_2 U_1 U)^\dagger \, U_2 U_1 U = \mathrm{I},$$
$$U_2 U_1 U \, (U_2 U_1 U)^\dagger = \mathrm{I}.$$

elements of the first row must be zero. It also follows that the first element must have modulus unity, and our construction ensures that it is real and positive:

$$U_2 U_1 U = \begin{pmatrix} 1 & 0 & 0 \\ 0 & e'' & f'' \\ 0 & h'' & j'' \end{pmatrix}. \tag{L.6}$$

The unitarity of this matrix means that its inverse is

$$U_3 = \begin{pmatrix} 1 & 0 & 0 \\ 0 & e''^* & h''^* \\ 0 & f''^* & j''^* \end{pmatrix}. \tag{L.7}$$

It follows that $U_3 U_2 U_1 U = \mathrm{I}$, so that

$$U = U_1^\dagger U_2^\dagger U_3^\dagger, \tag{L.8}$$

which is the required decomposition as a product of two-level unitary matrices.

If we had started with a 4×4, or two-qubit, unitary matrix, then pre-multiplying by a sequence of three suitable 2×2 matrices would leave the first entry equal to unity and all other elements in the first row and column zero. A further sequence of three two-level unitary matrices would complete the decomposition. Hence a general 4×4 unitary matrix can be decomposed into a product of six two-level unitary matrices. Extending this to larger matrices, we readily conclude that a $d \times d$ unitary matrix can be decomposed into a product of $d(d-1)/2$ two-level unitary matrices. The state space for a system of n qubits has dimension 2^n, and it follows that any unitary transformation can be decomposed into $2^{n-1}(2^n - 1)$ two-level unitary matrices.

Our second task is to show that any given two-level unitary matrix can be realized by the combined action of single-qubit and CNOT gates. It is helpful to label the states associated with the rows and columns of our matrix by the corresponding binary digit so that, for example, the rows of a three-qubit unitary matrix are associated with the states $|000\rangle, |001\rangle, |010\rangle, \cdots, |111\rangle$. A two-level unitary matrix couples, by design, only two states, labelled by the binary numbers s and t: $|s\rangle$ and $|t\rangle$. We can understand the required process by introducing Gray codes. A Gray code is a sequence of binary numbers, starting with s and ending with t, such that successive members of the sequence differ by exactly one bit. For example, given $s = 101001$ and $t = 110011$, a possible Gray code is

$$101001\,,$$
$$101011\,,$$
$$100011\,,$$
$$110011\,. \tag{L.9}$$

Let $g_1, g_2, \cdots, g_m$ denote our Gray code connecting s and t, with $g_1 = s$ and $g_m = t$. The idea is to construct a quantum circuit which induces he changes $|g_1\rangle \rightarrow |g_2\rangle \rightarrow \cdots \rightarrow |g_{m-1}\rangle$. We can then perform a ontrolled unitary transformation on the single qubit at which $|g_{m-1}\rangle$ nd $|g_m\rangle$ differ. Finally, we can undo the Gray code transformation $|g_{m-1}\rangle \rightarrow |g_{m-2}\rangle \rightarrow \cdots \rightarrow |g_1\rangle$.

It remains to show how the required Gray code transformation can be ealized as a quantum circuit. The first step is to swap the states $|g_1\rangle$ nd $|g_2\rangle$. If these two states differ (only) at the ith qubit then we can chieve the required transformation by applying a bit flip, or Pauli-X ate, to this qubit *constrained* on the other qubits being identical to nose in both $|g_1\rangle$ and $|g_2\rangle$. Next we perform a controlled operation to wap $|g_2\rangle$ and $|g_3\rangle$, and we continue in this way. After $m-2$ operations e shall have realized the Gray code transformation

$$|g_1\rangle \rightarrow |g_{m-1}\rangle,$$
$$|g_2\rangle \rightarrow |g_1\rangle,$$
$$|g_3\rangle \rightarrow |g_2\rangle,$$
$$\vdots$$
$$|g_{m-1}\rangle \rightarrow |g_{m-2}\rangle. \tag{L.10}$$

ll states that do not correspond to an element of the Gray code will be naffected by these operations. At this stage we can apply a controlled nitary operation which enacts the desired unitary transformation (orig-ally on $|s\rangle$ and $|t\rangle$) on the single qubit at which $|g_m\rangle$ and $|g_{m-1}\rangle$ differ. his unitary transformation needs to be conditioned on the states of all e other qubits being the same as in $|g_m\rangle$ and $|g_{m-1}\rangle$. The final step is invert the $m-2$ unitary operations that constituted the Gray code ansformation.

As an example, let us suppose that $|s\rangle$ and $|t\rangle$ differ by virtue of e states of just three qubits. It suffices to consider just these three ıbits, which we take to be in the state $|000\rangle$ or $|111\rangle$. We seek a rcuit to implement the two-level unitary matrix corresponding to the ansformation

$$|000\rangle \rightarrow a|000\rangle + b|111\rangle,$$
$$|111\rangle \rightarrow c|000\rangle + d|111\rangle, \tag{L.11}$$

nere the constraints of unitarity require that $|a|^2 + |b|^2 = 1 = |c|^2 + |d|^2$ d that $ac^* + bd^* = 0$. We start with a suitable Gray code,

$$000 ,$$
$$001 ,$$
$$011 ,$$
$$111 , \tag{L.12}$$

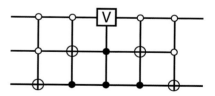

Fig. L.1 Quantum circuit for performing our example three-qubit transformation.

•m which we can read off the required quantum circuit, depicted in ʒ. L.1. The first two gates implement the Gray code transformation by anging the state $|000\rangle$ into the state $|011\rangle$. The third gate implements the unitary transformation

$$\hat{V} = \begin{pmatrix} a & b \\ c & d \end{pmatrix} \tag{L.13}$$

on the first qubit if the second and third qubits are both in the state $|1\rangle$. Finally, the last two gates undo the Gray code transformation. We have seen, in Section 6.3, how multiqubit gates such as those in Fig. L.1 can be formed from single-qubit and CNOT gates. It then follows that any desired multiqubit transformation can be implemented by a circuit formed only from single-qubit and CNOT gates. This establishes the desired result.

Nine- and five-qubit quantum codewords

Detection and correction of an arbitrary single-qubit error is possible using the Shor code, in which two nine-qubit states represent our logical 0 and 1:

$$|0_{\ell 9}\rangle = 2^{-3/2} \left(|0_{\ell 3}\rangle + |1_{\ell 3}\rangle\right) \otimes \left(|0_{\ell 3}\rangle + |1_{\ell 3}\rangle\right) \otimes \left(|0_{\ell 3}\rangle + |1_{\ell 3}\rangle\right),$$
$$|1_{\ell 9}\rangle = 2^{-3/2} \left(|0_{\ell 3}\rangle - |1_{\ell 3}\rangle\right) \otimes \left(|0_{\ell 3}\rangle - |1_{\ell 3}\rangle\right) \otimes \left(|0_{\ell 3}\rangle - |1_{\ell 3}\rangle\right).$$

$$(M.1)$$

Errors can arise from the effective action of any Pauli-X gate on any of the nine qubits, but it is apparent that only three of the nine possible Pauli-Z operators lead to different states. The actions of Z_1, Z_2, and Z_3, for example, introduce the same error for any superposition of $|0_{\ell 9}\rangle$ and $|1_{\ell 9}\rangle$. It follows that our error-correction protocol needs to distinguish between the original state and 21 possible distinct single-qubit errors ($9X + 9Y + 3Z$).

The two states $|0_{\ell 9}\rangle$ and $|1_{\ell 9}\rangle$ are simultaneous eigenstates of the eight mutually commuting operators

$$ZZIIIIIII,$$
$$IZZIIIIII,$$
$$IIIZZIIII,$$
$$IIIIZZIII,$$
$$IIIIIIZZI,$$
$$IIIIIIIZZ,$$
$$XXXXXXIII,$$
$$IIIXXXXXX,$$

$$(M.2)$$

with eigenvalue $+1$ in each case. It follows, of course, that any superposition state $\alpha|0_{\ell 9}\rangle + \beta|1_{\ell 9}\rangle$ will also be an eigenstate of these eight operators with eigenvalue $+1$. It is straightforward to show that any of the 21 distinct single-qubit errors can be identified, and so corrected, by their unique pattern of eigenvalues for the eight operators in eqn M.2.

It is reasonable to ask what are the shortest codewords that allow us to correct an arbitrary single-qubit error. We can arrive at a bound by a simple counting argument. If our codeword is formed from N qubits then we need to be able to identify X, Y, or Z errors affecting any of the qubits and, of course, the zero-error state. To do this we need each of the codewords representing the logical 0 and 1 to be associated with $3N + 1$ distinct possible states, one for the original state and $3N$ to account for the possible single-qubit errors. Hence we need at least $2(3N + 1)$ states in our 2^N-dimensional state space:

$$2^{N-1} \geq 3N + 1. \tag{M.3}$$

The smallest number of qubits satisfying this condition is $N = 5$, for which eqn M.3 is an equality, and there is indeed a five-qubit error-correcting code.

We can discuss the five-qubit error-correcting code using similar methods to those used for the Steane code in Section 6.4. To this end, we introduce the four mutually commuting operators

$$\mathbf{L}_1 = \text{IZXXZ},$$
$$\mathbf{L}_2 = \text{ZIZXX},$$
$$\mathbf{L}_3 = \text{XZIZX},$$
$$\mathbf{L}_4 = \text{XXZIZ}. \tag{M.4}$$

It is helpful to define our logical qubit states in terms of these operators and the five-qubit identity operator $\mathbf{I} = \text{IIIII}$ in the form

$$|0_{\ell 5}\rangle = \frac{1}{4} \left(\mathbf{I} + \mathbf{L}_1\right) \left(\mathbf{I} + \mathbf{L}_2\right) \left(\mathbf{I} + \mathbf{L}_3\right) \left(\mathbf{I} + \mathbf{L}_4\right) |00000\rangle,$$
$$|1_{\ell 5}\rangle = \frac{1}{4} \left(\mathbf{I} + \mathbf{L}_1\right) \left(\mathbf{I} + \mathbf{L}_2\right) \left(\mathbf{I} + \mathbf{L}_3\right) \left(\mathbf{I} + \mathbf{L}_4\right) |11111\rangle. \tag{M.5}$$

These two states, and all superpositions of them, are simultaneous eigenstates of the four operators in eqn M.4 with each eigenvalue being $+1$. Each of the possible operators has eigenvalues $+1$ and -1 and each of 15 arrangements of these, which include at least one value -1, corresponds to one of the 15 possible single-qubit errors. It follows that an error can be detected and then corrected by applying the relevant single-qubit Pauli operator.

In principle, the five-qubit codewords, being the shortest, should be the error-correcting code of choice. It turns out, however, that performing quantum information processing with these states is far from straightforward. The states in the seven-qubit Steane code are easier to manipulate and seem, for the present, a better candidate for practical quantum information processing.

Computational complexity

We have defined the difficulty of an algorithm by the way in which the resources required to realize it scale with the number of bits, n, in the input. It is both useful and important to be able to quantify this computational complexity. We do this by introducing three functions: (n), $\Omega(n)$, and $\Theta(n)$.

The function $O(n)$ sets an upper bound on the behaviour of a resource. The computing time $T(n)$ for a given algorithm is $O(g(n))$, for example, the function $g(n)$ bounds the large-n behaviour of $T(n)$. The precise statement is that $T(n)$ is $O(g(n))$ if there are (positive) constants c and such that

$$T(n) \leq cg(n), \qquad \forall \quad n > n_0. \tag{N.1}$$

ote that we are not interested in the values of the constants c and n_0, ly that they exist. As a simple example, suppose that

$$T(n) = 5n^3 + 6n + 4\log n. \tag{N.2}$$

r sufficiently large n, we can be sure that $T(n) < n^4$ and so conclude at $T(n)$ is $O(n^4)$, which we express simply as $T(n) = O(n^4)$. It is also ear that, for large n, $T(n) < 6n^3$ and hence that $T(n) = O(n^3)$.

It is also useful to be able to set a lower bound on the large-n value the resource. Our computing time is $\Omega(g(n))$ if there exist constants and n_0 such that

$$T(n) \geq cg(n) \qquad \forall \quad n > n_0. \tag{N.3}$$

r our example in eqn N.2, we can see that $T(n) > 5n^3$ and it follows at $T(n)$ is $\Omega(n^3)$ or $T(n) = \Omega(n^3)$.

The strongest statement we can make is when we find that a single nction $g(n)$ provides *both* an upper bound on $T(n)$ *and* also a lower und. When this is the case, we combine $T(n) = O(g(n))$ and $T(n) = n)$ in the single statement $T(n) = \Theta(n)$. Clearly, for our example in n N.2 we can write $T(n) = \Theta(n^3)$.

We noted in Section 7.1 that the familiar algorithm for adding two bit integers takes a time which scales like n. We can now make this tement more precise by stating that for this algorithm,

$$T(n) = \Theta(n). \tag{N.4}$$

r multiplication, the simplest and most familiar algorithm takes a time

$$T(n) = \Theta(n^2). \tag{N.5}$$

Now that we have the means to quantify complexity, we can show that there exists a more efficient algorithm. Let us denote the two n-bit numbers by x and y and divide the bit sequence for each into two parts of equal length by writing

$$x = 2^{n/2}a + b,$$
$$y = 2^{n/2}c + d, \tag{N.6}$$

We have implicitly assumed that n is even. For odd values, we can simply make a, b, c, and d into $(n+1)/2$-bit integers with the first bits in a and c being zero.

so that a, b, c, and d are all $n/2$-bit integers. Multiplying our integers x and y now gives

$$xy = 2^n ac + 2^{n/2}(ad + bc) + bd. \tag{N.7}$$

Multiplying by 2^n or $2^{n/2}$ corresponds simply to adding n or $n/2$ zeros to the bit string, so the only time-consuming parts are four multiplications of $n/2$-bit integers and the addition of the products ad and bc. The addition takes time $\Theta(n)$, and so we can write the total time taken using this method as

$$T(n) = 4T(n/2) + \Theta(n). \tag{N.8}$$

Solving this equation by recursion leads to the same scaling,

$$T(n) = \Theta(n^2). \tag{N.9}$$

There is a more efficient algorithm, however, and to see this we note that

$$(a + b)(c + d) = ac + bd + (ad + bc). \tag{N.10}$$

This means that we can compute $ad + bc$ by a single multiplication followed by subtracting ac and bd, so that only three multiplications are required and

$$T(n) = 3T(n/2) + \Theta(n), \tag{N.11}$$

the solution of which is

$$T(n) = \Theta(n^{\log 3}) \approx \Theta(n^{1.58}). \tag{N.12}$$

It is possible to do even better and to find an algorithm that is close to $\Theta(n)$.

The Bernstein–Vazirani algorithm

O

The Bernstein–Vazirani algorithm is a simple modification of that of Deutsch and Jozsa and, like the Deutsch–Jozsa algorithm, it solves a somewhat artificial problem involving an oracle.

Let b be an unknown n-bit string $B_n B_{n-1} \cdots B_1$ associated with the oracle and representing the number

$$b = \sum_{m=1}^{n} B_m 2^{m-1}. \tag{O.1}$$

If we input the string a then the oracle calculates the modulo-2 sum of the products of the corresponding bits of a and b. We write this single-bit function as

$$
\begin{aligned}
a \cdot b &= (A_n \cdot B_n) \, \text{XOR} \, (A_{n-1} \cdot B_{n-1}) \, \text{XOR} \cdots \text{XOR} \, (A_1 \cdot B_1) \\
&= A_n \cdot B_n \oplus A_{n-1} \cdot B_{n-1} \oplus \cdots \oplus A_1 \cdot B_1,
\end{aligned} \tag{O.2}
$$

where

$$a = \sum_{m=1}^{n} A_m 2^{m-1}. \tag{O.3}$$

Our task is to determine the value of b by addressing the oracle, and the challenge is to do this in the minimum possible number of trials.

It is not difficult to find the optimal classical algorithm. On each occasion that we address the oracle, we get a single bit of information in the form of the value of $f(a) = a \cdot b$. The unknown string b is n bits in length and so we need to perform at least n computations in order to solve the problem. The simplest way to achieve this is to input in turn the n values of a that have only a *single* 1 (and $n-1$ 0s). If $f(a) = 1$ then the corresponding bit in b is 1, but if $f(a) = 0$ then the corresponding bit in b is 0. After n trials, we have n bits and so know the value of b. A quantum computer allows us to solve the problem by addressing the oracle just once.

The oracle calculates the single-bit function $f(a) = a \cdot b$ and so, as with the Deutsch–Jozsa algorithm, we need a first string of n qubits and a second string of just a single qubit. We prepare the same input state as used in the Deutsch–Jozsa algorithm (eqn 7.21) and the oracle performs the transformation

$$2^{-(n+1)/2} \sum_{a=0}^{2^n-1} |a\rangle \otimes (|0\rangle - |1\rangle)$$

$$\rightarrow 2^{-(n+1)/2} \sum_{a=0}^{2^n-1} (-1)^{f(a)} |a\rangle \otimes (|0\rangle - |1\rangle)$$

$$= 2^{-(n+1)/2} \sum_{a=0}^{2^n-1} (-1)^{a \cdot b} |a\rangle \otimes (|0\rangle - |1\rangle). \tag{O.4}$$

We can see that, because of the superposition principle, the quantum processor has evaluated $a \cdot b$ in a single run for all values of a. It only remains to find a suitable measurement with which to extract the value of b. In order to see how this may be done, we first note that

$$a \cdot b = \sum_{m=1}^{n} A_m \cdot B_m \quad \mathrm{mod}\, 2, \tag{O.5}$$

which means that

$$(-1)^{a \cdot b} = (-1)^{\sum_{m=1}^{n} A_m \cdot B_m}. \tag{O.6}$$

This means that the output state in eqn O.4 is not an entangled state but rather an $n+1$-qubit product state of the form

$$2^{-(n+1)/2} \sum_{a=0}^{2^n-1} (-1)^{a \cdot b} |a\rangle \otimes (|0\rangle - |1\rangle)$$

$$= 2^{-(n+1)/2} \left(|0\rangle + (-1)^{B_n}\right) \otimes \left(|0\rangle + (-1)^{B_{n-1}} |1\rangle\right) \otimes \cdots$$

$$\cdots \otimes \left(|0\rangle + (-1)^{B_1} |1\rangle\right) \otimes (|0\rangle - |1\rangle). \tag{O.7}$$

The action of a Hadamard gate on a single qubit produces the transformation

$$\hat{\mathrm{H}} \frac{1}{\sqrt{2}} \left(|0\rangle + (-1)^{B_m} |1\rangle\right) = |B_m\rangle. \tag{O.8}$$

It follows that applying a Hadamard gate to each of the first n qubits allows us to read off the desired value of b by performing a measurement in the computational basis:

$$\hat{\mathrm{H}}^{\otimes n} \otimes \hat{\mathrm{I}}\, 2^{-(n+1)/2} \sum_{a=0}^{2^n-1} (-1)^{a \cdot b} |a\rangle \otimes (|0\rangle - |1\rangle) = |b\rangle \otimes \frac{1}{\sqrt{2}} (|0\rangle - |1\rangle). \tag{O.9}$$

(Alternatively, of course, we could simply measure the observable corresponding to the $\hat{\sigma}_x$ for each of the first n qubits.)

In spite of the similarities, the Bernstein–Vazirani algorithm has one important advantage over the Deutsch–Jozsa algorithm, and this is its ability to cope with errors. The Deutsch–Jozsa algorithm demonstrates a clear advantage over a classical algorithm only if we need to know for certain the nature of the function computed by the oracle. If a small obability of error is allowed then we can satisfy this requirement by a b-exponential number of classical trials. The output of the Bernstein–zirani algorithm is an n-bit string, which should be the hidden number The possibility, however, of a small number or errors in the compution may cause the measurement at the output to correspond to a mber other than b. If the error probability is sufficiently small, how-er, then the Hamming distance (see Section 1.4) between b and the mber given by the computation will also be small. In this case sim-y running the algorithm a small number of times and using majority ting for each of the bits generated in the output strings should suffice remove the errors. It is no longer possible to generate b in a single n, but the number of trials required will still be much less than that quired using a classical algorithm.

Discrete Fourier transforms

The discrete Fourier transform is, as its name suggests, a discrete form of the familiar Fourier transform in which the integration is replaced by a sum. We recall that the Fourier transform of a function $f(t)$ is defined to be

$$F(\omega) = \frac{1}{\sqrt{2\pi}} \int_{-\infty}^{\infty} f(t) e^{i\omega t} dt. \tag{P.1}$$

The Fourier transform can be inverted, and the inverse transform is

$$f(t) = \frac{1}{\sqrt{2\pi}} \int_{-\infty}^{\infty} F(\omega) e^{-i\omega t} dt. \tag{P.2}$$

The form of this inverse transform is intimately connected to the integral form of the delta function

$$\delta(t - t') = \frac{1}{2\pi} \int_{-\infty}^{\infty} e^{i\omega(t-t')} d\omega. \tag{P.3}$$

We should note that there is a conventional element in these definitions: you will often find the signs in the exponentials interchanged between eqns P.1 and P.2, and one of the prefactors is often chosen to be unity, with the other then being $1/(2\pi)$. It is important only that the product of these prefactors is $1/(2\pi)$.

The discrete Fourier transform acts to transform not a continuous function but rather a sequence of N complex numbers $x_0, x_1, \cdots, x_{N-1}$, into a sequence $X_0, X_1, \cdots, X_{N-1}$. It is defined by the summation

$$X_b = \frac{1}{\sqrt{N}} \sum_{a=0}^{N-1} x_a \exp\left(i \frac{2\pi ab}{N}\right). \tag{P.4}$$

The inverse discrete Fourier transform is

$$x_a = \frac{1}{\sqrt{N}} \sum_{b=0}^{N-1} X_b \exp\left(-i \frac{2\pi ab}{N}\right). \tag{P.5}$$

The form of the inverse relies on the simple identity

$$\frac{1}{N} \sum_{b=0}^{N-1} \exp\left(-i \frac{2\pi(a - a')b}{N}\right) = \frac{1}{N} \cdot \frac{1 - \exp\left(-i2\pi(a - a')\right)}{1 - \exp\left(-i2\pi(a - a')/N\right)}$$

$$= \delta_{a,a'}. \tag{P.6}$$

As with the Fourier transform, there is a conventional element in these definitions: you will often find the signs in the exponentials interchanged between eqns P.4 and P.5, and one of the prefactors is often chosen to be unity, with the other then being $1/N$. It is important only that the product of these prefactors is $1/N$.

We can consider the discrete Fourier transform as a unitary transformation. If we arrange the complex numbers x_a in a column vector, then multiplication by the unitary matrix

$$U = \frac{1}{\sqrt{N}} \begin{pmatrix} 1 & 1 & 1 & \cdots & 1 \\ 1 & \omega_N & \omega_N^2 & \cdots & \omega_N^{(N-1)} \\ 1 & \omega_N^2 & \omega_N^4 & \cdots & \omega_N^{2(N-1)} \\ 1 & \omega_N^3 & \omega_N^6 & \cdots & \omega_N^{3(N-1)} \\ \vdots & \vdots & \vdots & \ddots & \vdots \\ 1 & \omega_N^{(N-1)} & \omega_N^{2(N-1)} & \cdots & \omega_N^{(N-1)^2} \end{pmatrix}, \qquad \text{(P.7)}$$

where

$$\omega_N = \exp\left(i\frac{2\pi}{N}\right), \qquad \text{(P.8)}$$

gives a new column vector, the components of which are the transformed numbers X_b:

$$X_b = \sum_{a=0}^{N-1} U_{ba} x_a. \qquad \text{(P.9)}$$

The unitarity of the matrix in eqn P.7 is most simply expressed as the condition

$$\left(UU^\dagger\right)_{bc} = \sum_{a=0}^{N-1} U_{ba} U_{ac}^* = \delta_{bc}. \qquad \text{(P.10)}$$

It follows directly from the unitarity of U that

$$\sum_{b=0}^{N-1} X_b Y_b^* = \sum_{a=0}^{N-1} x_a y_a^*. \qquad \text{(P.11)}$$

The special case of this with $y = x$ is simply the discrete-Fourier-transform version of Parseval's theorem.

We conclude on a cautionary note by observing that eqn P.6 is not quite correct. If we add to a, or indeed to a', any integer multiple of N then the terms in the summation are unchanged. This means that we should write

$$\frac{1}{N} \sum_{b=0}^{N-1} \exp\left(-i\frac{2\pi(a-a')b}{N}\right) = \delta_{a,a'}^{\mathrm{mod}N}, \qquad \text{(P.12)}$$

where $\delta_{a,a'}^{\mathrm{mod}N} = 1$ if $a \equiv a' \bmod N$ and is zero otherwise. This is important, for example, in deriving the correct form of the state given in eqn 7.50.

An entropy inequality

e present in this appendix a proof of the inequality

$$S\left(\sum_i p_i \hat{\rho}_i\right) \leq -\sum_i p_i \log p_i + \sum_i p_i S(\hat{\rho}_i). \qquad \text{(Q.1)}$$

Let us start with a simpler problem in which the component density atrices $\hat{\rho}_i$ all represent pure states, so that

$$\hat{\rho} = \sum_i p_i |\psi_i\rangle\langle\psi_i| \qquad \text{(Q.2)}$$

d our inequality takes the simpler form

$$S\left(\sum_i p_i \hat{\rho}_i\right) \leq -\sum_i p_i \log p_i. \qquad \text{(Q.3)}$$

e can represent the density operator as a *pure* state of two entangled antum systems (a purification) of the form

$$|\psi\rangle_{AB} = \sum_i \sqrt{p_i} |\psi_i\rangle_A |\lambda_i\rangle_B, \qquad \text{(Q.4)}$$

ere the states $|\lambda_i\rangle_B$ are mutually orthonormal. Taking the trace over e B system gives, of course, the original density operator in eqn Q.2:

$$\text{Tr}_B\left(|\psi\rangle_{AB}\,_{AB}\langle\psi_i|\right) = \hat{\rho}_A. \qquad \text{(Q.5)}$$

e Schmidt decomposition (see Appendix D) of the state is necessarily the form

$$|\psi\rangle_{AB} = \sum_m \sqrt{\rho_m} |\rho_m\rangle_A |\phi_m\rangle_B, \qquad \text{(Q.6)}$$

ere the states $|\phi_m\rangle_B$ are mutually orthogonal and the ρ_m are the envalues of $\hat{\rho}_A$. The two reduced density operators $\hat{\rho}_A$ and $\hat{\rho}_B$ have e same eigenvalues and hence the same von Neumann entropy:

$$S(\hat{\rho}_A) = -\sum_m \rho_m \log \rho_m = S(\hat{\rho}_B). \qquad \text{(Q.7)}$$

e inequality in eqn 8.8 expresses the idea that performing a projective asurement can only increase the value of the von Neumann entropy. we perform a measurement on the B system in the $|\lambda_i\rangle_B$ basis then e probability for getting the result corresponding to the state $|\lambda_i\rangle_B$ d clearly be p_i, and hence the von Neumann entropy associated with the post-measurement state will be simply $-\sum_i p_i \log p_i$. It follows, therefore, that

$$S\left(\hat{\rho}_B\right) \leq -\sum_i p_i \log p_i. \tag{Q.8}$$

The equality of $S\left(\hat{\rho}_A\right)$ and $S\left(\hat{\rho}_B\right)$ then establishes the required inequality given in eqn Q.3.

It remains only to prove the more general inequality given in eqn Q.1. Let each of the component density operators $\hat{\rho}_i$ have the diagonal form

$$\hat{\rho}_i = \sum_j P_i^j |\rho_i^j\rangle\langle\rho_i^j|, \tag{Q.9}$$

so that

$$\hat{\rho} = \sum_{ij} p_i P_i^j |\rho_i^j\rangle\langle\rho_i^j|. \tag{Q.10}$$

It then follows from the inequality in eqn Q.3 that

$$
\begin{aligned}
S\left(\hat{\rho}\right) &\leq -\sum_{ij} p_i P_i^j \log(p_i P_i^j) \\
&= -\sum_i p_i \log p_i - \sum_i p_i \sum_j P_i^j \log P_i^j \\
&= -\sum_i p_i \log p_i - \sum_i p_i S\left(\hat{\rho}_j\right),
\end{aligned} \tag{Q.11}
$$

which is the inequality in eqn Q.1.

Quantum relative entropy

Section 8.1, we defined the quantum relative entropy to be

$$S(\hat{\sigma}\|\hat{\rho}) = \text{Tr}\left[\hat{\sigma}\left(\log \hat{\sigma} - \log \hat{\rho}\right)\right] \tag{R.1}$$

d made much use of the inequality

$$S(\hat{\sigma}\|\hat{\rho}) \geq 0. \tag{R.2}$$

this appendix, we provide a proof of this important inequality. We start by writing both of our density operators in their diagonal ms

$$\hat{\rho} = \sum_{m} \rho_m |\rho_m\rangle\langle\rho_m|,$$

$$\hat{\sigma} = \sum_{n} \sigma_n |\sigma_n\rangle\langle\sigma_n|, \tag{R.3}$$

ere the sets of states $\{|\rho_m\rangle\}$ and $\{|\sigma_n\rangle\}$ are the orthonormal eigen-tes of $\hat{\rho}$ and $\hat{\sigma}$, respectively. If we substitute these forms into eqn R.1 en we find

$$S(\hat{\sigma}\|\hat{\rho}) = \sum_{n} \sigma_n \log \sigma_n - \sum_{nm} \sigma_n |\langle\sigma_n|\rho_m\rangle|^2 \log \rho_m$$

$$= \sum_{nm} \sigma_n |\langle\sigma_n|\rho_m\rangle|^2 \left(\log \sigma_n - \log \rho_m\right), \tag{R.4}$$

ere we have used the completeness of the eigenstates of $\hat{\rho}$,

$$\sum_{m} |\langle\sigma_n|\rho_m\rangle|^2 = 1. \tag{R.5}$$

proceed by adding and subtracting the same term on the right-hand e of eqn R.4 to give

$$(\hat{\sigma}\|\hat{\rho}) = \sum_{nm} \sigma_n |\langle\sigma_n|\rho_m\rangle|^2 \left(\log \sigma_n + \log |\langle\sigma_n|\rho_m\rangle|^2\right.$$

$$\left. - \log \rho_m - \log |\langle\sigma_n|\rho_m\rangle|^2\right)$$

$$= \sum_{nm} \sigma_n |\langle\sigma_n|\rho_m\rangle|^2 \left[\log \left(\sigma_n |\langle\sigma_n|\rho_m\rangle|^2\right) - \log \left(\rho_m |\langle\sigma_n|\rho_m\rangle|^2\right)\right]$$

$$= \sum_{nm} P(n,m) \left[\log P(n,m) - \log Q(n,m)\right]. \tag{R.6}$$

We need only note that the functions $P(n, m) = \sigma_n |\langle \sigma_n | \rho_m \rangle|^2$ and $Q(n, m) = \rho_m |\langle \sigma_n | \rho_m \rangle|^2$ have the mathematical properties of probabilities, in that they are greater than or equal to zero and that

$$\sum_{nm} P(n, m) = 1 = \sum_{nm} Q(n, m). \tag{R.7}$$

It follows that the final line of eqn R.6 has the same form as a classical relative entropy and, as shown in Appendix B, this is necessarily greater than or equal to zero. This establishes the required inequality given in eqn R.2.

Our proof in Appendix B of the positivity of the relative entropy also showed that the relative entropy is zero if and only if the two probability distributions are identical. It follows, therefore, that the quantum relative entropy will be zero if and only if $P(n, m) = Q(n, m)$, so that

$$\sigma_n |\langle \sigma_n | \rho_m \rangle|^2 = \rho_m |\langle \sigma_n | \rho_m \rangle|^2, \quad \forall \ n, m. \tag{R.8}$$

This is true only if $\hat{\rho} = \hat{\sigma}$, and it follows that $S(\hat{\sigma} \| \hat{\rho}) = 0$ if and only if $\hat{\rho} = \hat{\sigma}$.

It is also possible to derive the positivity of the quantum relative entropy using the property of convexity (or of concavity). We recall that a continuous and differentiable convex function f satisfies the inequality

$$f(x) - f(y) - (x - y)f'(y) \geq 0, \tag{R.9}$$

where f' denotes the first derivative of f with respect to its argument. We make use of Klein's inequality for a convex function:

$$\text{Tr} \left[f(\hat{A}) - f(\hat{B}) - \left(\hat{A} - \hat{B} \right) f'(\hat{B}) \right] \geq 0, \tag{R.10}$$

where $\hat{A}$ and $\hat{B}$ are Hermitian (or, more strictly, self-adjoint) operators. We can prove Klein's inequality by introducing a complete orthonormal set of eigenvectors for $\hat{A}$ and $\hat{B}$ so that

$$\hat{A} = \sum_i a_i |a_i\rangle \langle a_i|,$$

$$\hat{B} = \sum_j b_j |b_j\rangle \langle b_j|. \tag{R.11}$$

This leads us to write the quantity in eqn R.10 as

$$\text{Tr} \left[f(\hat{A}) - f(\hat{B}) - \left(\hat{A} - \hat{B} \right) f'(\hat{B}) \right]$$
$$= \sum_i \langle a_i | \left[f(\hat{A}) - f(\hat{B}) - \left(\hat{A} - \hat{B} \right) f'(\hat{B}) \right] |a_i\rangle$$
$$= \sum_{ij} |\langle a_i | b_j \rangle|^2 \left[f(a_i) - f(b_j) - (a_i - b_j)f'(b_j) \right]$$
$$\geq 0, \tag{R.12}$$

where the final inequality follows from the fact that each term in the sum is, because of eqn R.9, greater than or equal to zero. The inequality in eqn R.2 follows directly from Klein's inequality on selecting $f(x) = x \log x$.

he Araki–Lieb inequality

e seek a proof of the Araki–Lieb inequality for von Neumann entropies, ich states that

$$|S(A) - S(B)| \leq S(A, B). \tag{S.1}$$

We consider a purification of $\hat{\rho}_{AB}$ formed by introducing a third quan- m system C. Let the orthonormal eigenstates of $\hat{\rho}_{AB}$ be $|\rho_m\rangle_{AB}$ with , the corresponding eigenvalues. It then follows that the pure state

$$|\psi\rangle = \sum_m \sqrt{\rho_m}|\rho_m\rangle_{AB}|\phi_m\rangle_C, \tag{S.2}$$

ere $\langle\phi_n|\phi_n\rangle = \delta_{nm}$, has the required reduced density operator for the and B systems:

$$\hat{\rho}_{AB} = \text{Tr}_C\left(|\psi\rangle\langle\psi|\right). \tag{S.3}$$

e fact that the combined state of the three systems is pure means at $S(ABC) = 0$ and it follows, therefore, that however we partition e state, the two resulting subsystems will have the same von Neumann tropy:

$$\begin{aligned} S(AB) &= S(C), \\ S(BC) &= S(A), \\ S(AC) &= S(B). \end{aligned} \tag{S.4}$$

badditivity (eqn 8.32) then requires that

$$\begin{aligned} S(BC) &\leq S(B) + S(C) \\ \Rightarrow S(A) &\leq S(B) + S(AB). \end{aligned} \tag{S.5}$$

we rewrite this condition for $S(AC)$ then we find that

$$\begin{aligned} S(AC) &\leq S(A) + S(C) \\ \Rightarrow S(B) &\leq S(A) + S(AB). \end{aligned} \tag{S.6}$$

mbining these two then gives the required inequality (eqn S.1).

Fidelity for mixed states

ur task is to generalize the formula for the fidelity for a pure state $|\psi\rangle$
d a mixed state $\hat{\rho}$,

$$F(|\psi\rangle\langle\psi|, \hat{\rho}) = \langle\psi|\hat{\rho}|\psi\rangle, \qquad (\text{T.1})$$

a form suitable for using with mixed states. We shall find that the
sired expression has the form of Uhlmann's transition probability:

$$F(\hat{\rho}, \hat{\sigma}) = \left(\text{Tr}\sqrt{\hat{\rho}^{1/2}\hat{\sigma}\hat{\rho}^{1/2}}\right)^2. \qquad (\text{T.2})$$

deriving this result, we follow the analyses of R. Jozsa, *Journal of*
odern Optics **41**, 2315 (1994) and of Nielsen and Chuang (see sugges-
ns for further reading in Chapter 8).
We recall that in Section 2.2 we found that we could write any mixed
te in terms of a pure state in a doubled state space. By this we mean
at if our density operator has the diagonal form

$$\hat{\rho} = \sum_m \rho_m |\rho_m\rangle\langle\rho_m|, \qquad (\text{T.3})$$

en the pure state

$$|\psi\rangle = \sum_m \sqrt{\rho_m}|\rho_m\rangle \otimes |\rho_m\rangle \qquad (\text{T.4})$$

l give precisely the same statistical properties for the first system
those associated with $\hat{\rho}$. The *purification* in eqn T.4 is not unique,
deed any state of the form

$$|\psi\rangle = \sum_m \sqrt{\rho_m}e^{i\theta_m}|\rho_m\rangle \otimes |\phi_m\rangle, \qquad (\text{T.5})$$

ere the states $|\phi_m\rangle$ are any orthonormal basis and the θ_m are any
ases, will be a purification of $\hat{\rho}$. It is helpful to write this general
rification of $\hat{\rho}$ in a different form. To do so, we first introduce the
normalized) maximally entangled state

$$|\mathcal{E}\rangle = \sum_m |m\rangle \otimes |m\rangle, \qquad (\text{T.6})$$

ere the states $|m\rangle$ form complete orthonormal bases over each of the
state spaces. It is straightforward to show that the state in eqn T.5

To do this, we need only ensure that
$\hat{U}$ transforms the states $|m\rangle$ into the
states $|\rho_m\rangle$.

and hence any purification of $\hat{\rho}$ can be written in terms of $|\mathcal{E}\rangle$ in the form

$$|\psi\rangle = \sqrt{\hat{\rho}}\,\hat{U} \otimes \hat{U}'|\mathcal{E}\rangle, \tag{T.7}$$

where $\hat{U}$ and $\hat{U}'$ are unitary operators.

The natural way to define the fidelity for mixed states is to apply the form for pure states to the purifications of the two density operators. To this end, we write the general purification of the second density operator, $\hat{\sigma}$, in the form

$$|\phi\rangle = \sqrt{\hat{\sigma}}\,\hat{V} \otimes \hat{V}'|\mathcal{E}\rangle. \tag{T.8}$$

The fidelity for the mixed states is then

$$F(\hat{\rho}, \hat{\sigma}) = \mathrm{Sup}_{|\psi\rangle, |\phi\rangle} |\langle \phi|\psi\rangle|^2, \tag{T.9}$$

where the maximization is carried out over all possible purifications $|\psi\rangle$ and $|\phi\rangle$. It follows that

$$F(\hat{\rho}, \hat{\sigma}) = \mathrm{Sup}_{\hat{U}, \hat{U}', \hat{V}.\hat{V}'} |\langle \mathcal{E}|\hat{V}^\dagger \sqrt{\hat{\sigma}}\sqrt{\hat{\rho}}\,\hat{U} \otimes \hat{V}'^\dagger \hat{U}'|\mathcal{E}\rangle|^2. \tag{T.10}$$

In order to proceed, we need to prove a simple result for expressions of the type given in eqn T.10. We note that

$$\begin{aligned}
\langle \mathcal{E}|\hat{A} \otimes \hat{B}|\mathcal{E}\rangle &= \sum_{m,m'} \langle m, m|\hat{A} \otimes \hat{B}|m', m'\rangle \\
&= \sum_{m,m'} \langle m|\hat{A}|m'\rangle \langle m|\hat{B}|m'\rangle \\
&= \sum_{m,m'} \langle m|\hat{A}|m'\rangle \langle m'|\hat{B}^\dagger|m\rangle \\
&= \mathrm{Tr}\left(\hat{A}\hat{B}^\dagger\right),
\end{aligned} \tag{T.11}$$

where the operators $\hat{A}$ and $\hat{B}^\dagger$ both operate on the original single state space in the final line. It follows, therefore, that our fidelity is

$$F(\hat{\rho}, \hat{\sigma}) = \mathrm{Sup}_{\hat{W}} \left|\mathrm{Tr}\left(\sqrt{\hat{\sigma}}\sqrt{\hat{\rho}}\,\hat{W}\right)\right|^2, \tag{T.12}$$

where $\hat{W} = \hat{U}\hat{U}'^\dagger\hat{V}'\hat{V}^\dagger$. The maximization over the possible unitary transformations may be found using the simple result that for any operator $\hat{C}$ and any unitary operator $\hat{W}$,

$$\left|\mathrm{Tr}\left(\hat{C}\hat{W}\right)\right| \leq \mathrm{Tr}\left|\hat{C}\right|. \tag{T.13}$$

The proof follows on writing

$$\hat{C} = \left|\hat{C}\right|\hat{S}, \tag{T.14}$$

where $\hat{S}$ is a unitary operator and $|\hat{C}|$ is a positive and therefore Hermitian operator. We can evaluate the trace in eqn T.13 using the basis of igenstates, $|c_n\rangle$, of $|\hat{C}|$:

$$
\begin{aligned}
\mathrm{Tr}\left(\hat{C}\hat{W}\right) &= \mathrm{Tr}\left(|\hat{C}|\hat{S}\hat{W}\right) \\
&= \sum_n \langle c_n||\hat{C}|\hat{S}\hat{W}|c_n\rangle \\
&= \sum_n |c_n|\langle c_n|\hat{S}\hat{W}|c_n\rangle.
\end{aligned} \tag{T.15}
$$

he operator $\hat{S}\hat{W}$ is unitary and it follows that

$$
\left|\langle c_n|\hat{S}\hat{W}|c_n\rangle\right| \le 1, \tag{T.16}
$$

ith equality for all $|c_n\rangle$ if $\hat{W} = \hat{S}^\dagger$. With this choice we find that the delity is

$$
\begin{aligned}
F(\hat{\rho}, \hat{\sigma}) &= \left(\mathrm{Tr}\left|\sqrt{\hat{\rho}}\sqrt{\hat{\sigma}}\right|\right)^2 \\
&= \left(\mathrm{Tr}\sqrt{\hat{\rho}^{1/2}\hat{\sigma}\hat{\rho}^{1/2}}\right)^2,
\end{aligned} \tag{T.17}
$$

hich is the required result.

Entanglement of formation for two qubits

e present in this appendix the form of the entanglement of formation two-qubit states as derived by W. K. Wootters in *Physical Review tters* **80**, 2245 (1998). We start by introducing the 'spin flip' trans- mation, which, for a single qubit, takes the form

$$\hat{\rho} \rightarrow \tilde{\hat{\rho}} = \hat{\sigma}_y \hat{\rho}^* \hat{\sigma}_y, \tag{U.1}$$

ere $\hat{\rho}^*$ is the complex conjugate of $\hat{\rho}$ expressed in the computational sis. This transformation has the effect of reversing the sign of the och vector:

$$\hat{\rho} = \frac{1}{2} \left(\hat{I} + \vec{r} \cdot \hat{\vec{\sigma}} \right)$$
$$\Rightarrow \tilde{\hat{\rho}} = \frac{1}{2} \left(\hat{I} - \vec{r} \cdot \hat{\vec{\sigma}} \right). \tag{U.2}$$

e spin-flipped state for two qubits is obtained, naturally enough, by plying the above transformation to both qubits:

$$\tilde{\hat{\rho}} = \hat{\sigma}_y \otimes \hat{\sigma}_y \, \hat{\rho}^* \, \hat{\sigma}_y \otimes \hat{\sigma}_y. \tag{U.3}$$

e form from $\hat{\rho}$ and $\tilde{\hat{\rho}}$ the positive operator

$$\hat{R}^2 = \sqrt{\hat{\rho}} \, \tilde{\hat{\rho}} \, \sqrt{\hat{\rho}}. \tag{U.4}$$

: the four positive eigenvalues of $\hat{R}$ be, in order of decreasing size, λ_1, λ_3, and λ_4. The concurrence $\mathcal{C}$ is then defined to be

$$\mathcal{C} \left(\hat{\rho} \right) = \mathrm{Sup} \left(0, \lambda_1 - \lambda_2 - \lambda_3 - \lambda_4 \right); \tag{U.5}$$

at is, $\lambda_1 - \lambda_2 - \lambda_3 - \lambda_4$ if this is positive and zero otherwise. It is zero he state is unentangled but takes a positive value if it is entangled. e concurrence is a measure of entanglement, albeit only for states of > qubits. The entanglement of formation is simply

$$E_{\mathrm{F}} \left(\hat{\rho} \right) = - \left(\frac{1 + \sqrt{1 - \mathcal{C}^2}}{2} \right) \log \left(\frac{1 + \sqrt{1 - \mathcal{C}^2}}{2} \right)$$
$$- \left(\frac{1 - \sqrt{1 - \mathcal{C}^2}}{2} \right) \log \left(\frac{1 - \sqrt{1 - \mathcal{C}^2}}{2} \right). \tag{U.6}$$

As a simple example, we can calculate the entanglement of formation for a pure entangled state. Consider the two-qubit state

$$|\psi\rangle = \cos\theta|00\rangle + \sin\theta|11\rangle. \qquad (\text{U.7})$$

The associated spin-flipped state is

$$|\bar{\psi}\rangle = \cos\theta|11\rangle + \sin\theta|00\rangle, \qquad (\text{U.8})$$

so that

$$\mathcal{C} = |\langle\psi|\bar{\psi}\rangle| = |\sin(2\theta)|, \qquad (\text{U.9})$$

which, on insertion into eqn U.6, gives the pure-state entanglement of formation $H(\cos^2\theta)$.